An Introduction to
Marine Ecology

Frontispiece Exposed rocky shore, Tsitsikama, South Africa.

An Introduction to
Marine Ecology

THIRD EDITION

R.S.K. Barnes
Department of Zoology
University of Cambridge
Cambridge CB2 3EJ

R.N. Hughes
School of Biological Sciences
University of Wales
Bangor, Gwynedd LL57 2UW

With contributions from
John Field, Dan Baird & Michel Kaiser

Blackwell
Science

For David, Andrew, Morvan & Annaëlle
and Helen, Ruth & Anne

© 1982, 1988, 1999 by
Blackwell Science Ltd
Editorial Offices:
Osney Mead, Oxford OX2 0EL
25 John Street, London WC1N 2BL
23 Ainslie Place, Edinburgh
 EH3 6AJ
350 Main Street, Malden
 MA 02148 5018, USA
54 University Street, Carlton
 Victoria 3053, Australia
10, rue Casimir Delavigne
 75006 Paris, France

Other Editorial Offices:
Blackwell Wissenschafts-Verlag
 GmbH
Kurfürstendamm 57
10707 Berlin, Germany

Blackwell Science KK
MG Kodenmacho Building
7–10 Kodenmacho Nihombashi
Chuo-ku, Tokyo 104, Japan

First published 1982
Reprinted 1986
Second edition 1988
Reprinted 1989, 1990, 1991, 1992,
 1993, 1995
Third edition 1999

Set by Graphicraft Limited,
Hong Kong
Printed and bound in Great Britain
at the University Press, Cambridge

The Blackwell Science logo is a
trade mark of Blackwell Science Ltd,
registered at the United Kingdom
Trade Marks Registry

A catalogue record for this title is
available from the British Library

ISBN 0-86542-834-4

Library of Congress
Cataloging-in-publication Data

Barnes, R.S.K. (Richard Stephen
Kent)
 An introduction to marine
ecology / Richard Barnes,
Roger Hughes.—3rd ed.
  p.   cm.
 Includes bibliographical
references
(p.   ) and index.
 ISBN 0-86542-834-4
 1. Marine ecology.
I. Hughes, R.N. II. Title.
QH541.5.S3B34 1999
577.7—dc21   98-38820
       CIP

DISTRIBUTORS

Marston Book Services Ltd
PO Box 269
Abingdon, Oxon OX14 4YN
(*Orders*: Tel: 01235 465500
     Fax: 01235 465555)

USA
Blackwell Science, Inc.
Commerce Place
350 Main Street
Malden, MA 02148 5018
(*Orders*: Tel: 800 759 6102
      781 388 8250
     Fax: 781 388 8255)

Canada
Login Brothers Book Company
324 Saulteaux Crescent
Winnipeg, Manitoba R3J 3T2
(*Orders*: Tel: 204 837-2987)

Australia
Blackwell Science Pty Ltd
54 University Street
Carlton, Victoria 3053
(*Orders*: Tel: 3 9347 0300
     Fax: 3 9347 5001)

For further information on
Blackwell Science, visit our website:
www.blackwell-science.com

1001711816

Contents

Preface

The first two editions of this book were aimed at students with some knowledge of ecology who were about to venture into the *mare incognitum* for the first time. The marine environment and its inhabitants are exciting and stimulating subjects for amateur naturalists, students and professionals, and continue to inspire each of the contributors to this book. The previous texts provided the reader with a broad introduction to distinct processes or subsystems of the ocean. Since then, the information base has continued to increase exponentially; much more is now known about ecological processes in the oceans than ever before. As in any branch of science, however, progress is spread unevenly among subdisciplines which is reflected in the content of the chapters. All have been revised in the light of new information, but in some areas growth has been so prodigious that we felt compelled to enlist specialist expertise. John Field deals with fisheries, an issue that has started wars and continues to cause political strife within and between nations. Dan Baird takes a more detailed look at ecosystems, the understanding of which are essential if we are to predict the effects of major changes such as global warming in our environment. Finally, Michel Kaiser has focused on the issues surrounding human interference and conservation that have become so prominent in recent years with the concept of biodiversity and its importance to mankind. As before, editorial co-operation among co-authors has ensured consistency and co-ordination. The result is a new book, inheriting parental character while offering the freshness of a new generation.

1 The nature and global distribution of marine organisms, habitats and productivity

1.1 Introduction

Homo sapiens is a terrestrial species and perhaps not surprisingly it tends to regard ecology as being mostly about the other terrestrial organisms with which it comes into daily contact. Ours is a planet largely covered by water, however; terrestrial habitats comprise less than one third of its surface; and of that liquid water, 99.7% is in the ocean. Some 99% of the planet's inhabited space is that ocean. To a first approximation, therefore, the ecology of the Earth is the ecology of its seas.

The ocean basins are not just 'drowned land': the sea floor and the land-masses are made of two different types of crustal materials: thin, dense oceanic crust and thick, light continental crust. Both float on the denser, upper layers of the mantle—the oceanic crust as a thin skin and the continental crust as large lumps—and both move in response to convection currents in that mantle. The oceanic crust is geologically young and is created continually along the mid-oceanic ridges; it then moves away from the ridges and is eventually resorbed into the mantle beneath the oceanic trenches. The continental crust, in contrast, is much older and it floats above, but is moved by, this sea-floor spreading.

The existence of these two forms of crust is reflected in the Earth's surface relief. Most of the ocean bed is a level expanse of sediment (with slopes of less than 1 in 1100) lying 3000–4000 m below sea-level. From this plain, the huge continental blocks rise steeply, with an average upward slope of some 1 in 14 but in some areas with an average slope of more than 1 in 3, up to a depth of between 20 and 500 m (with an average of 130 m) below sea-level. At this point, representing the angle between the side and top of the continental mass, the gradient usually changes dramatically, falling to about 1 in 600. The average height of the top surface of the continents is less than 1000 m above this point.

Thus, ignoring the water for a moment, we have a scenario of a level ocean bed from which arise sheer-sided, flattish-topped or gently domed blocks averaging just over 5300 m in height. Also arising from our level plain would be volcanoes and the mid-oceanic ridges. Of course this is a considerable oversimplification, not least because the movement of sediment from the continents to the oceans tends to blur the starkness of the relief. Major rivers, when they discharge into the sea, do not lose their integrity but flow in submarine canyons scoured out by sediment-laden water. The 'river water' is no longer fresh but a dense suspension of sediment in sea-water and this descends the sides of the continental mass, the sediments being discharged into the angle between that mass and the oceanic bed. Great fans of sediment may extend out up to 600 km from the continental base, resting at an average slope of about 1 in 60 and forming ramps leading from the ocean floor to the continental sides.

This surface topography spans a height of almost 20 km (from the highest point on any continental block to the lowest point in an oceanic trench) but nevertheless this is an insignificant fraction of the Earth's radius (0.3%). If, arbitrarily, we set the base of the rising continental blocks at a depth of 2000 m, then they would occupy 41% of the Earth's surface. This is not to say that the land occupies that proportion of the surface, however, for, as we have seen, the sides and parts of the rims of the continents are below sea-level. Today, some 73% of the continental surface area projects above the waves and so the sea in total accounts for 70% of the Earth's 510×10^6 km^2 surface—59% being that covering the ocean floor plus 11% over the submerged continental margins. The surface area of the continental masses appears more or less constant and so, therefore, is that of the ocean floor. However, the area of continental margin beneath the sea is, because of its shallow slope, subject to marked variation dependent on sea-level. A 100 m decrease in sea-level (such as might occur during the next glacial phase) would decrease the 11% of today to around 7%, whereas a 100 m rise in sea-level (which might result from the melting of all the Earth's ice) would increase it to nearly 20%. Changes of this magnitude have occurred in the past and have greatly affected the abundance and diversity of the shallow marine fauna.

The sea therefore covers the largest portion of the Earth's surface and it is even more important a habitat in terms of the total volume of the Earth regularly inhabited by living organisms. On land the inhabited zone usually extends only a few tens of metres above the ground and a metre or so below it; the oceans are inhabited from their surface right down to their greatest depths (in excess of 11 000 m): the sea therefore provides all but 1% of the living space on our planet. Although the largest, it is also the least known and least knowable portion, particularly with regard to its biology.

Sir Alister Hardy likened our attempts to investigate the ocean to a person in a hot-air balloon slowly drifting over a land hidden from view by dense fog. Every so often, the balloonist would let down a bucket on the end of a large rope, let it drag along the ground for a while, and then, after pulling it up, examine the contents. What sort of an impression and understanding of terrestrial biology would an observer gain using such a technique? We have perhaps progressed beyond this limitation—but not very far. Maintenance of a research ship at sea is also very much more expensive than operating a hot-air balloon. We know most about life in shallow, coastal waters and about the relatively slow-moving and small- to medium-sized organisms; we know least of life in the depths and of the smallest and largest, fastest-moving species. The reason for the depth limitation is self-evident. Note also, however, that fish population densities at depth are low— 1 per 1000 m^3 for example—and this imposes severe sampling problems. That relating to size of organism is not immediately obvious and is often overlooked. Most ecological information is still obtained from the sea by use of nets or by washing samples through a sieve, although the use of scuba techniques and minisubmersible craft has increased dramatically. Neither nets nor sieves can be used to retain the smallest or most delicate organisms; they either pass through or are fragmented beyond recognition or study. Small water or sediment samples have to be taken *in situ* with consequent problems of sampling accuracy and adequacy, and with respect to bacteria, problems of changes in the relative proportion of the individuals and species after culture. Neither can nets capture large, fast-swimming species: they simply avoid the net. Our knowledge of the largest squids, for example, is entirely derived from the occasional specimen cast up on a beach (and examined by a biologist before scavengers and decay render it useless) and from the hard parts (beaks) recovered from the whales which feed on them. Yet *Architeuthis* the giant squid may attain a total length of 17 m (and over 30 m has been claimed). Some whales (e.g. *Mesoplodon* and *Stenella* spp.) are also known only from a few specimens or fragments stranded on the shore, never having been seen alive. There is no scientific or probabilistic reason why Heuvelmans' (1968) thesis that several huge and as yet unknown fish and mammals occur in the oceans should not eventually be found to be correct. Indeed, the discovery in 1976 of the previously unknown megamouth shark (5 m long and 750 kg weight) is a case in point. Completely new types of animals are still being discovered: since 1980, two new phyla (the Loricifera and Cycliophora) and two new classes in well-known phyla (the Remipedia in the Crustacea, and the concentricycloid echinoderms) have been described.

There are, therefore, many areas of complete and almost complete ignorance, and there are several areas of controversy and doubt; but there is much that we do know and even more of which we are fairly certain. The following pages set out to introduce the reader to this body of knowledge and to present what currently appear to be the outlines of the ecology of the seas.

1.2 The nature of the ocean

We have already considered the basic shape of the crustal container housing the world's ocean and we must now put rather more flesh on these bones and describe those aspects of oceanic structure and those properties of sea-water that have a particular bearing on marine ecology. The study of marine science in general—oceanography—is, of course, a large field embracing physics, chemistry, geology and several other disciplines besides biology; here we must be very selective and many only marginally relevant topics cannot be covered. The reader is referred to Bearman (1989) for additional information.

As we will see several times later, marine organisms appear particularly to respond to and reflect three all-important environmental gradients: the latitudinal gradient in magnitude and seasonality of solar radiation from the poles to the equator (which will be deferred for detailed consideration below); the depth gradient from the sea surface to the abyssal sea bed; and the coastal to open water gradient, which often coincides with that in respect of depth. In fact, all three are interlinked and are often superimposed.

The most straightforward of the three is the depth gradient. Although a host of terms have been coined for specific sections of the 0–11 000 m gradient, the most important distinction is between the uppermost few metres of the water column which can be illuminated by sunlight and the remaining 97.5% which cannot. Light is exploited for different purposes by different organisms, and different intensities will limit different processes. Therefore 'illuminated' must be qualified by reference to the process concerned. The light intensity at the sea surface also varies in regular diurnal and seasonal patterns and in relation to cloud cover, and hence any illuminated zone will vary with the light intensity and with the translucency of the water. Much of the incident light is scattered at the surface and of that which does penetrate this barrier, most is very quickly absorbed so that light intensity decreases logarithmically with depth. In the Sargasso Sea, for example, where the water is particularly translucent and light penetration is greatest, only a maximum of 1% of the red light penetrating the surface remains by 55 m depth, only 1% of the yellow–green and violet light by 95 m, and only 1% of the blue by 150 m.

To photosynthesize, algae require light (particularly of the shorter wavelengths) and one can calculate the depth down to which the light is sufficient to permit their growth. In the most translucent, oceanic water and under conditions of full sunlight, the limiting depth for photosynthetic production is of the order of 250 m; in clear, coastal waters this reduces to

about 50 m; and in highly turbid waters it is to be measured only in centimetres. Clearly, therefore, all primary fixation of organic compounds by photosynthetic organisms must be a phenomenon confined to the surface waters. The depth zone in which this is possible, the 'photic' (or 'euphotic') zone, averages some 30 m deep in coastal waters and some 150 m in the open ocean; the remainder of the depth gradient (and at night the whole ocean) is 'aphotic'. Below about 1250 m, there is insufficient sunlight for any biological process and hence, except for light produced by the organisms themselves, the ocean is thereafter lightless.

Light is one form of energy arriving from the Sun; the second of great ecological consequence is heat. It is no accident that the element of a domestic kettle is situated at the bottom of the water mass enclosed within this heating appliance. As the water in contact with the element is warmed so it becomes less dense and rises, thereby allowing more, cooler water to replace it and be heated in turn. The heating process operates on convection currents which would not form if the element was positioned near the surface of the water mass. But, discounting geothermal sources, this is precisely the situation with respect to the source of heat and the oceans. The surface waters of the sea receive heat from the Sun; therefore they become less dense and float at the surface; therefore they receive yet more heat; and so on. The end result is a body of hot, less dense water floating on top of a much larger mass of cold, dense water; the interface between the two, or more strictly the zone of rapid change in water temperature (Fig. 1.1), is termed the 'thermocline'. As with the photic zone, the position and magnitude of the thermocline are variable, but as water has a high specific heat it can absorb much heat with relatively little change in temperature and it will retain its heat for a long time in the presence of a temperature gradient. Diurnal changes in temperature are confined to the uppermost few metres and, even there, are rarely more than 0.3°C in the open ocean or more than 3°C in coastal areas.

The thermocline is therefore a feature of the upper 1000 m, below which the temperature of the ocean falls from a maximum of 5°C down to between 0.5 and 2.0°C. In contrast, at the surface, temperature may vary from −2°C to more than 28°C dependent on latitude (in contrast to fresh water, the density

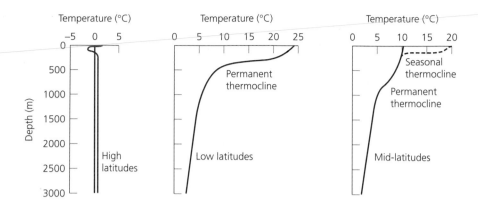

Fig. 1.1 Characteristic temperature profiles at different latitudes in the open ocean. (The presence of cold, low-salinity water near the surface in polar latitudes disturbs the otherwise vertical profile there.) (After Wright 1977–78.)

of sea-water increases uniformly with a decrease in temperature down to near −2°C). Thermoclines are permanent features of the oceanic depth gradient in all but the highest latitudes, their magnitude depending on the temperature differential between surface and bottom waters. In regions experiencing an alternation of hot and cold seasons, a marked, though shallow and temporary, seasonal thermocline is superimposed on the relatively weak, permanent thermocline during the hot season (Fig. 1.1). The importance of this surface water–bottom water density difference is that it produces a barrier to mixing of these two water masses. Those dissolved substances taken out of the water in the photic zone and incorporated into living tissue which sink through the thermocline (as a result of gravitational forces) cannot be replaced by local mixing. Waters above a thermocline may therefore become depleted in these essential dissolved nutrients whereas the bottom waters hold large, untappable stocks (Fig. 1.2). (It should be noted that it is thought that before mid-Miocene time deep-ocean water was some 10°C warmer than at present. This would have had considerable repercussions on nutrient availability.)

The third and final feature associated only with the surface layers of the sea which must be mentioned here is wind-induced mixing. Winds blowing over the surface of the ocean impart some of their energy to the water, causing waves to form and inducing turbulent mixing of the surface layers down

to maximum depths of the order of 200 m. This potential zone of mixing is therefore within the same depth range as the potential photic zone and the temporary seasonal thermocline of temperate latitudes; the precise relationships between these three depths at any one time are of great importance with respect to the potentiality of photosynthetic production. If,

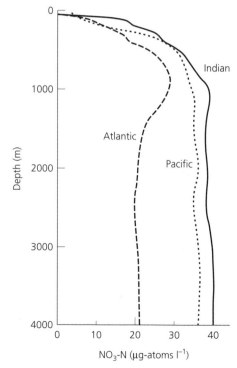

Fig. 1.2 The vertical distribution of nitrate in tropical and subtropical regions of the Atlantic, Indian and Pacific Oceans. (After Sverdrup et al. 1942.)

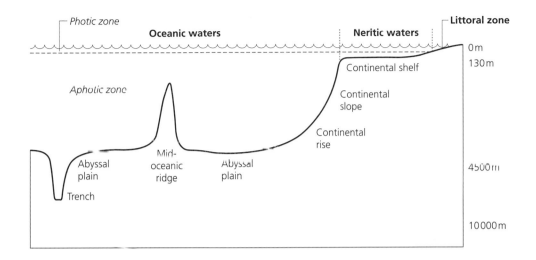

Fig. 1.3 Section of the ocean showing the major habitat subdivisions.

for example, mixing extends well below the photic zone, photosynthetic organisms may spend much more time below their threshold light intensity than above it and be unable to achieve sufficient production to balance their own energetic requirements (see Sections 2.2.1 and 2.2.2). Moreover, if mixing does not extend down to the thermocline, then photosynthetic production may also be reduced to low levels because of exhaustion of the nutrient supplies (Fig. 1.2).

The second major gradient is that stretching outwards from the coast into the open ocean, and it also involves variation in nutrients, depth and mixing. Several important subdivisions of the marine habitat can be made on this basis:

1 The immediate coastal or 'littoral' region from the upper limit of sea-water cover down to some 30 m depth.

2 The areas of submerged continental margins—the so-called 'neritic' water and the underlying 'continental shelf'.

3 The rapidly descending sides of the continental masses—the 'continental slope' with the more gently sloping 'continental rise' at the base of the slope.

4 The oceanic floor, usually termed the 'abyssal plain'.

5 The mid-oceanic ridges—vast mountain chains rising from the abyssal plain to within 2000 m or so

of the surface (and occasionally breaking surface in the form of mid-oceanic islands).

6 The 'hadal regions' of the deep-ocean trenches—chasms in the abyssal plain descending from 6000 m to, in several cases, below 10 000 m.

The waters cradled within the continental slopes and the deep ocean floor are differentiated from the coastal neritic waters by being termed 'oceanic' (see Fig. 1.3). For present purposes, the three basic sections of the coastal to open water gradient are: (a) the littoral; (b) the neritic–continental shelf; and (c) the oceanic–abyssal (the latter including the continental slopes and rises, the mid-oceanic ridges and the deep trenches).

The essential feature leading to the separation of the littoral zone as a distinct part of the marine ecosystem is the extreme shallowness of the water. Light may penetrate to the sea bed and, indeed, in areas with tidal water fluctuations part of the bottom may become exposed temporarily to the air and receive solar radiation directly, diurnally or semi-diurnally. Plants and algae associated with the sea bed can survive there and add their contribution to the total marine production. As we shall see later, the photosynthetic organisms occupying this littoral fringe of insignificant area on a world basis may contribute quite markedly to this total. The transitional position of the littoral zone between land and sea has many other repercussions on the ecology of its characteristic organisms. Marine species, for example, may diminish interspecific competition by colonizing high intertidal levels; and the upper parts of marine shores may support luxuriant stands of semi-aquatic vegetation of basically terrestrial ancestry. Such salt-marshes, mangrove-swamps and, at slightly lower levels, sea-grass meadows (Chapter 4) also contribute significantly to essentially marine food webs.

Light more rarely penetrates to the bed of the somewhat deeper shelf seas (from 30 m to an average 130 m depth), and bottom-living, photosynthetic organisms are here much less significant. Nevertheless, shelf seas differ in many respects from open oceanic waters. As wind action can keep surface waters mixed down to depths of some 200 m, shelf seas are well mixed and the impoverishment which can result from the loss through gravity of nutrients incorporated into tissue in the surface waters, is much less marked than in deeper systems. Coastal waters also receive the discharge of ground waters and rivers draining 100×10^6 km^2 of land, and river water on average contains twice as much nutrient per unit volume as does sea-water (organic and inorganic materials dissolving in the water which percolates through rocks and the soil before being discharged by river and groundwater flow). For both these reasons (and in some areas also as a result of the upwelling considered below), shelf seas are particularly productive.

Many bottom-living, coastal animals have evolved larval stages which swim for a time in the water, in part perhaps to exploit the productive, nutrient-enriched waters (Section 9.3.2.2). Hence neritic waters may also be characterized by the abundance of these larvae. Once again, therefore, although shelf seas comprise only some 3% of the ocean's area, they contribute much more than their proportional share to its total productivity. They are also locally very important and extend out from the coast for distances of up to 1500 km: such seas as the North and Baltic, Yellow and East China, Chukchi and Bering, Hudson Bay, much of the South China, Java, Arafura and Timor and even the Arctic Ocean are shelf seas.

Most of the world's ocean is the open sea cradled between the continental masses (Table 1.1). From what has been said above it will be apparent that the surface waters of this region are relatively nutrient poor, stable and overlie the cold, dark ocean depths. A marked change in the sediments also occurs. On the continental shelf, the sediments are predominantly sands, silts and clays of terrestrial origin (material eroded from the land by wave action,

Table 1.1 Percentages of the world ocean comprised by various habitat types and depth intervals.

A. Habitat types	
Littoral zones	negligible
Continental shelf	3
Oceanic areas	97
Continental slopes	12
Continental rises	5
Mid-oceanic ridges, mountains, etc.	36
Abyssal plains	42
Ocean trenches	2

B. Depth intervals	
0–1000 m	12
1–2000 m	4
2–3000 m	7
3–4000 m	20
4–5000 m	33
5–6000 m	23
> 6000 m	2

discharged by rivers or deposited by glaciers), locally with high percentages of the pulverized remains of molluscs, corals, etc., and with exposed bedrock in areas of rapid water movement. The 40 000 km^3 per annum of river discharge into the world's coastal seas contains 13×10^6 t of continental sediments (together with $> 140 \times 10^6$ t of organic carbon in the form of humic acids, plant debris, etc.; this is some 15 times the amount of organic carbon removed annually from the ocean by man through exploitation of its living resources). Over much of the abyssal plain, however, calcareous oozes formed from the skeletal remains of minute protists suspended in the water mass (e.g. coccolithophores and foraminiferans) dominate down to depths of 4500 m. The solubility of the carbonate ion (CO_3^{2-}) varies with temperature and pressure, and below about 5000 m calcium carbonate goes into solution; hence below this depth there are no calcareous oozes. Other minute protists of the water column possess siliceous hardparts (e.g. diatoms and radiolarians), and oozes formed largely of their remains are locally common in high latitudes, especially around Antarctica, and are also an important deep-sea sediment in low latitudes between depths of 4000 and 6000 m. In the deepest areas of the ocean (below 6000 m) the dominant sediment is a fine, inert red clay (Fig. 1.4). The open ocean as a whole is very poorly known; only of the surface

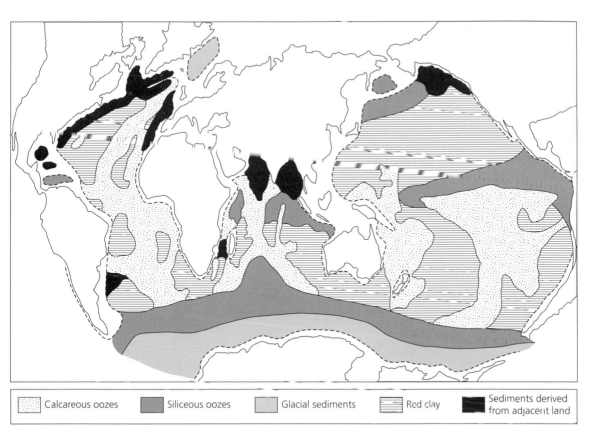

Calcareous oozes	Siliceous oozes	Glacial sediments	Red clay	Sediments derived from adjacent land

Fig. 1.4 Distribution of ocean-floor sediments. (After Wright 1977–78.)

photic zone do we have a reasonable biological understanding.

Thus far we have neglected the properties of the water itself and to these we now turn. The water forming the world ocean is, of course, salty. In fact it approximates a 3.5%—or 35‰—solution (by weight) in which the dominant anion is chloride (19‰ by weight, making up 86.8% of the total anions) and the dominant cation is sodium (10.5‰ by weight and 83.6% of the total cations). The concentration of salts in sea-water is remarkably constant; below 1000 m depth it varies only between 34.5‰ and 35.2‰, although clearly surface waters will be diluted in areas of freshwater discharge and concentrated in regions with a marked excess of evaporation over precipitation—to over 36‰ in the tropical open oceans and to even higher values in semi-enclosed areas such as the Red Sea.

With the exception of the nutrient salts required by photosynthetic organisms, however, the chemical composition of sea-water and variation in this composition appear to play a very small role in marine ecology. Indeed, it has been argued (Barnes & Mann 1980, 1991) that the basic form taken by the ecology of aquatic systems is almost completely unrelated to the nature of the aquatic medium and is essentially similar in fresh, brackish and marine waters. Nevertheless, the abundance of such nutrients as nitrate and phosphate is often of very great significance. Both are only minor constituents of sea-water, nitrate averaging 0.5 p.p.m. and phosphate an order of magnitude less; both may reach potentially limiting concentrations in surface waters. Their distribution, abundance and flux, and the consequences for marine productivity, will reoccur frequently in the following pages.

Although, nutrients excepted, details of the composition of sea-water are of little ecological importance (but of great interest in marine physiology), the movement of the water is critical and underlies many ecological processes and distributions, being particularly important in the circulation of nutrients and oxygen. Discounting tidally induced mixing, which generally is of significance only in very shallow waters, the two main categories of large-scale water movement are density-driven currents and the various processes responsible for upwelling (and its converse, downwelling).

Upwelling, as its name suggests, is the movement of water from relatively deep in the ocean into the photic zone, i.e. movements parallel to the depth gradient and perpendicular to the surface. Its importance is that it is one of the few mechanisms by which the nutrient stocks of the aphotic regions can be introduced into the surface waters. Three processes may induce upwelling. First, deep currents when they meet such an obstacle as a mid-ocean ridge will be deflected upwards and may gush forth into surface waters. Secondly, when two contiguous water masses are moved apart, as for example when water immediately to the north of the equator moves northwards under the influence of Coriolis force and similarly water to the south of the equator is moved southwards, a 'hole' is left between them and water upwells to fill it. The depth from which water upwells is then dependent on the quantities of surface water moved laterally and on their current velocity. This is upwelling due to areas of divergence. Thirdly, and generally most importantly, when water is driven away from a coastline by wind action, an equivalent conceptual 'hole' is left which is filled by upwelling. In areas in which the continental shelf is of little extent, the upwelled water must be sucked straight from relatively deep zones (Fig. 1.5). We will look at the distribution of centres of upwelling below (Section 1.4). Conversely, of course, areas of downwelling will occur in regions of convergence of water masses and when water is driven onshore: in each case, the water has nowhere to go but down, and thereby surface waters are entrained into the depths.

The main currents in surface oceanic waters are driven by the prevailing winds (Fig. 1.6) and move along paths determined by the topography of the ocean basin and by Coriolis force (Fig. 1.7). These play an important role in the distribution of surface-living organisms, but, as we noted earlier, wind action does not penetrate below some 200 m depth

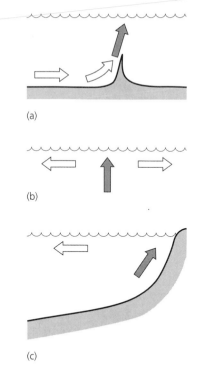

(a)

(b)

(c)

Fig. 1.5 Upwelling mechanisms consequent on: (a) an underwater ridge; (b) divergent surface currents; (c) the movement of water away from a coastline.

and these surface currents have little influence on the large-scale movement and mixing of sea-water. The major currents in the sea as a whole are caused by changes in the density of certain surface water masses, and the currents generated, because of the shape of the sea bed, then downwell and cause compensatory upwelling. The density of sea-water changes in relation to variation in temperature and/or salinity: cold sea-water is denser than warm, and full-strength sea-water is denser than that diluted by fresh water. Over the full ranges of temperature and salinity encountered in the oceans, salinity is a more potent factor in altering density than temperature: an increase in temperature from 7°C to 20°C decreases the density of sea-water by 0.002, a similar decrease to that achieved by diluting it from 36‰ to 33.5‰. In the oceans, changes in temperature and salinity usually occur together, and both can be illustrated by considering briefly the generation of the most important density-driven current in the ocean—that emanating from the south polar seas.

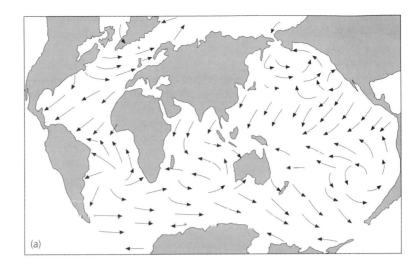

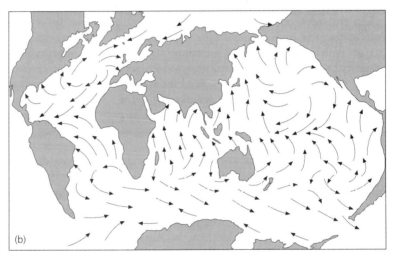

Fig. 1.6 The general distribution of winds across the world ocean in (a) January and (b) June.

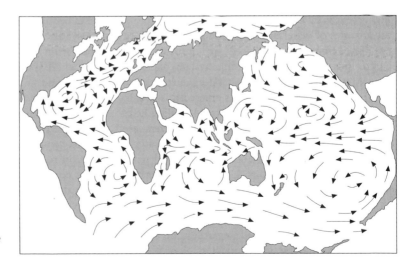

Fig. 1.7 The general pattern of surface currents in the world ocean.

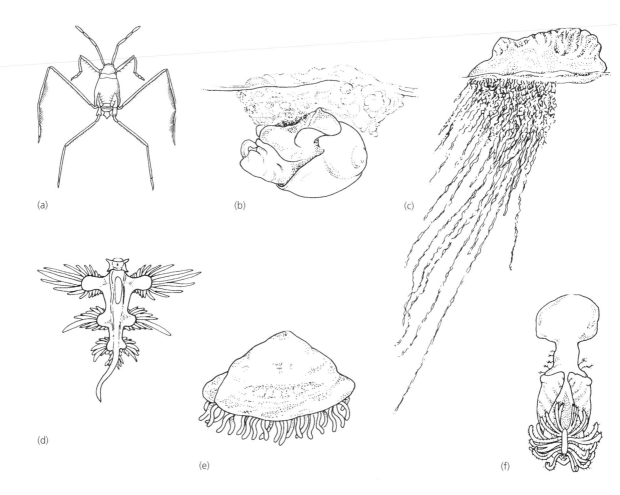

(a)

(b)

(c)

(d)

(e)

(f)

Fig. 1.10 Representative members of the pleuston:
(a) *Halobates*; (b) *Ianthina*; (c) *Physalia*; (d) *Glaucus*;
(e) *Velella*; (f) *Lepas fascicularis*.

barriers are present *par excellence* on land, but the ocean is continuous and, relatively, stable and uniform. These conclusions are reinforced by the observation that a small minority of marine species (about 2%) inhabit the water mass itself; most species are associated with the more diverse and more fragmented bottom sediments.

Degree of diversification into species is one measure of success but it is not the only one nor, indeed, one of the better. Abundance of a group (whether measured as number of individuals, biomass or productivity) and length of time for which a basic body plan has been able to persist are other measures. In these terms, many marine groups are highly successful.

Regardless of their phylogenetic position, marine organisms can be placed in two large categories dependent on whether they live in the water mass (pelagic) or on or in the bottom sediments or rock (benthic). A minor, third category is required for those organisms that straddle the air–water interface (pleustic). These categories, although extremely useful, are by no means mutually exclusive or rigidly definable. We have already seen (Section 1.2), for example, that some species are benthic as adults but pelagic as larvae, and a number of pelagic organisms may spend much time resting on or feeding at the sediment–water interface (these are often termed bentho-pelagic).

If organisms are positively buoyant they will float at the air–water interface like corks. In some animals, seabirds such as auks, petrels and shearwaters for example, flotation serves as a resting phase from

Table 1.2 Systematic list of the main groups of planktonic organisms (parasites excluded).

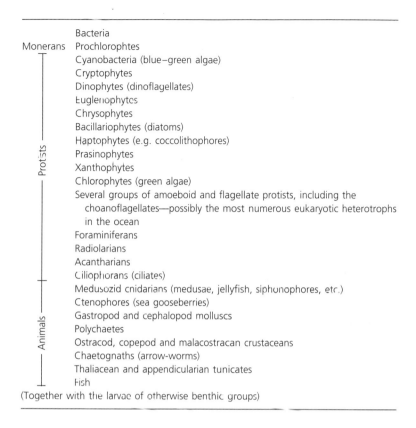

Monerans	Bacteria
	Prochlorophtes
	Cyanobacteria (blue–green algae)
	Cryptophytes
	Dinophytes (dinoflagellates)
	Euglenophytes
	Chrysophytes
	Bacillariophytes (diatoms)
Protists	Haptophytes (e.g. coccolithophores)
	Prasinophytes
	Xanthophytes
	Chlorophytes (green algae)
	Several groups of amoeboid and flagellate protists, including the choanoflagellates—possibly the most numerous eukaryotic heterotrophs in the ocean
	Foraminiferans
	Radiolarians
	Acantharians
	Ciliophorans (ciliates)
	Medusozid cnidarians (medusae, jellyfish, siphonophores, etc.)
	Ctenophores (sea gooseberries)
	Gastropod and cephalopod molluscs
Animals	Polychaetes
	Ostracod, copepod and malacostracan crustaceans
	Chaetognaths (arrow-worms)
	Thaliacean and appendicularian tunicates
	Fish

(Together with the larvae of otherwise benthic groups)

which to depart either into the water for food or into the air for longer distance dispersal. In others, their association with the boundary layer is permanent. This may be achieved by the possession of a float, as in some cnidarians (the Portuguese man of war, *Physalia*, and by-the-wind-sailor, *Velella*), crustaceans (the stalked barnacle *Lepas fascicularis*) and molluscs (the gastropods *Ianthina* which secretes a frothy raft and *Glaucus* which has bubbles of air in its gut), or by utilizing the surface tension of the interface. The latter property is exploited by the sea-skaters, e.g. *Halobates*, hemipteran insects which stride across the surface film in the same fashion as the more numerous pond-skaters of fresh-waters (Fig. 1.10). Intuitively one might think that an existence tied to this interface, being distributed at the mercy of surface current and wind direction, and without any mechanism for changing depth, etc., would have more than its fair share of difficulties, and indeed the number of pleustic species is very small—all are largely confined to the tropics. Most are carnivores (in the broadest sense of the word), either

passively dangling a food-catching apparatus down into the water or being dependent on other species of the pleuston.

The pelagic category is very much larger and it includes all those species of the water column which are distributed at the mercy of currents (the plankton), in that their powers of locomotion are insufficient to enable them to make headway against current action, and those, usually larger, species which can swim more powerfully (the nekton). Swimming ability is usually related to size and all gradations are found—from minute organisms without any means of propulsion to large animals capable of migrating from the Arctic to the Antarctic Ocean and back again at will. Like pelagic and benthic, therefore, plankton and nekton are mainly terms of convenience.

The plankton (Plate 1, facing p. 64) suspended in sea-water range in size and phylogenetic position from bacteria of < 0.1 μm in diameter to relatively large jellyfish > 0.5 m across the umbrella (Table 1.2 and Figs 1.11 and 1.12). As our knowledge of them is

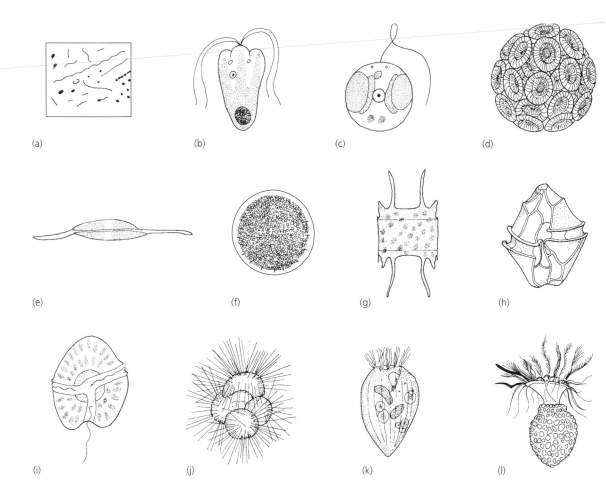

Fig. 1.11 Representative members of the ultra-, nano- and microplankton: (a) bacteria; (b and c) flagellates; (d) coccolithophore; (e–g) diatoms; (h and i) dinoflagellates; (j) foraminiferan; (k) ciliate, (l) tintinnid.

mainly derived from the use of nets and filters of standard pore size, it is usual to refer to different (logarithmic) size classes:

ultraplankton $\begin{cases} \text{femtoplankton, 0.02–< 0.2 μm;} \\ \text{picoplankton, 0.2–< 2 μm;} \end{cases}$
nanoplankton, 2–< 20 μm;
microplankton, 20–< 200 μm;
macroplankton, 200–< 2000 μm;
megaplankton, 2000 μm or more.

This supplements and in part corresponds to a systematic classification in that the ultraplankton are chiefly monerans, the nano- and microplankton are protists, and the macro- and megaplankton are animal. It is normal practice to refer collectively to the photosynthetic members of the pico- to microplankton as phytoplankton (and, together with their larger seaweed relatives of the littoral zone, as algae), and to the heterotrophic species of micro- to megaplankton as the zooplankton. The abundance of different size classes in one water mass in the northern Pacific Ocean is shown by way of example in Fig. 1.13.

The dilute suspension of life in the water that is the plankton—only some 0.1–1 mm³ of living organisms per litre even in the rich photic zone—displays a number of special features. It is evident that the phytoplankton, which, with the exception of the littoral algae and a few regions of significant chemosynthesis, are the sole source of the food energy that fuels the entire oceanic food web, are between 0.2 and 200 μm in size. Most are also *very* small: pico-, nano- and microphytoplankton occur in

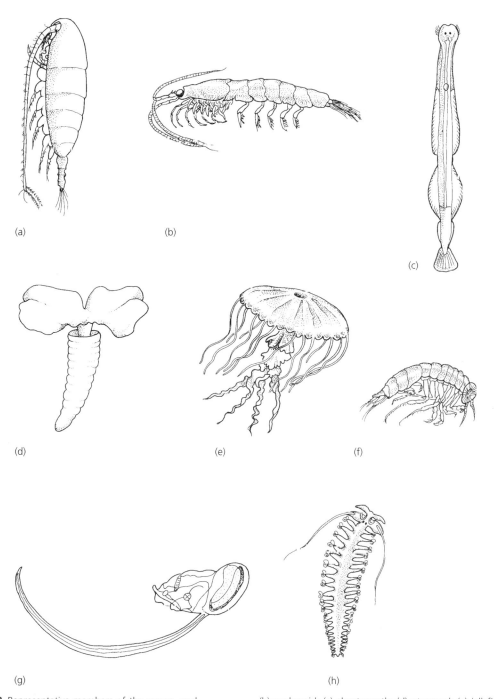

Fig. 1.12 Representative members of the macro- and megaplankton: (a) copepods (copepods make up nearly 80% of all individual planktonic animals caught by nets); (b) euphausid; (c) chaetognath; (d) pteropod; (e) jellyfish; (f) hyperiid amphipod; (g) appendicularian; (h) polychaete.

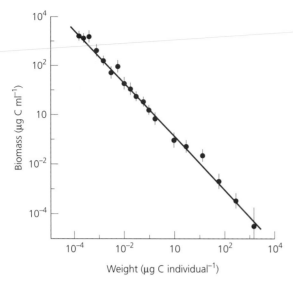

Fig. 1.13 The abundance of planktonic organisms in the surface waters of the North Pacific Gyre in relation to individual body size. (After Rodriguez & Mullin 1986.)

the general ratios of $10^{11-15} : < 10^4 : < 10$ cells ml^{-1}, with prochlorophytes, cyanobacteria and prasinophytes dominating the picoplanktonic fraction. On land the dominant photosynthetic organisms are many orders of magnitude larger, and are often larger than the herbivores which graze them. Why, then, are the marine autotrophs so small? There are several answers to the question. Plants are large on land in part because they need supporting tissue—woody material enabling the photosynthetic apparatus to be raised nearer the light source than their competitors. Tree-shaped plants in the sea would need, on average, to be 4000 m tall, and for 3800 of those metres would have to grow without external supplies of energy: this strategy is clearly a non-starter. But why have the protists not produced a large, floating, multicellular organism which could intercept more light than smaller, potentially competing species? One of the problems facing all species which need to remain in the photic zone is the gravitational attraction exerted on all living tissue which, because of the included salts and organic molecules, is denser than sea-water. Here, being small helps. A particle 2 µm in diameter will sink only 10 cm through distilled water at 20°C in a time of over 8 h; whereas, under the same conditions, a 62 µm diameter particle will sink 10 cm in less than 30 s. Weight is proportional

to volume, and frictional resistance to surface area, so small objects with a large surface area per unit volume will sink relatively slowly—small phytoplankton cells will effectively be suspended in the water and their rate of fall will be dwarfed by water movements resulting from wind-induced turbulence.

These turbulent water movements may carry many photosynthetic organisms out of the photic zone, however, and herbivores may consume many algae per unit time. Hence, to make good these and other losses and to maintain populations of their descendants in the photic zone, the phytoplankton must be able to multiply rapidly. Here again, small size is important; available energy must be devoted to multiplication rather than put into individual growth. The intrinsic rate of population growth, r_{max}, is in living organisms roughly related to weight (W g) by the expression, $r_{max} = 0.025 \, W^{-0.26}$. An organism 10 µm long has a generation time of 1 h; one 10 m long has a generation time of 10 years. A third and final consideration is nutrient uptake. Terrestrial plants take up nutrients through their root systems, which have a very large surface area in relation to the plant's volume. Planktonic algae cannot afford to increase their mass, so in order to have a large enough surface area to take in sufficient nutrients to support their mass/volume-dependent metabolism they must be small. It is even generally true that within the size range of the phytoplankton, the smallest species are characteristic of areas with a premium on efficient nutrient uptake and on multiplication to offset herbivore pressure.

Shape as well as size is relevant: phytoplankton are rarely spheres (see Section 9.2.2). Many have long spinous processes and others are completely spine shaped. This may help to slow the sinking rate even further, but more importantly it increases the effective size of an organism from the point of view of a potential consumer. A sphere of given volume is easier to handle and consume than a needle of equivalent volume.

The non-photosynthetic protists and animals of the plankton face exactly the same problems and have solved them in a similar fashion. Members of the macro- and megaplankton do possess locomotory systems more effective than flagella and cilia at enabling them to counteract gravity, but nevertheless they also possess adaptations lowering their specific gravity. These include gelatinous tissue, the presence of oil droplets and the replacement of heavy bivalent cations by light monovalent ones. They too may also bear long spines or setae increasing their effective body size.

One problem characteristically facing the consumers within the plankton is that of finding more food when that in their immediate vicinity has been exhausted. A terrestrial animal could merely walk or fly to another patch, i.e. change its location within the habitat, but because by definition plankton move with their environment and cannot overcome current action, this is not nearly so easily achieved. Many terrestrial animals, however, can move only in a horizontal plane, whereas the plankton have a large vertical component in their habitat and this makes all the difference. If one is travelling in a fast-moving ship or bus, one cannot effectively alter one's movement with respect to the Earth's surface by moving along the vehicle—its motion is much greater than one's own—but it is a simple matter to move in a vertical plane and climb from one deck or floor to another. In the sea, water masses at different depths move at different speeds and in different directions (see e.g. Fig. 1.8), and these different currents are analogous to the various decks of our hypothetical ship. By moving downwards, a planktonic organism will eventually enter a different current and on ascending back near the surface again a new patch of surface water will be encountered (Fig. 1.14). In effect, the only way for a marine organism with limited powers of locomotion to change its horizontal position in the photic zone is to move vertically: one migration downwards and then back up to the surface again (see Section 2.5).

Plankton occur throughout the oceans from the uppermost few centimetres of the water layer (where a characteristic assemblage of species—the neuston —are permanently associated with the undersurface of the air–water interface) down to the abyssal depths. Greatest development, in terms of abundance and productivity, not surprisingly is located in the photic zone and there the plankton forms a self-sufficient community in all but complete nutrient regeneration. Algae fix carbon photosynthetically and are consumed by herbivores, which in turn are taken by carnivorous zooplankton (both herbivores and carnivores releasing inorganic nutrients back into the water). The bacterioplankton subsist on soluble organic compounds released by the phytoplankton,

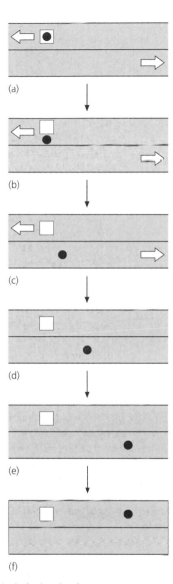

Fig. 1.14 Vertical migration between two water masses flowing in different directions and/or at different speeds can effect a change in the surface water mass inhabited by the migrating organism.

on the faeces of the zooplankton and on dead organisms of all types. Bacteria are engulfed by the protistan and some animal zooplankton, and these protists, other bacteria, and the detrital substrate are eaten by omnivorous and detritus-feeding members of zooplankton. In addition to these permanently planktonic organisms (the 'holoplankton'), other species acting as consumers occur in the form of the

Fig. 1.15 Representative members of the nekton: (a) squid; (b) shark; (c) deep-sea fish; (d) tunny; (e) flatfish; (f) turtle; (g) sea snake; (h) seal; (i) penguin; (j) whale.

larval stages (the 'meroplankton') of, for example, nektonic fish and, especially in neritic waters, of benthic invertebrates. Although self-sufficient, the plankton is not closed. Except in the littoral zone, it entirely supports the nekton and benthos as well.

The nekton (Fig. 1.15) are differentiated from the plankton only on the basis of swimming ability (Plate 2, facing p. 64). The large cephalopod molluscs (squid), crustaceans (e.g. prawns) and vertebrates (fish, turtles, sea snakes, seals, sea-cows, whales, etc.) making up this category of pelagic organisms are the terminal consumers of the sea, mostly being carnivorous although a few are herbivores and even fewer take detritus. Like most top predators, they need to range widely in search of food concentrations—and hence the swimming ability which characterizes them. Swimming is not only of use in moving relative to the Earth's surface; as anyone who has watched fish in a river will know, a fish also needs to swim against a current to stay in the same place. In this manner nektonic organisms may establish territories in such areas as coral reefs and remain there against the water flow. Active swimming requires the development of muscular systems and often of relatively large size. The nekton are therefore those pelagic species with the greatest energy requirement simply to counteract gravity. Some fish and cephalopods have evolved buoyancy devices to achieve neutral density but a concomitant of this is a lessened ability to change depth in the water column rapidly. Nekton which chase prey actively have generally dispensed with buoyancy aids and maintain themselves in the water solely through constant locomotion. Others have abandoned a perpetually pelagic life and rest on the bottom between active forays for food or even whilst feeding (the bentho-pelagic groups).

The problem is most intense for the deep-sea fish, which live in a habitat with low food availability. Sufficient energy to maintain an active swimming regime to achieve suspension is not available, yet they are pelagic feeders and must also maximize the chances of prey capture. The solution has been a bizarre suite of adaptations. Body tissue is largely reduced to a gelatinous consistency (so that a 1.5 m long fish may weigh only as many kilograms); this reduces both gravitational attraction and maintenance metabolic requirements. Prey is often lured towards the floating fish-trap by bioluminescent organs and, when in range, it is engulfed by a cavernous mouth, the predator retaining only sufficient muscular ability to make the short, sharp lunge required. As prey density is low, response to the lure is infrequent, and any 'bite' must be seized almost regardless of size. Jaws can be dislocated and stomachs distended to accommodate prey items almost twice the length of the predator.

In spite of the efficiency of food uptake engendered by food shortage, some of the potential food available in the water column escapes utilization by pelagic consumers and settles on the sea bed, largely in the form of faecal pellets and other items low in nutritive value. This then forms the basic food resource of the benthos, and in general it will be the more abundant the smaller the distance between sea bed and photic zone (although anomalies may occur where deep water lies immediately adjacent to land and can receive an input of river-borne debris, etc.).

Except in the littoral zone, the benthos is lacking in photosynthetic organisms. Chemosynthetic bacteria (nitrifying, sulphur, hydrogen, methane, carbon monoxide, and iron bacteria) occur in benthic sediments, however, and can there fix carbon dioxide by oxidizing ammonia, hydrogen sulphide, methane, etc., in some cases using dissolved oxygen but in others by utilizing that bound in, for example, sulphates. In many cases, the inorganic substrates used are produced by the decomposition of organic matter produced earlier and elsewhere by photosynthesis, and organic compounds themselves may be used as carbon sources: in these cases, chemosynthesis is only regenerating organic matter. But some bacteria can produce organic materials quite independently of photosynthesis. Reduced sulphur compounds issuing from submarine volcanic vents—fumaroles—are used as energy sources by some sulphur bacteria in their fixation of dissolved carbon dioxide. Such chemosynthetic bacteria are also symbiotic in various gutless nematodes, oligochaetes, pogonophorans and bivalve molluscs associated with sulphur-rich marine microhabitats, and provide their only known source of food materials (some 50% of the carbon fixed chemosynthetically passing to the host). Equivalent chemosynthesis-dependent communities occur around seeps of unheated water and of petroleum.

As for the plankton, benthic organisms are isolated for study using pores of known size (in sieves and filters) and hence a size-based series of categories is also used: microbenthos ($< 100\ \mu m$); meiobenthos ($100–500\ \mu m$); and macrobenthos ($> 500\ \mu m$). The microbenthos comprises bacteria and protists; the meiobenthos, animals together with a few large protists (particularly foraminiferans); and the macrobenthos category is, except in the shallowest waters, entirely animal. And again as for the plankton, it is customary to differentiate between the photosynthetic and animal members of the benthos by terming them, respectively, the phytobenthos and zoobenthos (see Plates 3 and 4, facing p. 64) (Figs 1.16 and 1.17).

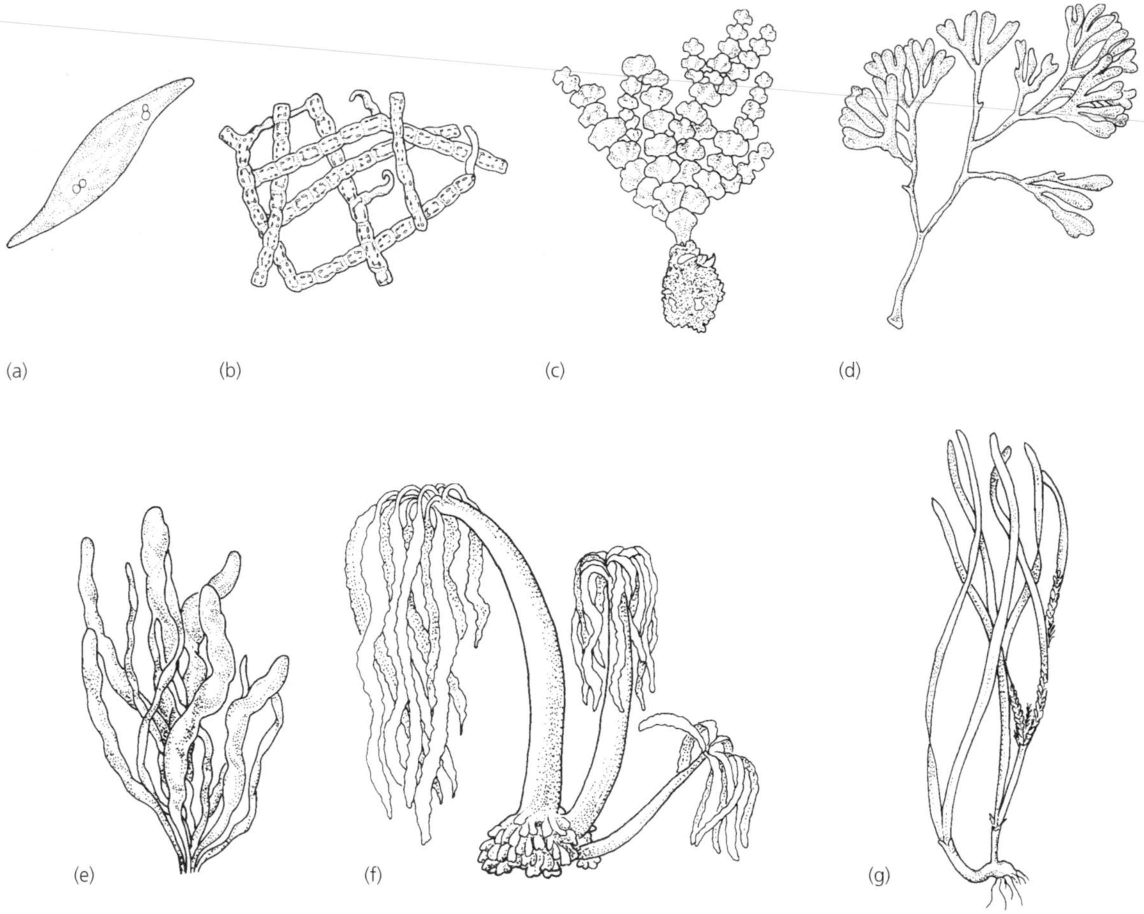

(a)   (b)   (c)   (d)

(e)   (f)   (g)

Fig. 1.16 Representative members of the phytobenthos:
(a) benthic diatom; (b) mat-forming alga; (c) coralline alga;
(d) wrack; (e) *Enteromorpha*; (f) the kelp *Postelsia*;
(g) sea-grass.

Figure 1.18 illustrates the relative abundance of the different size categories. The benthos includes a much more diverse collection of organisms than the plankton, especially of animals (Table 1.3) of which all phyla with marine examples are represented. A distinction between 'epifauna' and 'infauna' is almost universal, the epifauna living on the sediment–water interface rather as *Halobates* lives on the air–water interface, and the infauna living within the sediment itself. In contrast with the pleuston, however, several animals may span the interface by extending a feeding system into the overlying water whilst the rest of the body is hidden within the substratum.

As one might expect, on hard substrata such as exposed bedrock most organisms are epifaunal, whereas most animals are infaunal in soft sediments; but in the deeper areas of the abyssal plain and in the ocean trenches, many species in groups which are characteristically infaunal are to be found on the surface of the ooze. This change of living station is probably to be associated with decrease in predation intensity. Several researchers (e.g. Clark 1964; Mangum 1976) have advanced the thesis that the earliest meio- and macrobenthos were epifaunal and remained thus until the advent of large predatory species which could roam over the sediment surface. These predators exerted a strong selection pressure in favour of a life spent in hiding beneath the surface. In coarse sediments, meiobenthos can move between the grains without displacing them, but macrobenthos must burrow. Clark suggested

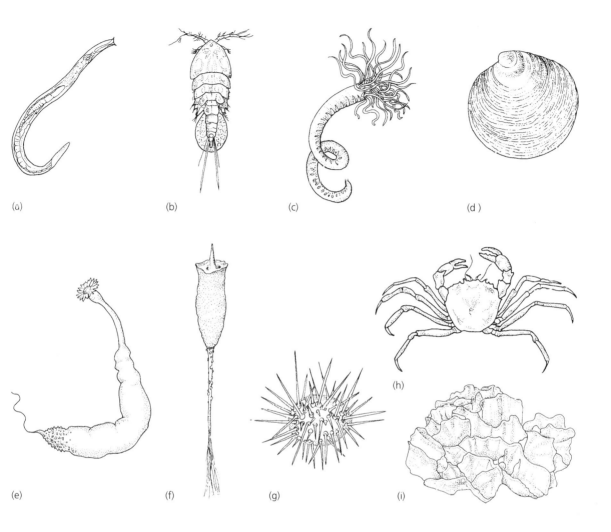

Fig. 1.17 Representative members of the zoobenthos:
(a) nematode; (b) harpacticoid; (c) polychaete; (d) bivalve;
(e) sipunculan; (f) sponge; (g) sea-urchin; (h) crab;
(i) bryozoan

that fluid-filled body cavities first evolved in this connection—as hydrostatic skeletons permitting burrowing—and Mangum argued that burrowing would have resulted in problems in obtaining oxygen and that the evolution of respiratory pigments by the annelids went hand-in-hand with the invasion of the sediments. As the density of potential prey falls in the deep sea (Section 7.2), so life as a predator becomes progressively more difficult and this mode of nutrition is poorly represented below 6000 m (members of otherwise largely predatory

groups, e.g. the starfish, turning to ooze-eating instead). The absence or scarcity of predators has then permitted animals to return to the ancient epi-faunal existence.

Over the whole of the continental shelf and abyssal plain, two feeding types predominate: suspension feeding on particles carried by or settling out of the water; and deposit feeding on material on or in the sediment (see Lopez & Levinton 1987; Jørgensen 1990; Ward *et al.* 1993). In both cases, the ultimate food organisms are probably bacteria subsisting on the rain of refractory debris from the pelagic system. The bacteria may be captured and consumed directly or only after passage along a bacteria → protist → meiofauna chain, but below about 30 m (or whatever the depth of the photic zone may be)

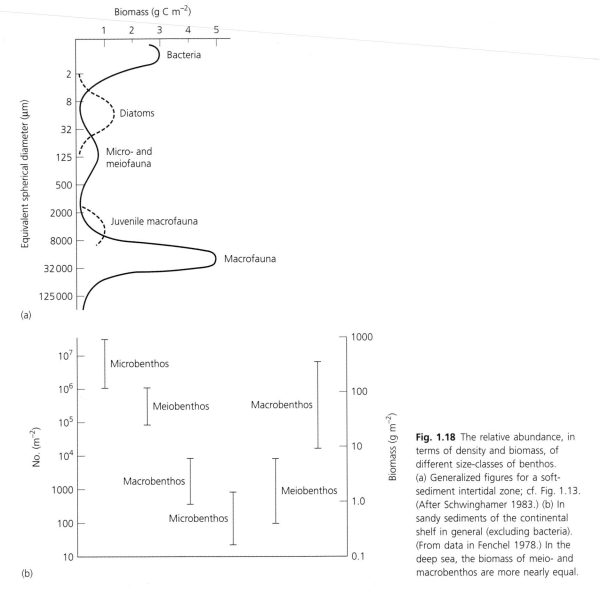

Fig. 1.18 The relative abundance, in terms of density and biomass, of different size-classes of benthos. (a) Generalized figures for a soft-sediment intertidal zone; cf. Fig. 1.13. (After Schwinghamer 1983.) (b) In sandy sediments of the continental shelf in general (excluding bacteria). (From data in Fenchel 1978.) In the deep sea, the biomass of meio- and macrobenthos are more nearly equal.

the benthos will be dependent on bacterial activity. Deposit feeding is viable on all types of sediment, although large quantities may have to be sorted and/or consumed, but suspension feeding will be successful only if there are sufficient particles of potential food suspended in the water. Therefore, suspension feeders are most characteristic of shallow regions and of areas of relatively rapid water movement into which an animal may extend a passive filter and at intervals wipe off particles retained. In practice, this means that they are typical of harder substrata than those occupied by deposit feeders (Fig. 1.19).

The benthos of the littoral zone has been reserved for comment separate from that of the shelf and abyssal plain, because it is the zone to which benthic photosynthesis is confined and in which the latter may exceed that achieved in the water column. Three categories of photosynthetic organisms occur: (1) microalgae essentially similar to those of phytoplankton but here associated with soft sediments,

Table 1.3 Systematic list of the main groups of benthic organisms (parasites excluded).

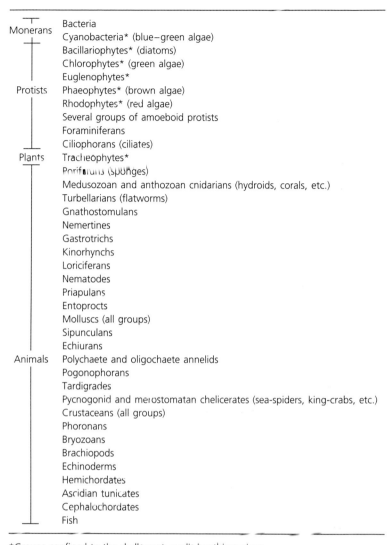

Monerans	Bacteria
	Cyanobacteria* (blue–green algae)
	Bacillariophytes* (diatoms)
	Chlorophytes* (green algae)
	Euglenophytes*
Protists	Phaeophytes* (brown algae)
	Rhodophytes* (red algae)
	Several groups of amoeboid protists
	Foraminiferans
	Ciliophorans (ciliates)
Plants	Tracheophytes*
	Poriferans (sponges)
	Medusozoan and anthozoan cnidarians (hydroids, corals, etc.)
	Turbellarians (flatworms)
	Gnathostomulans
	Nemertines
	Gastrotrichs
	Kinorhynchs
	Loriciferans
	Nematodes
	Priapulans
	Entoprocts
	Molluscs (all groups)
	Sipunculans
	Echiurans
Animals	Polychaete and oligochaete annelids
	Pogonophorans
	Tardigrades
	Pycnogonid and merostomatan chelicerates (sea-spiders, king-crabs, etc.)
	Crustaceans (all groups)
	Phoronans
	Bryozoans
	Brachiopods
	Echinoderms
	Hemichordates
	Ascidian tunicates
	Cephalochordates
	Fish

*Groups confined to the shallowest, sunlit benthic regions.

symbiotic within littoral animals such as reef corals, and occurring in various other microhabitats; (2) larger, multicellular and macroscopic algae in a variety of forms but especially as the large, leathery seaweeds of rocky outcrops and the finer, more filamentous species growing on the surfaces of coarser seaweeds or on rock; (3) the sole examples of marine tracheophyte communities, the sea-grasses, salt-marsh herbs, and mangrove-swamp shrubs and trees. The mat-forming or interstitial protists (and moneran cyanobacteria = blue–green algae), the fine, epiphytic seaweeds and the sporeling stages of the coarse wracks and kelps provide a direct source of food for many herbivores, especially gastropod molluscs, whereas the larger seaweeds and the tracheophytes on their decay form abundant detritus ultimately consumed by suspension and deposit feeders.

The constraints on size of individual primary producers do not apply in the littoral fringe of the seas and, for example, giant kelps such as *Macrocystis* can achieve total lengths in excess of 50 m, can grow at a rate of 50 cm per day, and can lead to extensive areas of coast being visibly dominated by photosynthetic organisms. The kelps and other large

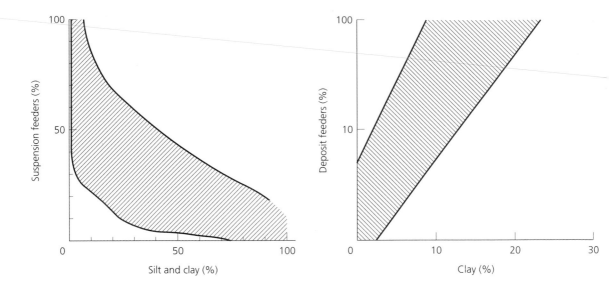

Fig. 1.19 The relative importance of suspension and deposit feeders in substrata of differing degrees of fineness. (After Menzies *et al.* 1973. Reprinted by permission of John Wiley & Sons, Inc.)

seaweeds although growing from the substratum are not rooted in it; their holdfast system serves only as an anchorage, not a nutritive function. They, and the tracheophytes, are consumed by relatively few species, most notably by some sea-urchins and opisthobranch sea-hares. Thorson (1971) has given a graphic description of the consumption of the giant kelp *Nereocystis* by a Californian sea-hare, *Aplysia*: 'If one gives them a stem . . . the thickness of a broomstick one can hear their jaws crunch away at the crisp algal tissue, rather like a child chewing a raw carrot; half a metre of stem is probably regarded as a reasonable meal every 24 hours.' Kelps also release copious dissolved organic substances.

The pelagic and benthic regions of the sea are most closely and most complexly interlinked in the littoral zone, and here all the rich plankton of neritic seas can be filtered from the water by benthic suspension feeders. It is therefore in the littoral zone that the benthos reaches its apogee in systems such as coral reefs, sheltered estuaries and lagoons, and rocky intertidal regions. Suspension feeders dominate. They display two features rarely seen in terrestrial animals. First, they are sedentary or completely sessile. To find food, there must be relative motion

between an animal and its food-bearing environment. For suspension feeders the food is contained within the water column and this may move, through tide or current, past the stationary animal, or if natural water movement will not suffice the suspension feeder can set up its own current by the beating of cilia, setae or other appendages. By living in a tube within the sediment and creating a current of water through that tube, an animal may derive the benefits of an infaunal existence whilst retaining respiratory and feeding access to water. Sessile suspension feeders are amongst the most inanimate of animals, at least as adults: the skeleton in animals is a structure facilitating locomotion, but in sponges it serves the converse function—it prevents the sponge from changing shape whilst the internal water current moves relative to the cells.

Secondly, suspension feeders or more generally small-particle feeders are often colonial in the sense that different individual modular units, zooids or polyps, together form a much larger group-structure of characteristic shape. The colony is produced asexually, the founding polyp budding off new clonal polyps which themselves bud, and so on, individual polyps often retaining tissue contacts with those nearest them. This is an analogous process to the creation of a metazoan body by the repeated asexual division of a founding zygote. As in the metazoan body, division of labour amongst the asexual division products may occur. This reaches its height in bryozoan and hydrozoan colonies (especially the pelagic siphonophores), in which feeding, reproduction, defence and, in the siphonophores, movement and flotation are carried out by different modular units or aggregations of such units. Such colonies have a considerable claim to be considered as individual organisms in which there are two hierarchies of organizational levels: polyps formed of differentiated cells; and the colony/individual formed of a differentiated series of polyps (see Hughes 1989).

1.4 Latitudinal gradients and global patterns

The third of the three great gradients affecting marine organisms was omitted from consideration in Section 1.2 and can conveniently be discussed now. Solar energy is not distributed evenly across the surface of the globe. Regions within the tropics experience per day 12 h of daylight and 12 h of darkness, the light being intense and the heat input being large. In higher, temperate latitudes, summer daylight exceeds a duration of 12 h per day and winter daylight periods are less than 12 h, the light received is of lower intensity and the overall heat input is lower. Finally, in polar regions, a 6-month period of continuous, low-level daylight alternates with 6 months of darkness, temperatures being low throughout the year. This global pattern of light and heat receipt is of profound importance to the photosynthetic potential of the photic zone, both via its direct effects on photosynthesis and via the establishment of thermoclines.

If the light energy received by surface waters is sufficient for the phytoplankton to fix more carbon per day in photosynthesis than they need to respire in that period, then the heat received at the same time will be of the order of magnitude required to generate a thermocline. Thus, in open tropical waters, light intensities will permit phytoplanktonic production to last throughout the year but a deep and permanent thermocline will also persist throughout all months. Light energy input may be favourable but nutrient availability will be unfavourable, and a low continuous level of production will result, totalling only some 20–150 mg C fixed m^{-2} of surface daily.

Moving further from the equator into temperate zones, seasonality becomes evident. During the winter months, the light intensity is so low that the photic zone is extremely shallow and well above the depth to which wind-induced mixing occurs. Therefore, although in the absence of a marked thermocline nutrients are plentiful, there is little or no production during the winter. At the start of spring, nutrients are still abundant in surface waters and as soon as the relationship between the depths of mixing and the photic zone becomes favourable, an outburst of production can take place. As this is happening, however, so is a seasonal thermocline being established and by sometime in summer the trapped stocks of nutrients will have been diminished and/or grazing pressure built up with the result that the outburst is ended. At the end of the summer the seasonal thermocline breaks up, allowing the ingress of more nutrients, and if the light and mixing regimes remain favourable for a time before the onset of winter conditions, a second autumnal outburst of production may take place. This pattern of phytoplanktonic production need not, however, be reflected by a similar pattern of algal biomass (see Section 2.2.4).

In even higher latitudes, nutrients are always abundant in surface waters but only during 5 months of the year is sufficient light available, and for the first and last of these months the relative depth of mixing is unfavourably large. Therefore the period of photosynthetic production is limited to only some 3 months of the year, but when for a brief time light, nutrients and surface mixing are all favourable an outburst of massive proportions can occur. The magnitude of the summer period of production in high latitudes is sufficiently large to yield a higher daily average productivity (150–300 mg C m^{-2} daily) than in tropical open waters, in spite of the occurrence of unproductive months.

This analysis has been based solely on the distribution of solar energy to the open waters of the ocean, but the proximity of coasts and of areas of upwelling is also relevant, and they are not distributed evenly across the globe either. Coastal waters, for example, occupy an increasing percentage of the sea as one progresses northwards from the equator, and we have already seen that nutrients are more plentiful over continental shelves. These enable neritic waters to support productivities in the range 125–1000 mg C m^{-2} daily, with an average value of the order of 500 mg C m^{-2} daily, or some three times the productivity of the adjacent areas of oceanic

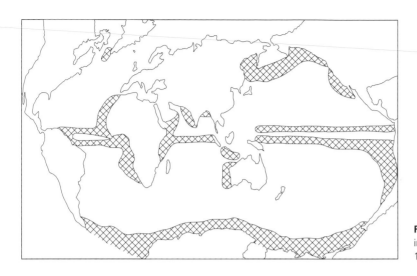

Fig. 1.20 The main areas of upwelling in the world ocean. (After Wright 1977–78.)

water. Tropical neritic waters receiving more light are, of course, more productive than those in higher latitudes (all things being equal).

The second factor impinging is upwelling, and in Section 1.2 it was indicated that upwelling is particularly marked where water is driven away from a coastline and where surface water masses diverge. The force moving water away from a coast is wind, and the globe is subject to two broad belts of persistent winds which cause upwelling: the northeast trades in the Northern Hemisphere and the southeast trades in the Southern (Fig. 1.6). These have the effect of driving water away from the western coasts of continents in their path (and onto eastern coasts). Water therefore upwells along the western coast of northwest and southwest Africa, the western coast of the USA and of central South America, and the coast of Western Australia; whereas the major zone of upwelling as a consequence of divergence is along the equator (see Section 1.2). In addition, we have seen that upwelling is associated with the generation of density-driven current systems in high latitudes, so that water upwells around Antarctica and, to a lesser extent, in the Arctic. Thus the distribution of upwelling in the world ocean is as shown in Fig. 1.20. The abundance of nutrients in these systems is sufficient to permit an average daily productivity in the range 500–1250 mg C m^{-2}, with an overall average of some 625 mg C m^{-2} daily. It should be noted, however, that upwelling is rarely a continuous phenomenon; it may be seasonal in occurrence or only

sporadic, and the magnitude of primary production varies accordingly. Failure of upwelling to occur is often referred to as 'El Niño'. Commonly, it results from an oscillation in the atmospheric processes that bring about the trade winds: easterly trades become replaced temporarily by westerly equatorial ones (Fig. 1.21). This has dramatic effects, not only on productivity patterns, but on reef-coral mortality and even on the frequency of occurrence of icebergs off eastern Canada (Glynn 1988; Philander 1990).

Combining the latitudinal gradient, coastal effects and zones of upwelling, a global distribution of pelagic primary production of distinctive pattern results (Fig. 1.22). A map of the world is on too small a scale to show the influence of the littoral zone on global productivity, but for the purposes of comparison with the figures given above, the productivity of benthic algae and plants lies in the range 500–25 000 mg C m^{-2} daily (more usually within 2500–5500 mg C m^{-2} daily): benthic production per unit surface area exceeds pelagic production by a factor of ten. Although they can only inhabit less than 0.5% of the surface area of the oceans, by virtue of their large size, coastal macrophytes also account for two-thirds of the total biomass of marine photosynthetic organisms. The existence of productive coral atolls in the middle of the barren tropical oceans is a paradox which will receive special consideration in Chapter 6.

It is not only overall productivity that varies with latitude, so do the nature of the organisms

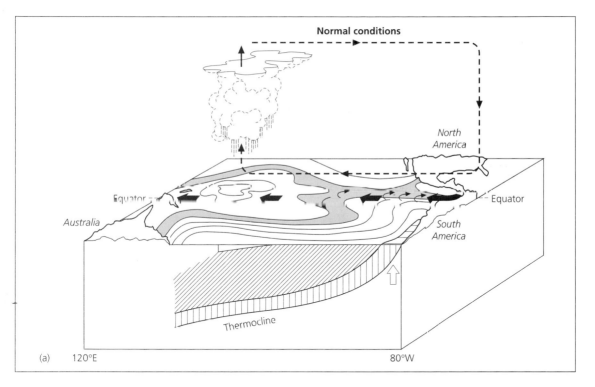

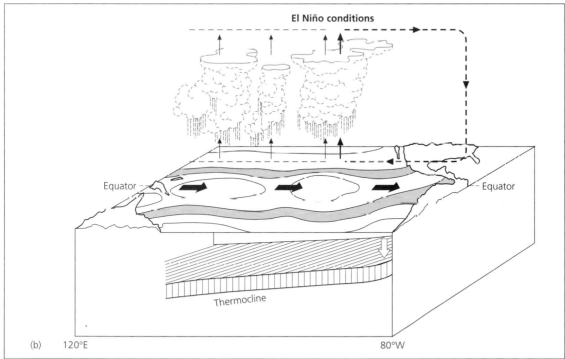

Fig. 1.21 El Niño. (a) Normal atmospheric and hydrographic conditions under the influence of the trade winds that result in upwelling of water along the western margin of South America. (b) El Niño conditions in which higher atmospheric pressure in the western Pacific Ocean causes offshore winds to be weaker. Upwelling ceases and warm water extends to the South American coast. (After Cromie 1988.) El Niño occurs once every 5 years on average and particularly severe cases occur once every 20 years on average (see Enfield 1989).

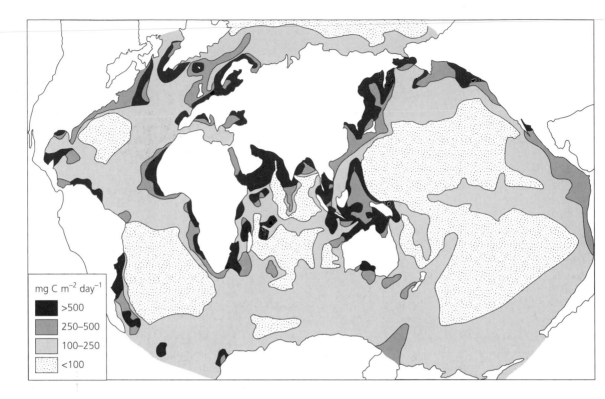

Fig. 1.22 The distribution of phytoplanktonic primary productivity in the world ocean. (After Koblentz-Mishke *et al.* 1970.)

achieving this production and the characteristics of the assemblages of species supported. Amongst the pelagic algae, for example, diatoms make up a significant fraction of the total only in high latitudes, along the equator and off western continental margins, i.e. in zones of abundant nutrients. Conversely, dinoflagellates, coccolithophores, cyanobacterial blue–green algae and several other pico- and nanoplanktonic groups are most important in tropical, open-ocean waters. The pico- and nanoplankton form up to 90% of the biomass and of the production of tropical oceanic phytoplankton; in some areas, picoplanktonic cyanobacteria alone may be responsible for 90% of the total productivity. Other distribution patterns can be correlated directly with nutrient levels—that of the larger benthic kelps centres on coastal regions of upwelling—but not all latitudinal patterns are so easily explained. Whole habitat types may show a tropical vs. high-latitude division: coral reefs and mangrove-swamps are tropical whereas salt-marshes replace mangrove-swamps in temperate and boreal zones. The same phenomenon is seen in individual systematic groups. The ocypodid and grapsid crabs which dominate many intertidal habitats in the tropics are absent from high latitudes, for example, even though apparently comparable habitat-types are still available. Attempts have been made to correlate such distribution patterns with a single environmental variable, temperature in the case of reef corals and rainfall and/or frost for the mangroves, but the detailed patterns still remain largely unexplained.

Although not an explanation in itself, Thorson (1957) showed many years ago that if one compares similar coastal regions in different latitudes, the number of infaunal species present per unit area is almost constant, but the species richness of the epifauna increases with decrease in latitude (at least in respect of the Northern Hemisphere). The observed pattern of species diversity in the coastal macrobenthos (Fig. 1.23) is largely, if not entirely, attributable to the increase in the epifaunal component alone. Coral reefs and mangrove-swamps are two of the spatially complex tropical systems which (other things

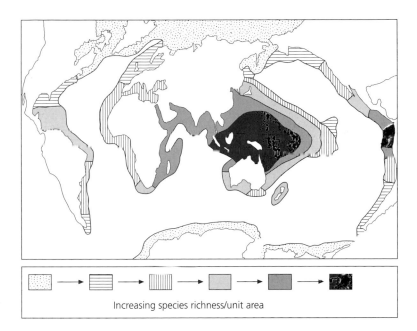

Fig. 1.23 The relative species richness of different shallow-water regions of the world ocean. (After Valentine & Moores 1974.)

Increasing species richness/unit area

being equal) have permitted large numbers of epifaunal species to adapt to small microhabitats and to coexist, but this does not account for the absence of reefs and mangrove-swamps from more temperate conditions. Both are, however, mainly epifaunal phenomena and it is probably no accident that the infauna which appear insensitive to these latitudinal effects are the component least exposed to the extremes and fluctuations of the shelf–littoral environment. The infauna are cushioned by living within the sediments and the less labile interstitial habitat. This story will be pursued further in Chapter 3 (and see Ormond et al. 1998).

This chapter has set out to introduce the sea and its organisms, and to provide a very broad, overall picture of marine primary production. In the following seven chapters we will concentrate on the more detailed patterns and processes characterizing the major subdivisions of the marine environment outlined above. The world ocean, however, forms one large ecosystem in that all its component habitats interlink to form a three-compartment system —pelagic, benthic, fringing—typical of all aquatic environments (Barnes & Mann 1991). Chapter 11 will examine the nature of these interactions and their magnitudes.

It was emphasized repeatedly in Chapter 1 that several processes are confined to the surface waters of the seas, although the depth down to which each penetrates varies with the nature of the process. In this chapter we will consider the water column above the arbitrary depth of 1000 m so as to include all potential surface effects; deeper waters will be covered in Chapter 7.

2.1 The nature of pelagic photosynthesis

Light energy in the sea is used by organisms to synthesize complex organic molecules in three rather different processes, of which two are confined to special habitat types. Where simple, dissolved organic compounds (e.g. glucose) are already abundant in the water—as a result of previous photosynthetic activity—these can be converted into more complex substances by the process of 'photo-assimilation'. In this, glucose, acetate, etc. replace carbon dioxide as the carbon source, but otherwise the process is similar to the photosynthetic pathways described below. True photosynthesis involves the introduction of hydrogen into the carbon dioxide molecule to form compounds with the empirical formula of $n(CH_2O)$, i.e. carbon dioxide is reduced or hydrogenated. In regions containing little oxygen, several bacteria, and in some circumstances various protists, use compounds such as hydrogen sulphide as the hydrogen donor and photosynthesize according to the equation

$$CO_2 + 2H_2X \rightarrow CH_2O + 2X + H_2O \qquad (2.1)$$

where H_2X is any reduced molecule other than water. The most frequent process in the oceans as a whole, however, is the photosynthetic fixation of carbon by the phytoplanktonic prochlorophytes, cyanobacteria and protists using water as the hydrogen donor, in which the X of eqn (2.1) is replaced by O:

$$CO_2 + 2H_2O \rightarrow CH_2O + O_2 + H_2O. \qquad (2.2)$$

This process liberates oxygen, which is not the case in bacterial photosynthesis.

Algal photosynthesis is therefore basically the same process as that in terrestrial plants and botanical texts may be consulted for the biochemical details (Parsons *et al.* 1984). (It should be noted, however, that eqns (2.1) and (2.2) are empirical simplifications of what, in reality, are a complex series of biochemical reactions, only one of which, that dissociating the hydrogen donor to release hydrogen ions and free electrons, requires light energy.) The light of 400–720 nm wavelength which powers the essential first stage of photosynthesis is absorbed by various photosynthetic pigments which are responsible for the characteristic colours of the phytoplankton. In the dominant prochlorophytes, cyanobacteria and protists (though not in the photosynthetic bacteria), the most important of these pigments is chlorophyll *a*, which exhibits peak absorption of light of 670–695 nm wavelength. However, the phytoplankton possess a host of other accessory pigments (Table 2.1) which absorb light of shorter wavelengths, in part using it directly to dissociate water molecules and in part passing the light energy to chlorophyll *a*.

The wide range of light-absorbing pigments present in planktonic (and benthic) algae allows a broad band of wavelengths to be used in photosynthesis, and it has also permitted different protist groups to specialize in the use of different wavelengths and therefore to trap light not absorbed by other species. In particular, phytoplankton species inhabiting relatively deep sections of the photic zone are adapted to the prevailing wavelengths by the possession of the most appropriate pigments. The photosynthetic abilities of a given species can be encapsulated in an 'action spectrum' which combines information on the wavelengths which can be absorbed and on the

Table 2.1 Photosynthetic pigments of marine phytoplankton.

	Prochlorophyta	Cyanobacteria	Chlorophyta	Xanthophyta	Chrysophyta	Bacillariophyta	Cryptophyta	Dinophyta	Euglenophyta	Prasinophyta	Haptophyta
Chlorophyll *a*	+	+	+	+	+	+	+	+	+	+	+
Chlorophyll *b*	+		+						+	+	
Chlorophyll *c*	+			?	+	+	+	+			+
α carotene		+						+		+	
β carotene	+	+	+	+	+			+	+	+	+
γ carotene			+							+	
ε carotene							+	+			
Fucoxanthin				+	+	+		+			+
Neofucoxanthin					+	+					+
Diadinoxanthin					+	+		+			+
Diatoxanthin					+	+					+
Dinoxanthin								+			
Peridinin								+			
Neoperidinin								+			
Lutein		+	+		+					+	
Zeaxanthin		+								+	
Flavoxanthin		+									
Violaxanthin		+		+	+					+	
Neoxanthin		+		+					+		
Alloxanthin							+				
Monodoxanthin							+				
Crocoxanthin							+				
Siphonoxanthin										+	
Myxoxanthophyll		+									
Myxoxanthin		+									
Anthraxanthin		?									
Astaxanthin				+					+		
Oscilloxanthin		+									
Echinenone		+							+		
Phycocyanins		+					+				
Phycoerythrins		+					+				
Others			+	+				+	+	many	+

amount of carbon fixed by unit quantity of the light captured by each pigment (the efficiency of energy transfer to photosynthesis varies with the nature and functional role of the various pigments). Representative action spectra are shown in Fig. 2.1.

This photosynthetic fixation is responsible for the primary generation of organic compounds in the sea. Carbohydrates, fats and proteins are all synthesized and the total quantity of carbon or energy fixed forms the 'gross primary production'. The latter suffers three immediate fates. Some of the gross production will be broken down by the respiratory metabolism of the photosynthetic organism itself. Some will be incorporated into its tissues and fluids, and thereby constitute growth. Some will leak from the alga into the surrounding water and there join a pool of dissolved organic substances. 'Net primary production' is the term usually given to that proportion incorporated into the protist's body and therefore available to herbivores (mainly because of the importance which historically has been attached to the primary producer → herbivore food chain). However, net production should also include the loss of organic substances to the water as they are also utilized by consuming species, especially by bacteria in the microbial loop (see Section 2.4) but also by

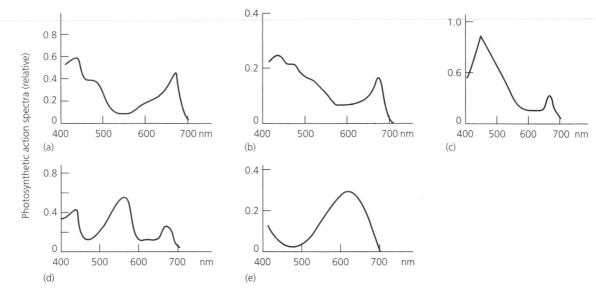

Fig. 2.1 Photosynthetic action spectra of some marine phytoplankton: (a) *Chlorella*; (b) *Gonyaulax*; (c) *Coccolithus*; (d) *Phormidium*; (e) *Cyanidium*. (After Parsons *et al.* 1977.)

some zooplankton. All phytoplankton are leaky—a phenomenon utilized, and magnified, in various symbiotic relationships (see Chapter 6)—and in nutrient-poor waters the loss may be equivalent to more than 40% of the gross primary production. Release of glycolic acid may result in concentrations of this substance of 0.1 mg l^{-1} in sea-water.

2.2 Factors limiting primary production

Chapter 1 identified four factors as impinging on the magnitude of primary production achieved in the sea—light, mixing, nutrient availability and grazing—and we will now investigate these in some detail.

2.2.1 Light

The relationship of photosynthetic production to the rate of supply of light energy ('irradiance' or 'intensity' as measured, for example, in watts per unit area of langleys per unit time) takes the general form shown in Fig. 2.2. With increase in light intensity from zero there is first a linear phase of increasing photosynthesis, then a plateau is attained, and finally photosynthesis decreases at high intensities. During the linear phase, light is limiting and production is proportional to light intensity; the photosynthetic mechanism is saturated along the plateau; and harmful effects of too intense a supply of radiation set in at the start of the decline and increase thereafter. These harmful effects are both real and apparent. The real ones are due to the destructive action of ultraviolet radiation and to an overflow of light energy into oxidation processes (photo-oxidation); in bright sunlight these account for most of the decline. To

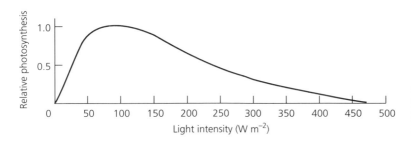

Fig. 2.2 The relationship between light intensity and photosynthetic production in planktonic diatoms (mean of five species). (After Ryther 1956.)

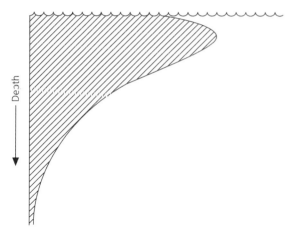

Fig. 2.3 The relationship between depth and photosynthetic production in the surface waters of the ocean.

tailed by self-shading if dense congregations of algae are present near the surface. Nevertheless, constancy in the general shape of this curve (especially when plotted as percentage of the maximum production in the water column against a measure of light penetration) has permitted the use of formulae in the calculation of the total net photosynthetic production (ΣP) achieved beneath each square metre of sea surface. One such is

$$\Sigma P = n P_{max} d \qquad (2.3)$$

where $n P_{max}$ is the rate of maximum photosynthesis (photosynthetic rate per unit population at saturation multiplied by phytoplankton density) and d is the depth at which the intensity of the most deeply penetrating wavelength is reduced to 10% of its surface value. It is therefore customary to express production per unit area of surface, not per unit volume.

In contrast to photosynthesis, respiration rate does not vary with depth (except in the uppermost layers, where it may be enhanced by photorespiration) and it is usually considered to dissipate some 10% of the gross primary production achieved at saturation. Hence, if allowance for respiration is made on the curve relating photosynthesis to depth (Fig. 2.4), a point will occur at which photosynthetic fixation and respiratory dissipation of carbon are in balance. This is known as the 'compensation point' or depth. Clearly it will vary in the water column with changes in light intensity, so it is usual to express the compensation point as the depth at which the carbon fixed (or oxygen produced) in photosynthesis over a period of 24 h is equal to the carbon dissipated (or oxygen consumed) in respiration over that same period. The depth, of course, may still vary seasonally. Above the compensation point, phytoplankton may grow and multiply; below it, they must subsist on accumulated reserves, form inactive resting bodies, or starve. Although expressed as a depth, the real variable involved is light intensity and hence the corresponding level of irradiance is known as the compensation light intensity. In practice, rules of thumb are used to establish the compensation depth, rather than precise measurements of photosynthesis and respiration. One such rule places it at the depth to which 1% of surface light penetrates and the same depth then separates the photic and aphotic zones (see Section 1.2).

some extent, however, the decrease in photosynthesis may only be apparent in that: (a) a light-stimulated increase in respiration, 'photorespiration', may occur, resulting in an increased metabolism of fixed compounds; and (b) if high light intensity is coupled with shortage of the nutrients also required in photosynthesis (see Section 2.2.3), the rate of leakage of fixed products from the alga is increased, sometimes equalling 90% of the total net production.

If we apply this production–light-intensity relationship to the surface waters of the sea (through which light decreases exponentially with depth), the pattern displayed in Fig. 2.3 would be expected. Near the surface, excess light may result in photoinhibition whereas at depth, light intensity will quickly decline to limiting values. Average values for the critical points on this distribution would lie in the general area of 170 W m^{-2} for the onset of photoinhibition and 120 W m^{-2} for saturation. Different species and different algal groups, however, display differing characteristic values; in diatoms, for example, both saturation and inhibition occur at lower light intensities than in dinoflagellates.

A distribution of photosynthesis with depth in the sea similar to that of Fig. 2.3 is indeed found in nature, with minor modifications resulting from the distribution in the water column of the photosynthetic organisms (Fig. 2.3 assumes a uniform distribution). On relatively dull days, of course, there will be no photoinhibition; and light penetration may be cur-

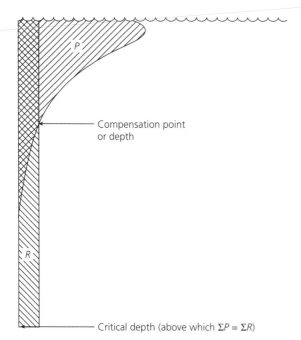

Fig. 2.4 The relationship between depth, phytoplanktonic photosynthesis and phytoplanktonic respiration in the surface waters of the ocean, showing the compensation and critical depths.

2.2.2 Turbulence

At any one time, the compensation depth is a real balance point in the metabolic physiology of a phytoplanktonic cell. A second point which can be derived from graphs of respiration and photosynthesis against depth is a purely notional one but it is nevertheless vital in the relationship between light penetration and the zone of surface mixing. A depth will occur in any water mass above which the gross primary production of carbon in that water column is equal to the total respiratory release in the same column (Fig. 2.4); i.e. the excess of production over respiration above the compensation point is balanced by the excess of respiration over production below it. This second point is known as the 'critical depth' and in terms of irradiance it is the depth above which the average light intensity is equal to the compensation light intensity. It can be calculated from a knowledge of the compensation light intensity, the light intensity at the surface and the extent of light penetration.

On average, a photosynthetic organism can cir-culate through the water column above the critical depth whilst still being in photosynthetic–respiratory balance over 24 h. Wind-induced turbulence may extend down to depths of 200 m yet the photic zone may be much more shallow (Section 1.2). It follows from the above that net photosynthetic production will be possible only if the depth to which mixing takes place is higher in the water column than the critical depth (Fig. 2.5). If mixing extends below the critical depth, as it frequently does during the winter in relatively high latitudes, the average light intensity experienced by a photosynthetic organism will be less than the compensation light intensity and production will be negative until such time as cir-cumstances change; for example, when the depth of mixing decreases with abatement of wind velocity, or when light intensity increases on the approach of spring or summer.

2.2.3 Nutrients

In the discussion in Section 2.2.1, photosynthesis was described in terms of the production of $(CH_2O)_n$ from CO_2 and H_2O (eqn (2.2)), but other elements are required by even the most completely autotrophic of organisms in that they form parts of proteins, enzymes, energy stores, energy carriers and other molecules. Hence, to grow and metabolize, both energy and the necessary building blocks and energy carriers must be available. Most of these material requirements are available to excess in sea-water, but concentrations of nitrogen and phosphorus in par-ticular can at times and in certain areas be very low. In some tropical and Southern Ocean regions, trace elements may also possibly be in limited supply. A more complete photosynthetic equation is therefore

$$5.44 \text{ MJ light energy} + 106 \text{ mol CO}_2 +$$
$$90 \text{ mol H}_2\text{O} + 16 \text{ mol NO}_3^- + 1 \text{ mol PO}_4^{2-} + \quad (2.4)$$
$$\text{trace elements} \rightarrow 3.3 \text{ kg organic matter} +$$
$$150 \text{ mol O}_2 + 5.39 \text{ MJ heat.}$$

In addition, a number of photosynthetic organ-isms are not completely autotrophic. Dinoflagellates and various other algae (particularly members of the Chrysophyta, Cryptophyta and Haptophyta), and some species in all algal groups, require external sources of certain organic compounds, especially vit-amins, as they are unable to synthesize them them-selves. It has been known for many years that, to

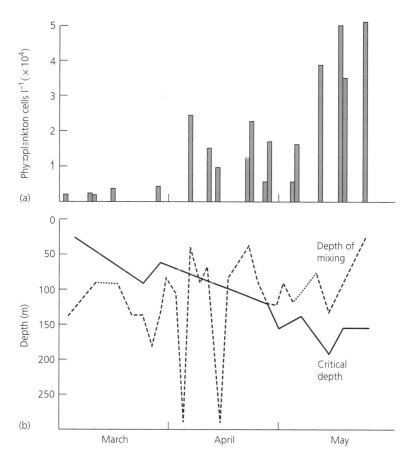

Fig. 2.5 The relationship between phytoplanktonic abundance and (1) the depth to which the surface layers are mixed, and (2) the critical depth. (After Sverdrup 1953.) (Note that phytoplankton are abundant only when the depth of mixing is less than the critical depth.)

culture these 'auxotrophs', addition of such materials as humic acids must be made to the culture medium.

Analysis of the extent to which nutrient shortage limits marine photosynthesis might appear, in theory, to be a relatively simple matter. In fact, it is fraught with conceptual and investigative difficulty and before presenting the evidence it will be helpful to review the nature of the problems. One major difficulty centres on the observation that different species have differing abilities with which they can take up environmental nutrients and, indeed, have differing requirements for these nutrients. Uptake of nutrients is an active process that can operate against a concentration gradient but is nevertheless dependent on the external concentration. Rather as with light absorption, at low external concentrations of a nutrient, uptake is dependent on concentration, but at a certain (higher) level the uptake mechanisms saturate and a plateau is attained (Fig. 2.6). The rate of uptake (U) can therefore be expressed as

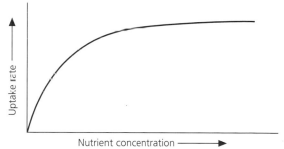

Fig. 2.6 The general relationship between the external nutrient concentration and the rate of uptake of that nutrient by algae.

$$U = \frac{U_{\max} C}{C + K_{c}} \tag{2.5}$$

where $U_{\max}$ is the rate at saturation, C is the concentration of the nutrient in question, and K_{c} is a constant equal to the nutrient concentration at which U equals $0.5U_{\max}$. The half-saturation constant, K_{c},

Table 2.2 Half-saturation constants for nitrate and ammonium uptake. (From data of Eppley et al. (1969) and MacIsaac & Dugdale (1969).)

Phytoplankton from	K_c nitrate (µg-at. l^{-1})	K_c ammonia (µg-at. l^{-1})
Nutrient-poor areas of the tropical Pacific	0.01–0.21	0.10–0.62
Nutrient-rich areas of the tropical Pacific	0.98	
Nutrient-rich areas of the polar Pacific	4.21	1.30
Oceanic in general	0.1–0.7	0.1–0.4
Neritic (diatoms)	0.4–5.1	0.5–9.3
Neritic and littoral (flagellates)	0.1–10.3	0.1–5.7

is an indication of the ability of a given photosynthetic organism to take up that particular nutrient from a range of external concentrations. (In fact, K_c is not truly constant; it varies with temperature and with the internal nutrient concentration of the alga.)

The point at issue, however, is that different members of the phytoplankton have widely varying values of K_c (Table 2.2), such that some species can take up nutrients only from high external concentrations whereas others can do so from progressively lower concentrations. Thus although it is easy to see how falling nutrient concentrations (resulting, for example, from uptake from a finite pool trapped above a thermocline (see below)) could limit the photosynthetic production of a *given* phytoplanktonic *species* with a high nutrient demand, it is less easy to see how this could limit phytoplanktonic production *in total*. What happens in such circumstances is that local selection pressures now favour species which can take up nutrients from lower external concentrations, and as concentrations drop still further so yet different species will be at an advantage, and so on. This process is reflected by the fact that the phytoplankton species dominating a given water mass at a given time have K_c values appropriate to the ambient nutrient concentrations (Fogg 1980); indeed, different races within a single species characteristically have individual uptake rates suited to the external concentrations in their preferred habitat conditions. It is also evidenced by seasonal successions of phytoplankton (Section 2.3). Therefore it is only by demonstrating that the phytoplankton

characterizing waters of low nutrient status are, and must be, inherently less productive than those of more nutrient-rich areas, that a nutrient limitation on phytoplanktonic production in total can be established. This does in fact appear to be the case. It must be emphasized, however, that demonstrations of nutrient limitation (whether mineral or organic) of individual species are not strictly relevant to nutrient limitation of production in general.

A second factor complicating any straightforward relationship between nutrient concentration and productivity is that growth and multiplication rates are related to the concentration of nutrients within the primary producer and not to those in the surrounding water. Phytoplankton can store nutrients taken up at times of relative plenty and use them for subsequent production even in the absence of external supplies. From two to more than five further generations may be fuelled from stored sources.

Such complications notwithstanding, there is abundant evidence of a general relationship between external nutrient supplies and phytoplankton biomass and productivity, largely in the form of correlations between the two in both space and time. We saw in Chapter 1 that the global pattern of productivity correlated well with nutrient levels, being high in the nutrient-rich shelf waters, in areas of upwelling and during the absence of a thermocline (solar radiation permitting), and low in the nutrient-poor, tropical oceanic regions. A more detailed example of such a spatial correlation is shown in Fig. 2.7. Similarly, in some areas with seasonal climates nutrients are known to be depleted during the presence of the warm-season thermocline and production also decreases; both nutrient levels and productivity increase again when the thermocline disappears (Fig. 2.8). Finally, it is known that when sea-water is enriched artificially with nutrients—as when sewage is discharged into semi-enclosed bays—primary production is stimulated, suggesting that nutrients were limiting hitherto. It should be noted, however, that this relationship between nutrient stocks and phytoplankton productivity is only a general one: significant exceptions are known. There are, for instance, some regions of high nitrogen and phosphorus status that have much lower than expected productivity, e.g. in the subarctic Pacific and near the equator. Here shortage of trace elements, such as iron, may limit production.

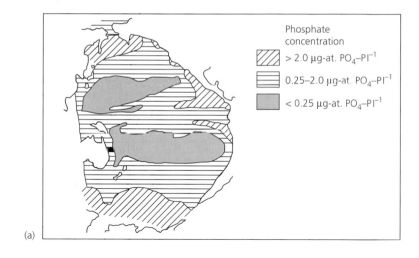

(a)

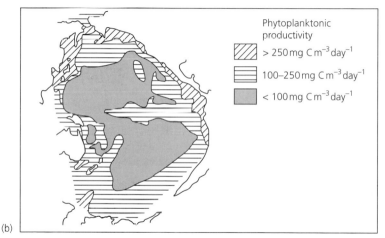

(b)

Fig. 2.7 The general correspondence between (a) the distribution of nutrients, in this case of phosphates at 100 m depth (after Reid 1962), and (b) phytoplanktonic productivity in the same ocean (the Pacific). (From data of Koblentz-Mishke *et al.* 1970.)

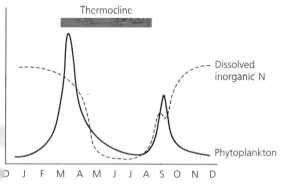

Fig. 2.8 An idealized relationship between phytoplanktonic biomass in the North Atlantic Ocean and (1) the external concentration of nitrogen, and (2) the duration of the seasonal thermocline.

Early work concentrated on the potential limiting role of phosphorus, as inorganic phosphates are less abundant in sea-water than ammonia and nitrates, the forms in which most phytoplankton require their nitrogen. Phosphorus and nitrogen atoms are usually present in the ratio of $1 : 15$. In part, the bias in favour of phosphorus was also a reflection of the availability and sensitivity of techniques for the routine determination of the two elements. However, with the discovery that several algae possess alkaline phosphatases and could obtain their phosphorus from the more abundant stocks of dissolved organic phosphates, attention switched to nitrogen, and this element was until recently generally accepted as that associated with most cases of nutrient limitation (see Hecky & Kilham 1988; Howarth 1988) (see also Section 3.2.4). It may well be significant in this context that some of the most successful primary producers in areas or at times of extreme nutrient poverty are the cyanobacteria (Plates 5 and 6, facing p. 64), some of which can obtain their nitrogen from dissolved nitrogen gas, of which there is no shortage in the sea. Indeed, the recent discovery of the widespread importance of picoplanktonic cyanobacteria (Chapter 1) has brought the wheel full circle: as they are capable of boosting the nitrogen status of the water, the role of phosphorus as *the* limiting nutrient has been returning to favour. (Nitrogen fixation by cyanobacteria is a feature of great ecological importance in other nutrient-poor waters, including on coral reefs (see Section 6.10.1).)

Phytoplankton are small and effectively can take up nutrients only from the thin layer of water surrounding them. Without relative motion, this thin layer would soon be impoverished regardless of the overall concentrations of nutrients in the larger water mass in which they are suspended. Motile flagellates can change water masses on a microscale and dinoflagellates can even maintain a movement of water over their surface by the beating of the transverse flagellum. This may be the main selective advantage of the possession of flagella, the latter often being arranged such that the protist spirals through the water, maximizing the flux of water past its surface. The immotile diatoms, however, have no active means of maintaining conditions favourable to nutrient uptake over their surface. The tendency to sink as a result of gravitational forces, rather than being completely disadvantageous, appears necessary to the maintenance of suitable nutrient-uptake gradients in these species, and the shapes of some diatoms cause rotational movements of the cell to occur as it falls through the water column, achieving the same end as the possession of a flagellum. Indeed, it seems likely that phytoplankton cells can adjust their buoyancy so as *not* to be neutrally buoyant.

How then can we explain the mode of action of nutrient shortage in limiting overall production? As suggested above, the solution to the problem is most likely to be found in the nature of the phytoplankton species that are at a selective advantage under different nutrient regimes, and we can best illustrate this by examining the two extreme cases. In areas or at times of nutrient abundance, there will be no premium on efficiency of nutrient uptake and species with high half-saturation constants will not be at any selective disadvantage. Except in areas of permanent nutrient abundance in the tropics, however, conditions conducive to photosynthesis will not persist: concentrations of nutrients may be reduced by utilization or the input of light energy will decline. Hence the period of nutrient abundance is a temporary bonanza which, as in all similar circumstances, is best exploited by an *r*-strategy (i.e. one that maximizes the potential rate of population increase (see Section 9.1). Under these conditions, it is often light which is limiting as a result of self-shading. Therefore there will be selective advantages associated with a 'bloom life style': species will be favoured which multiply very rapidly to pre-empt the light and utilize the nutrients whilst they are available; and anti-herbivore defences will not be required as herbivore build-up will either be slow in relation to the rate of increase of the blooming phytoplankton (Section 2.2.4) or may even be non-existent (Section 2.4). In other words, individuals which devote a large proportion of their available energy to rapid growth and multiplication and 'make hay while the sun shines' (i.e. are highly productive) will not be at a selective disadvantage; indeed, they will probably be at an advantage in that they can out-compete more slowly increasing forms for the available light.

Turning to the other end of the spectrum, productive, opportunistic species will be unable to survive in a system in which there is a premium on efficiency or on quality rather than quantity. High half-saturation constants may allow large quantities of nutrients to be taken up rapidly to fuel a high nutrient demand economy, but such an economy cannot be maintained in areas of resource shortage. In the nutrient-poor surface waters of the tropics light may be abundant but competition for nutrient will be intense, as will be competition amongst consuming species for their food (Section 2.7). Here there will be a premium on the efficient utilization of resources and on avoidance of being consumed. Selection will, at this end of the spectrum, favour species which devote a smaller proportion of their energy to production: it is the species that can survive best on least which will be the victors of competitive battles. The result is an assemblage of light-demanding, *K*-strategists efficient at removing nutrients from low concentrations and using them with maximum metabolic efficiency (K being the symbol for 'carrying capacity' and indicating that when resources are scarce the ability to compete effectively is more advantageous than a high value of r).

2.2.4 Grazing

The final potential check on phytoplanktonic production is the removal of algal biomass by herbivores. It is clear that herbivores can have a marked effect on the biomass of the larger microphytoplanktonic species. If, using known figures for the multiplication rates of algae or estimates of the quantities of nutrients taken up (as evidenced by decreases in environmental nutrient concentrations), one calculates the quantity of algal biomass that should have been produced during unit time, one finds that the actual biomass present is very low in comparison—perhaps as little as 0.5% of that expected (Fig. 2.9). In relatively stable areas, herbivores may remove microphytoplanktonic production as fast as it is formed. This is not always the case however. In high latitudes in which there is a summer bloom of the larger members of the phytoplankton, build-up of herbivore numbers may lag some 6 weeks behind that of their food (Fig. 2.10). The herbivorous zooplankton of markedly seasonal areas often overwinter in resting stages and do not begin their annual breeding cycle until they have fed on the algae available once the

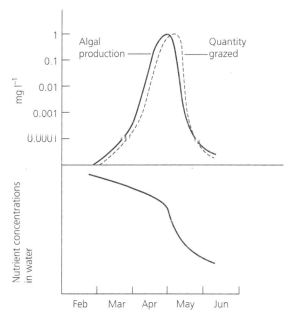

Fig. 2.9 Algal production during the spring phytoplankton bloom in the temperate North Atlantic Ocean in relation to the decline in the external nutrient levels caused, and the grazing pressure exerted by herbivorous zooplankton. (After Cushing 1959.)

bloom has started. The delay in increase of herbivore densities and the large difference between the potential growth rates of algae and herbivores ensure the

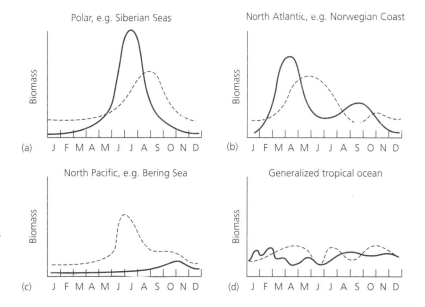

Fig 2.10 Seasonal patterns in the biomass of phytoplankton (continuous line) and zooplankton (dashed line) in different latitudes and oceans. ((a–c) after Heinrich 1962; (d) after Parsons et al. 1977.)

production of large quantities of algal tissues which are not taken by consumers. No hard-and-fast latitudinal rule can be applied, however, because in some areas, the surface waters of the North Pacific northwards of 40°N, for example, the dominant herbivores (the copepod *Neocalanus*) can 'predict' the onset of the bloom, breed in advance of its start using accumulated reserves, and make greater inroads into the biomass when it is produced (see Parsons & Lalli (1988) for a detailed treatment of North Pacific and North Atlantic contrasts). (It is necessary here to distinguish clearly between *removal* of algal productivity and its *utilization* by the consuming herbivores. Although the phytoplankton are relatively digestible, much may in fact remain undigested (Section 2.4), and further losses occur between capture and ingestion, especially in respect of crustaceans. One-fifth of the weight of algae collected may be lost on breakage of the cells during handling, and many small fragments produced during comminution are never ingested.)

The trophic interrelations between algae and herbivores are by no means entirely one way. By consuming algal biomass and metabolizing their tissues, herbivores release nutrients from their organic binding and excrete them into the environment. Equivalently, by feeding on bacteria which have themselves absorbed the dissolved organic material released by or from the phytoplankton, the protistan microplankton release further stocks of nitrogen and phosphorus back into the water. These consumers thus regenerate incorporated nutrients and make them available to the phytoplankton again, thereby stimulating productivity. Indeed, it is likely that in nutrient-poor areas, primary productivity would cease were the herbivores and bacterivores to cease regenerating nutrients *in situ*: they may stimulate as much productivity as they consume. The question therefore arises: although herbivores may remove much, and in some cases the vast majority, of phytoplankton production, do they limit algal productivity as well as algal biomass? There is no general agreement on this question, nor indeed is there any reason to expect a single general answer. It is relatively easy to see that when a phytoplankton bloom has reached a limit set by decreasing light intensity or decreasing nutrient concentrations, the delays inherent in the herbivore response pattern can result in an eventual bias in favour of the herbivore popu-

lations and a crash of the algal populations. Here herbivores graze down a bloom which had already been stopped by other agencies, their consumption of phytoplanktonic production increasing from maybe less than 10% to over 50%. Whether, over large areas, herbivores can limit production which is not already at a ceiling as a result of unfavourable conditions of light, turbulence or nutrients is not so simple a matter. At some times, herbivore populations follow those of their prey in a typical Lotka–Volterra fashion; at other times, herbivore decline may actually precede that of their food source, partly because they may have 'internal clocks' that trigger a resting state and/or a move to greater depths even though the stocks of phytoplankton are still high. The position can be rendered especially complex by the effects of carnivores. In some areas, for example, it seems possible that the herbivorous zooplankton would frequently overexploit the phytoplankton were they not themselves being controlled by predators such as ctenophores, although 'top-down controls' are generally less prevalent in the sea than they are in fresh-waters (Valiela 1991). Further aspects of these interactions will be considered later in this chapter, but it is certainly safe to conclude here that herbivore pressure is one element in a kaleidoscope of factors that influence marine photosynthetic populations: it is often very difficult to determine exactly which of the several factors is limiting production at any one time.

2.3 Distribution of phytoplankton in space and time

Global maps such as that of Fig. 1.23 are useful in summarizing broad latitudinal and regional trends, but they may give a most inaccurate impression of uniformity over large areas. The distribution of phytoplankton species and productivity on a small scale is, in fact, very patchy in both the temporal and spatial planes.

Mention of seasonal changes in phytoplankton has already been made and we can start by amplifying these patterns. The general form of seasonal successions of the dominant larger species is shown in Fig. 2.11. Most of the available information relates to the microplanktonic species and the extent to which the pico- and nanoplankton mirror their patterns is largely unknown: those few studies which have

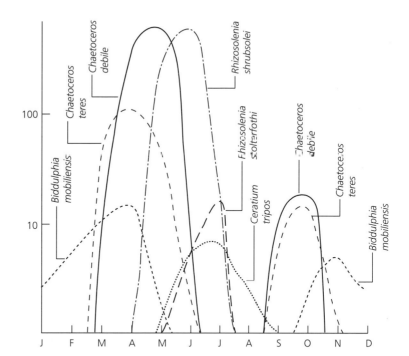

Fig. 2.11 Seasonal succession of dominant microplanktonic algae in the Irish Sea (approximate numbers in 1000s per m³ averaged over 14 years). (From data of Johnstone *et al.* 1924.)

been carried out indicate that they may not. The succession of microplanktonic species appears largely to be a result of changes in the quantity and quality of nutrients and of organic vitamins, etc. Auxotrophs, in particular, require previous conditioning of the water by other species before they can bloom. In the Sargasso Sea, for example, vitamins play an important role in determining the successional sequence. Diatom species requiring external sources of vitamin B_{12} dominate during the spring and greatly reduce its environmental concentration. They release vitamin B_1 into the water, however, and this is required by a coccolithophore, *Coccolithus*, which blooms when vitamin B_1 concentrations pass its threshold value. In turn, this alga releases vitamin B_{12} and so conditions the water for the succeeding diatoms. (Vitamin B_{12} is also released by several planktonic bacteria.)

Not all external metabolites are beneficial or neutral to other organisms. Dinoflagellate species are notably associated with the release of toxins into the water (see Section 3.3.5) and several algae appear to release substances with bactericidal properties. The flagellate *Olisthodiscus* produces a tannin-like compound which stimulates the growth of a diatom when present in low concentrations but inhibits it in high concentrations.

Productivity also varies within single periods of daylight in all but polar latitudes. One might predict that levels of primary production would vary with the diurnal change in light intensity and therefore be symmetrical about midday, but this is not the case. Photosynthetic production is usually greater before midday than after, by a factor of ten in tropical waters decreasing to zero near the poles. Many factors may interact to produce this pattern, but one important contributory cause is the tendency of the phytoplankton to divide at night. In the morning, the phytoplankton is dominated by young, growing, relatively productive individuals whereas by the afternoon the algae are older, slower growing and senescent.

In addition to this temporal variability, the phytoplankton is also distributed very variably in space (Angel 1994; Denman 1994): distinct patches occur, separated by intervening, relatively barren areas. The scale of the patchiness extends from the order of 100 km or more down to a few decimetres, large patches comprising aggregations of smaller patches. Several types of process appear to be responsible: physical, reproductive and feeding. Amongst the physical processes are those resulting in local turbulence, including the special case of Langmuir circulation. Weak

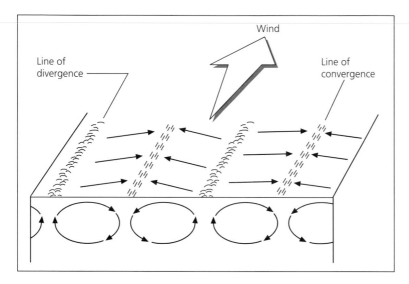

Fig. 2.12 The Langmuir circulation. Each vortex is c. 5 m in the shorter diameter and c. 15–30 m in the longer.

to moderate winds blowing persistently across the sea establish long, parallel, rotating cylinders of water, with adjacent cylinders rotating in opposite directions (Fig. 2.12). This Langmuir circulation therefore forms alternate streaks of downwelling and upwelling which may extend for tens of kilometres and be separated by some tens of metres. Buoyant particles will aggregate in the downwellings and sinking particles in the upwellings; the aggregations representing concentrations of over 100 times the background levels. Not only phytoplankton but zooplankton and pleuston become concentrated in these 'wind rows'. More generally, any purely local turbulence can raise nutrient concentrations over limited areas and there lead to increased phytoplankton production.

On an even larger scale, major hydrographical systems can engender patchiness. The large boundary currents found off the eastern margins of the continents, for example, may spin off cylinders of water with diameters of up to 250 km, depths of 1000 m and life spans of up to 3 years. Such self-contained volumes of water may have properties differing from the surrounding ocean, and these may be reflected in different productivities, faunal and floral diversities, and so on. Equivalently, areas of the ocean occur where two water bodies with differing characteristics abut, for example where shallow, well-mixed, coastal water meets deeper, offshore, stratified waters: these are termed 'fronts'

(Fig. 2.13a). Productivity is often enhanced at such fronts as a consequence of convergence or divergence, or as a result of the mixture of waters with complementary properties (Fig. 2.13b) (see Le Fèvre 1986; Mann & Lazier 1996).

These physical processes are mainly responsible for large-scale patchiness, whereas the two biological mechanisms operate over smaller distances. On the smallest of scales, the division products of a single phytoplanktonic individual will tend to remain together and so magnify any original random unevenness of horizontal distribution. The other biological cause of patchiness is a product of the interaction between the grazing of herbivores and the biomass of their phytoplankton food (Fig. 2.14). Quite simply, in areas of greater than average herbivore density (in relation to the amount of food present) phytoplankton will be grazed down, whereas in areas of lower than average grazing pressure, phytoplankton biomass can increase. This will lead to patches dominated by zooplankton interspersed with ones dominated by algae. The system may well be dynamic, however, in that having grazed down an area the herbivores can, through vertical migration (see Section 1.3 and Section 2.5), move to another patch and proceed to graze that down too, perhaps in a manner described by the 'marginal value theorem' (see Krebs & McCleery 1984). Meanwhile, in the area vacated the phytoplankton can recover. In part, it is this cycle of grazing down and recovery,

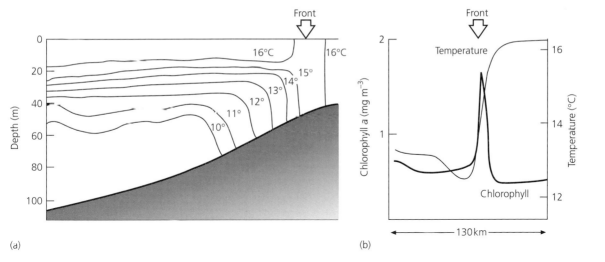

Fig. 2.13 Phytoplankton and fronts. (a) A diagrammatic section through some 280 km of the Celtic Sea in August, showing the boundary ('front') between the stratified and vertically mixed water masses. (Simplified from Simpson 1976.) (b) Phytoplankton abundance across such a Celtic Sea front. (After Savidge 1976.)

represented in space as a series of discrete patches along a phytoplankton-dominated to zooplankton-dominated continuum, which makes analysis of this, as of other such shifting aggregations, so difficult and so locally variable.

It must be emphasized that most of our discussion above has been based on the more easily studied microphytoplankton (the diatoms, etc.) and herbivorous copepods. It is becoming increasingly more obvious, however, that in many areas the smaller pico- and nanophytoplankton are likely to be much the more important size fractions in productivity, nutrient uptake and energy flow, although so far difficulties of sampling have prevented quantification of their general importance. 'Guestimates' of some two-thirds of average photosynthetic biomass and productivity are probably not far off the mark. The picophytoplankton are thought to be responsible for effectively all the productivity of oceanic waters off Brazil! Much the same argument applies to the role of the gelatinous pelagic siphonophores and tunicates, and the protistan ciliates and heterotrophic flagellates, amongst consumers (see below).

2.4 Zooplanktonic production

Because the phytoplankton can multiply rapidly, it is relatively easy to measure their potential productivity by confining them for short periods in small experimental containers. The measurement of potential productivity of the zooplankton, however, poses much greater problems by virtue of their slower individual growth rates and longer intergeneration times. It has been achieved in 'mesocosms' (Grice & Reeve 1982)—large *in situ* volumes of water enclosed within fine mesh netting suspended from a floating platform—but the database from this source remains small and biased towards a few individual species. The evidence, however, clearly indicates that secondary productivity generally reflects the magnitude of primary productivity. Thus, in the most coastal regions of neritic water and in some areas of upwelling, daily zooplanktonic production may average some 75 mg C m^{-2}, decreasing to 40–60 mg C m^{-2} in shelf waters generally, and to 12–16 mg C m^{-2} in tropical oceanic waters. (This positive correlation between primary and secondary production suggests that the abundance of grazing species does depend on, rather than control, the abundance of the phytoplankton.) The action of the consuming species leads to some 75% of pelagic primary production being dissipated in the top 300 m of the water column and 95% in the uppermost 1000 m.

Most estimates of zooplanktonic production have in fact been obtained by multiplying values of

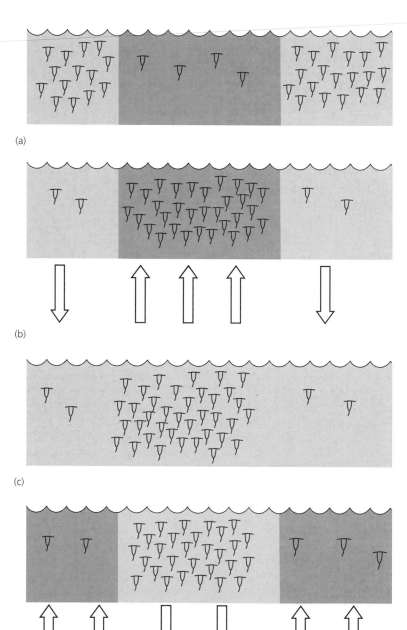

(a)

(b)

(c)

(d)

Fig. 2.14 A cyclic generation of patches dominated by zooplankton interspersed with patches dominated by phytoplankton. Patchiness is a consequence of the feeding behaviour and aggregation of the zooplanktonic herbivores. (After Bainbridge 1953.)

primary production by one or more conversion factors. That most frequently used is the 'ecological efficiency', defined as the amount of energy extracted from a given trophic level divided by that supplied to the trophic level. The ecological efficiency is itself the product of two other coefficients, the 'ecotrophic efficiency' (the proportion of the annual production of a trophic level taken by consumers) and the 'growth efficiency' (the annual weight increment of the consumers divided by the weight of food consumed). The production of a given tropic level (P) is therefore given by

$P = BE^n$

or

$$P = B(E_c G)^n \qquad (2.6)$$

where B is the annual primary production fuelling the system, E is the ecological efficiency, E_c the ecotrophic efficiency, G is the growth efficiency, and n is the number of trophic levels through which energy must pass to reach the trophic level of interest. One other efficiency may be mentioned at this point. The 'transfer efficiency', namely, the ratio of the production of one trophic level to that of the next, is a reasonable estimate of the ecological efficiency (on the assumption that the energy extracted from a given trophic level is proportional to its production.)

Unfortunately, several factors combine to render these methods of calculating secondary production open to question: no general agreement exists on the values to be assigned to E, and efficiencies can vary dramatically from area to area and from time to time; the concept of a trophic level is not one which can readily be applied to the marine ecosystem; and phytoplanktonic biomass is not the only food source available to the zooplankton. Therefore, all estimates of zooplanktonic production (and of secondary production generally) are still not much more than guesses based on certain assumptions. However, because the assumptions, and possible criticisms of them are, if anything, more instructive than are the estimates of production derived from them, we shall now investigate each in turn.

It is most convenient to start by considering the concept of a trophic level. This concept arose in terrestrial ecology to describe stages in the plant → herbivore → carnivore chain, and even in terrestrial ecology is of dubious validity. This stems from two observations. First, food webs are almost everywhere extremely complex (Fig. 2.15) and, for example, what was identified as a sixth-trophic-level species in one marine study, because it fed in part on a fifth-level organism, also took animals in the fourth, third and second trophic levels. Secondly, in many ecosystems—and, it has been claimed, in all —some 90% of the energy flow does not derive directly from living photosynthetic organisms but passes along the dissolved or particulate detritus food chain in which the trophic level concept is even less applicable (as the primary detritus is itself an aggregate including organisms which could conceivably

be placed in many trophic levels). A trophic level is therefore an abstraction bearing little relation to anything existing in the real world, and the marine system is no exception.

Values which different workers consider should be assigned to the efficiency of energy flow from consumed to consumer span a wide range, and the individual 'efficiencies' are themselves variable in nature for sound ecological reasons. This partly results from dissipation of energy by trophic interactions within a single notional trophic level, as a consequence of omnivorous diets or even cannibalism, but there are many other causes. Cushing (1971), for example, found that values of the transfer efficiency between phytoplankton and herbivores varied in a number of upwelling areas with the magnitude of the primary production. Efficiency was greatest (c. 24%) in regions of relative food scarcity and least (c. 3%) in areas of food abundance: a similar premium on efficiency to that noted in Section 2.2.3. Growth efficiency may vary within the herbivorous zooplankton from 7% to 50%, and ecotrophic efficiency between phytoplankton and herbivores may vary from only a few per cent (< 5%) to over 90% (see Table 2.3).

This introduces one exception to the general relationship between phytoplanktonic and zooplanktonic productivity. In some very rich coastal areas, primary production may be of such large magnitude that, paradoxically, it is not consumed by pelagic organisms; it sediments out of the water column creating a large benthic oxygen demand and, occasionally, anoxia. Such examples are exactly comparable with the more familiar eutrophic ponds and lakes: zooplankton may actually be asphyxiated by the thick algal soup if they are not killed by lack of dissolved oxygen first. Even in tropical waters in general, maybe only 15% of the phytoplanktonic productivity is consumed by planktonic herbivores.

What little is certain in the field of trophic efficiencies is that the ecological efficiency of energy flow in the sea is generally higher than the 10% often quoted in terrestrial ecology. At the phytoplankton–herbivore interface, it may lie in the vicinity of 20% (partly as a consequence of the greater digestibility of marine photosynthesizers), decreasing to between 10 and 15% further up the food web.

The third general assumption of eqn (2.6) is that living phytoplankton are the main food source of pelagic consumers. There is no question that

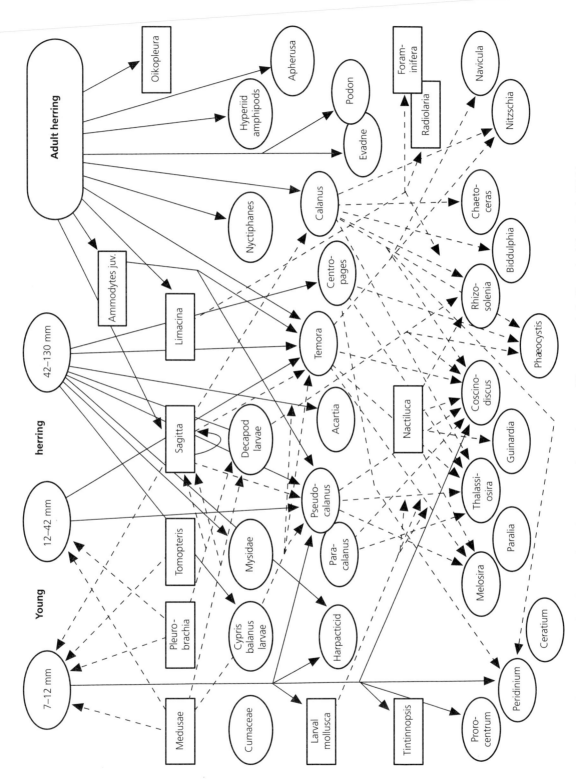

Fig. 2.15 The feeding relations of the herring, *Clupea harengus*, in the North Sea at different stages in its life cycle. (After Hardy 1924.)

Table 2.3 Gross growth efficiencies (energy from the food converted into new tissues in the consumer) of marine organisms, based on field and laboratory results. (After Pomeroy 1979.) Perhaps an average of 40% of the ingested carbon is incorporated into consumer tissue, whereas 30% is lost as particulate or dissolved organic matter and the other 30% is remineralized.

Species	% Efficiency
Daphnia	30–80
Oikopleura (pelagic tunicate)	4–50
Sagitta hispida (chaetognath)	25–50
Sagitta setosa (chaetognath)	9–18
Calanus hyperboreus, adult (copepod)	13–30
Calanus hyperboreus, stage IV–V copepodite larvae	15–50
Calanus finmarchicus, adult	14
Calanus finmarchicus, subadult	25–50
Calanus helgolandicus, nauplius–adult	18–72
Rhincalanus nasutus, nauplius–adult (copepod)	25–55
Euphausia pacifica (krill)	30
Palaemon adspersus (prawn)	1–10
Nereis diversicolor, adult (polychaete)	14–43
Mactra, 1–2 years old (bivalve)	55
Asterias (starfish)	55
Pleuronectes platessa, 0–1 year old (fish)	30
Pleuronectes platessa, over whole life	15

photosynthetic production is the base of the food web; the debate revolves around the quantities and proportion of living primary producers grazed. The alternatives are detritus and the bacteria subsisting on dissolved organic compounds released by the phytoplankton. Bacteria are known to be consumed by heterotrophic protists, which may themselves be taken by small zooplankton, and some zooplanktonic species, e.g. appendicularian tunicates, may obtain up to 50% of their requirements by feeding directly on bacteria. Similarly, faecal pellets, the gelatinous coverings of various planktonic organisms, and dead algal cells are colonized by a variety of bacteria, protists, and small metazoans (nematodes, etc.), and the whole detrital aggregates are eaten by many larger planktonic animals, some possibly subsisting entirely on this diet (Section 3.2.2). Faecal pellets, in particular, are often very rich in food materials. Some phytoplankton appear able to pass through copepod guts without being digested, but, being encased in a heavy faecal pellet, they sink rapidly (at some 100 m per day) and join the detrital pool. A

few even emerge alive and can gain freedom from the pellet and continue photosynthetic life.

Further, many of the living phytoplankton consumed may not be taken by organisms usually considered to be the grazers: it has been estimated that up to 70% of net photosynthetic production in the open oceans may be ingested by heterotrophic flagellates, ciliates and other protists constituting the so-called 'microbial loop'. This 'loop' may really be an almost closed circuit, with little energy passing to the larger copepods, etc. (see Section 2.6). Finally, there is the matter of which zooplankton really consume most phytoplankton. Reliance on plankton nets to sample the consumers gave the impression that copepods and other solid-bodied animals were numerically dominant, but more recent use of scuba techniques and minisubmersibles has shown that the hitherto neglected gelatinous plankton (which break up in nets and become unrecognizable) may really be the dominant individuals in several areas.

Although these alternative pelagic pathways to the grazing food chain are known, there is no concensus on their relative importance: estimates vary from more than 90% to an insignificantly small percentage of the total energy flow. Until such time as these and the other problems are resolved, there can be no certainty of the magnitude of secondary production nor of the detailed factors influencing it. All estimates must be treated with a modicum of suspicion.

A second important role performed by the consumers in the pelagic system has been appreciated only relatively recently. It was believed for many years that bacteria were the agents responsible for releasing from their binding those inorganic nutrients incorporated into organic tissues and thereby making them available again to the phytoplankton. Recently, however, it has been appreciated that bacteria themselves have a very high demand for phosphates and nitrates; indeed, as high as or higher than, weight for weight, that by the phytoplankton because bacterial tissues contain as many as or more phosphorus and nitrogen atoms for every carbon atom, than do those of the photosynthetic algae (Table 2.4). Bacteria may be a sink for nutrients, not a source (although, nevertheless, bacteria are extremely important in changing the form in which elements such as nitrogen occur). It now appears more likely that it is the protist and animal consumers of phytoplankton and bacteria that are

Table 2.4 Carbon : nitrogen ratios in various organisms.
(After Valiela 1984.)

Terrestrial tracheophytes	> 100 : 1
Marine tracheophytes	17–70 : 1
Macroalgae	10–60 : 1
Fungi	10 : 1
Phytoplanktonic algae	6–10 : 1
Bacteria	< 6 : 1

responsible for nutrient regeneration. Zooplankton excrete between 2 and 10% of their body nitrogen and 5–25% of their body phosphorus each day, with higher values when food is relatively abundant. Recycling of these nutrients is therefore rapid; in temperate seas, for example, the turnover time of phosphate is only 1.5 days. Nitrogen turns over somewhat more slowly, but even so each atom is recycled up to 20 times before it passes out of the photic zone. Even in upwelling areas, in which nutrients are plentiful, recycling supplies some 20% of the requirements for incorporation into tissues.

2.5 The spatial distribution of zooplankton

Zooplankton, like (and partly as a consequence of) the phytoplankton, are distributed patchily. Horizontal patchiness (Fig. 2.16) has already been discussed (Section 2.3), but it is also relevant to the main subject of this section—a phenomenon that forms 'one of the most striking and characteristic aspects of the behaviour of marine zooplankton' (Raymont 1983). Most, though not all, members of the zooplankton undertake diurnal vertical migrations through the water column, of somewhat less than 400 m on average in the smaller species and of over 600 m (up to some 1000 m) in the larger. These vertical movements, which may involve sustained upward swimming speeds of 12–200 m h^{-1} dependent on size, and downward speeds some three times faster, may be undertaken twice each day.

Most of these migrations form a similar pattern. During the daylight hours, the zooplankton are relatively deep in the water column, but during

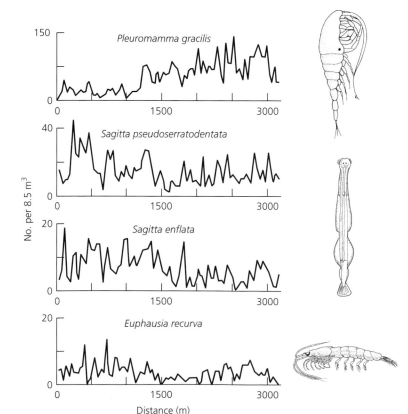

Fig. 2.16 Horizontal patchiness in four species of zooplankton. (After Wiebe 1970.)

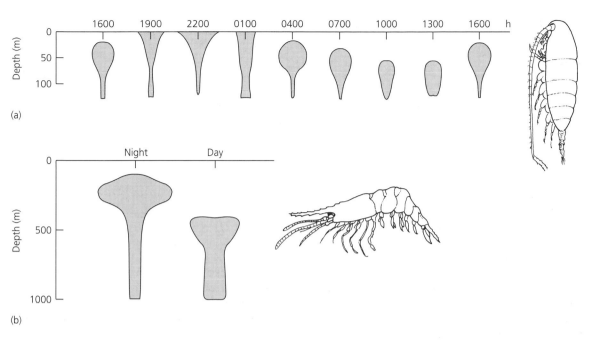

Fig. 2.17 Characteristic vertical movements undertaken by zooplanktonic populations: (a) a surface-dwelling species such as the adult female of the copepod *Calanus*; (b) a deeper-living species, e.g. an acanthephyrid prawn.

the evening dusk they ascend to the surface. They then disperse somewhat through the water column in the middle of the night, reaggregate at the surface before dawn, and then descend again to the day depth (by active downward swimming) as light intensity increases (Fig. 2.17). These movements correspond to those of the once enigmatic 'deep scattering layers' of the sea discovered during the Second World War and subsequently shown to comprise aggregations of migrating planktonic and nektonic animals.

Although many phylogenetically unrelated animals do show this or a similar pattern, it is by no means universal or stereotyped. It may be suppressed in certain life-history stages, at certain times, or in whole areas. In polar latitudes, for example, migrations are predominantly seasonal in that the zooplankton remain at the surface during the polar summer and at depth during the long polar winter. Further, some species display (in some areas and/or at some times) reversed migrations, being near the surface during the day and at depth at night; other species migrate under some conditions but not under others, and many migrate to different day depths in different regions. A number of species do not carry out any vertical migrations at all.

Interpretation of this widespread and important activity pattern has been bedevilled by failure to distinguish between causes, consequences and timing mechanisms, and very many hypotheses have been advanced in putative explanation. There seems little doubt that changes in light intensity provide the triggering and timing mechanisms for the various phases of the pattern: the zooplankton follow bands of constant light intensity as these bands move through the surface waters with regular day–night alternation. But there seems no reason to suppose that restriction to certain ambient light intensities provides the reason for the movements. Here many of the exceptions to the standard pattern are helpful in proving the rule. The exceptions listed above share the common factor of food availability (or lack thereof). The zooplankton remain in the immediate surface waters during the phytoplankton bloom of the polar summer, and remain at constant depth during the winter when no food is available and they are living on their food reserves. Stages in the life history which do not migrate are normally the non-feeding stages, whereas the stages which show most marked vertical migrations are those requiring most food (e.g. adult females). Migratory movements are suppressed in the presence of abundant food (see, e.g. Metaxas & Young 1998) and are stimulated in its absence. Non-migrants are either tied to a specific microhabitat (e.g. the neuston) or are predators. Therefore, as outlined in Section 1.3 and as originally suggested by Hardy over 30 years ago, vertical migrations are most plausibly regarded as the only effective way in which an animal capable of limited lateral movement (relative to its water mass) may change its position in the surface waters when its immediate environment has reached near-threshold concentrations of food. It is optimal foraging in a three-dimensional environment.

Therefore, vertical movement on the part of the herbivores is a response to the horizontal patchiness of the phytoplankton (itself in part herbivore created), and the herbivores are followed by many of the carnivores which depend on them. Changes in light intensity provide a dependable timing mechanism and, all things once again being equal, it is probably selectively advantageous to move at such times as to keep within areas of relatively low light intensity, to minimize the impact of visually hunting predators, i.e. to be at the surface at night and at depth during the day. In general, however, predators can be expected to be adapted to the prevailing environmental conditions in the habitats occupied by their prey.

Granted that these movements occur, many and varied will be the consequences. Populations will be well mixed, increasing rates of gene flow and decreasing the incidence of speciation. Consumption of the phytoplankton will decline at threshold concentrations providing the algae with an opportunity of recovery. Consumption of food in the relatively warm surface waters and its digestion at depth in cooler surroundings when vertically migrating may lead to the gain of an 'energy bonus'. Metabolic rate and the energy required to sustain it are proportional to temperature and so activity is high in the surface waters, leading to the capture of many algae; but by migrating into cooler waters, less of the energy from the meal will have to be devoted to maintenance of a high metabolic rate (i.e. respired) and a greater proportion can then be put to growth, reproduction or storage.

2.6 The bacterioplankton

The pelagic zone abounds in various types of Gram-negative bacteria, many of them motile, which subsist on a diet of the dissolved organic compounds released by living phytoplankton and leached from dead tissues. Bacterioplanktonic productivity is very difficult to measure as any attempt to enclose them in small volumes of water or to culture them almost immediately alters the nature of the system. One series of samples taken from within the phytoplankton-maximum zone indicated that biomasses of some 6 mg C m^{-3} produced 42 mg C m^{-3} daily, but it is not yet known to what extent such values are widespread. About 0.5–5 mg C m^{-3} day^{-1} in open water, rising maybe to 0.5 g C m^{-3} day^{-1} coastally, seems possible. Bacterial numbers appear to approximate 10^5–10^7 ml^{-1} of oceanic water, but higher densities are found in the water adjacent to large masses of organic debris, for example that deriving from salt-marshes or in association with faecal pellets. More than 10^9 bacteria ml^{-1} and more than 10^5 of the largely bacteria-dependent microplankton ml^{-1} have been recorded from such microenvironments. Clearly, bacterially mediated food webs may prove to be much more important than their frequency of mention in classic reviews and standard texts might indicate, with perhaps up to 60% of primary production passing along a dissolved organic compounds → bacteria → choanoflagellate chain (part of the microbial loop; see Section 2.4 and Fenchel (1988)). In tropical waters, the production of bacterioplankton may exceed that of the phytoplankton and it forms just as important a food source. More usually, the range would be of the order of 20–50% of phytoplanktonic production (Fig. 2.18). The nature of the planktonic food web, including the microbial loop and bacterioplankton, is summarized in Figs 2.19 and 2.20.

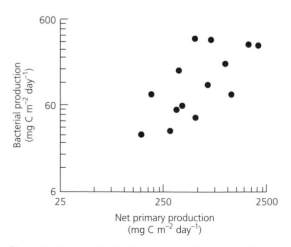

Fig. 2.18 The magnitude of bacterial productivity in relation to that of the phytoplankton. (After Cole *et al.* 1988, with permission.)

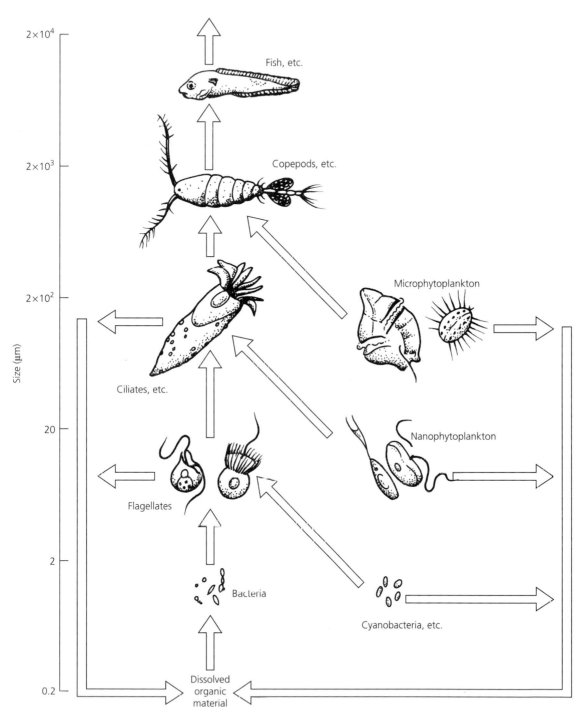

Fig. 2.19 Links in the planktonic food web, showing the sizes of the various organisms and the importance of dissolved organic matter and the microbial loop. (After Fenchel 1988.) Heterotrophs are on the left-hand side and photosynthetic autotrophs on the right. With permission from the *Annual Review of Ecology and Systematics*, Volume 19, © 1988, by Annual Reviews.

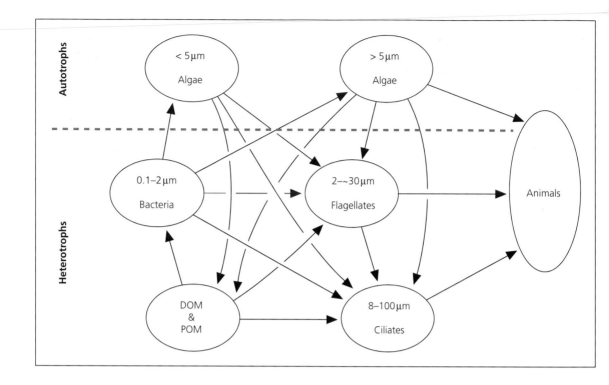

Fig. 2.20 The complexity of trophic relations within the microbial loop system and between it and larger organisms. (After Sherr & Sherr 1988.) It should be noted that much energy may be dissipated by predatory interactions within the flagellate and ciliate compartments, and that bacteria and < 5 μm algae (mainly prochlorophytes and cyanobacteria) are too small to be taken directly by most animals.

2.7 Diversity and other community characteristics

The plankton of highly productive areas (zones of upwelling and some inshore regions) show clear differences from those characterizing open, unproductive tropical waters, and moderately productive systems are, as one might expect, intermediate in these respects. Several of the differences have already been mentioned, in passing, earlier in this chapter. Species in productive waters are typically large and inefficient, and generate a large annual production in relation to their own average biomass. The plankton as a whole is species poor and has relatively little total community biomass supported by each unit of primary production. Carnivorous species are not well represented, anti-consumer adaptations are not prevalent, and food chains may be relatively short and simple.

In contrast, planktonic species in unproductive waters are typically very small, efficient, show a smaller annual production in relation to their biomass, and possess anti-consumer adaptations. Carnivores are an important element in the species-rich oceanic communities, food chains are long and relatively complex, and each unit of primary production supports a relatively large community biomass. Classic pyramids of biomass are often inverted. Kiørboe (1993) has provided a review of the relationship between phytoplankton cell size and food-web structure.

Most of these properties revolve around the efficiency, which is at a premium when resources are in short supply relative to the needs of the consumers (though not necessarily in absolute terms), whether the consumers be phytoplankton or zooplankton. The high diversity of tropical oceanic plankton, however, may appear to be something of a paradox in that one might expect a highly competitive system to lead, through competitive exclusion, to species poverty. This expectation rests on an assumption that conditions remain favourable to one particular competitor for a sufficient length of time to allow ousting of other species to occur, and this assumption is probably not justified in the pelagic environment. Neither does it allow for the evolution of dietary specialization. On a small scale the open ocean is not a uniform environment; changing patterns of turbulence, changing concentrations of inorganic and organic nutrients, etc., lead to a spatial and temporal mosaic of microhabitats, and balances of competitive advantage probably change continually. The pressure from consuming species, checking the potential build-up of any one prey species, may also have the effect of preventing a given species from monopolizing resources and achieving dominance, thereby acting as a counterbalance to any temporary superiority.

In these community characteristics, the oceanic–upwelling contrasts are almost exactly paralleled by other ecosystem dichotomies. Thus coral reefs, the deep sea, oligotrophic lakes and rainforest parallel the oceanic system, whereas mangrove-swamps, shallow lagoons, eutrophic lakes and boreal forests parallel the upwelling zones.

Infaunal species are of interest not only to biologists but also to geologists, and our knowledge of the daily lives of many benthic animals has been derived more from geological work than from the activities of biologists (the pioneering work of Schäfer (1972) is still a mine of useful information). The study of palaeoecology is treated as a branch of geology, and to interpret palaeoecological data it is necessary to know how organisms become fossilized, with what biases and under what conditions. Hence organism–sediment relationships, such as burial and burrowing, have been studied for many years, especially by German 'actuopalaeontologists'. Burrowing activities are also of interest to sedimentological geologists because bedding layers are disturbed as animals move through them (causing 'bioturbation'). Dense populations of polychaetes may circulate all the upper 10 cm of sediment through their guts in less than 2 years; and a population of the long, thin polychaete *Heteromastus* (8–15 cm long, 1 mm diameter) can move sediment from 10–30 cm depth to the surface during the four months of autumn at a rate of 2.5 l m^{-2} daily—this is equivalent to a layer 20 cm deep! For a long period, benthic biologists were preoccupied with descriptive and classificatory exercises (see Section 3.5) and it is only relatively recently that detailed and experimental ecological information has become available (for reviews of this work, see Fenchel 1978; Gray 1981; Reise 1985; Wilson 1990; Barnes 1994).

3.2 Trophic relations

3.2.1 Sources of food

In essence there are two forms of input of organic matter into shelf–littoral benthic systems: detritus and living plankton from the overlying water mass; and living and non-living organic matter associated with the sediment itself. In addition, dissolved organic compounds are certainly important in nourishing bacteria and they may be usable by some benthic animals, although this is controversial and has been so for 75 years. If the quantity of living organisms beneath a given square metre of sea surface is accorded a value of unity, then that of particulate organic detritus in the water would lie in the range 1–10, and dissolved organic material in the order of 50–500. The dissolved organic compounds clearly constitute a vast pool of potential food (equivalent to the total primary production of the ocean for 30 years or more); much of the pool, however, appears to be highly resistant to metabolic processes, even those of bacteria, and it is probably effectively inert (as the large size of the pool might suggest) with a turnover time of up to 3500 years. The current consensus is that although dissolved organic compounds may be of importance to some worm-like animals, they are unlikely to be of wide significance to metazoans (but see *American Zoologist* 1982).

The pelagic or benthic provenance of the primary food sources is reflected by the division of benthic organisms into suspension and deposit feeders, respectively (Section 1.3). The division, as might be expected, is by no means absolute, as the surface deposits may be put into suspension by water movements and thereby be made available to suspension feeders, and several organisms can feed in both modes. The division is nevertheless a useful one. With the exception of the microbenthic algae of littoral sediments, however, the truly benthic food sources are obtained, by sedimentation, from the water column—representing potential food materials which have not been intercepted by pelagic organisms. Hence it is most convenient to consider food inputs under two headings: microbenthic algae, and rates of fall-out onto the sea bed (regardless of whether they are taken by suspension or deposit feeders).

Unicellular and filamentous algae can live only on or in the surface layers of the sediment in the littoral zone, where the photic zone extends down to the substratum. Here, productivities of the order of 0.2–1.3 g C m^{-2} daily can be achieved, dependent mainly on water clarity. Mat-forming species are confined to the relatively stable muds, whereas sands are characterized by small species living attached to, or between the grains. There, benthic microalgae can photosynthesize over a much wider range of light intensity than can the planktonic species. Little or no photoinhibition occurs (Fig. 3.3), adapting littoral species to full sunlight at low tide, and they can also utilize very low intensities such as may be experienced at high tide. (In contrast to the water column, blue light is absorbed first in sediments and red light penetrates furthest.)

Over the larger, continental shelf areas, this *in situ* photosynthesis is missing and the benthos is

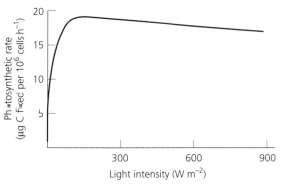

Fig. 3.3 The relationship between light intensity and photosynthetic production in benthic diatoms exposed to natural sunlight. (After Taylor 1964.) It should be noted that full, midday sunlight (870 W m⁻²) results in only a 10% inhibition of the maximum rate. (Compare with Fig. 2.2.)

supported solely by pelagic production and by detrital materials emanating from the coastal fringe. The measurement of fall-out rate is difficult, not least because to estimate the quantity of material actually reaching the bottom, a cup or sediment collector must be sited near to the sea bed, but there it is also likely to collect material resuspended from the sediment surface, leading to an overestimation of deposition rate. Published rates are therefore variable within wide limits, but they do generally reflect: (a) the depth of water through which material must travel; (b) the magnitude of pelagic production; (c) the proximity of additional sources of detritus. The relationship between pelagic photosynthesis and fall-out rate is

not linear because a higher percentage of the production of nutrient-rich areas may reach a given depth than that in unproductive systems as a consequence of the relative inefficiency of utilization discussed earlier (Section 2.2.3 and Fig. 3.4). As an approximate indication of magnitude, the annual input of organic carbon onto the continental shelf probably averages some 100 g C m⁻², with somewhat less than 25 g C m⁻² yearly near the edge of the shelf, and more than 300 g C m⁻² yearly in favourable littoral areas. Analysis of the contents of sediment traps suggests that most of this input is in the form of the faecal pellets of herbivorous zooplankton, in which much material remains undigested (Section 2.4). Salp faeces have been recorded as exporting up to 23 mg C m⁻² per day from the surface waters of the North Pacific. Faecal pellets are proportional in size to that of the animal producing them, and large pellets sink faster than small ones and are therefore more likely still to contain utilizable material by the time they reach the sea bed. Areas of shelf beneath waters dominated by small zooplankton, e.g. those off the southeastern USA, receive a relatively small input and correspondingly support a sparse fauna.

The planktonic food available to suspension feeders is naturally related to the same three variables as is fall-out rate, and, in addition, to the extent of water movement near the sea bed, which can bring about renewal of suspended supplies. Suspension feeders may be of great importance in translocating food from the water column to the sediment surface. Much of

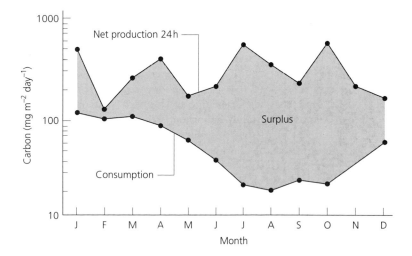

Fig. 3.4 The relationship between net phytoplanktonic production and zooplanktonic consumption in a lagoonal area of Kerala, India, showing a large, ungrazed 'surplus' of phytoplankton. (After Qasim 1970.)

Table 3.1 Inputs of food materials into the benthic systems of two shallow-water, marine environments. (From W.J. Wolff 1977 and Day *et al*. 1973.)

Grevelingen Estuary, Netherlands (g C m^{-2} yearly)	
Detritus from fringing vegetation:	
salt-marshes	0.3–7
sea-grass meadows	5–30
Detritus from adjacent sea	155–225
Run-off from adjacent land	2
Phytoplankton	130
Benthic microflora	25–57
Total	317–451
Barataria Bay, Louisiana (g C m^{-2} yearly)	
Detritus from fringing salt-marshes	297
Phytoplankton	209
Benthic microflora	244
Total	750

the material removed from suspension may not be suitable food for the organism concerned or may not be digested; materials which are not ingested are rejected as 'pseudofaecal' pellets, often bound in mucus, and both the faeces and pseudofaeces are deposited on the substratum and are there available to deposit feeders. In oysters, for example, this biodeposition may amount to over 1 kg C m^{-2} yearly.

Budgets for two, shallow-water, benthic systems are given in Table 3.1 and these indicate the variation that occurs in the relative importance of different forms of food input.

3.2.2 Utilization of food sources

Microbenthic algae and dead phytoplankton cells may be consumed and digested by benthic animals as easily and as efficiently as planktonic herbivores obtain their food, and the same is true of the living planktonic component of the diet of suspension feeders. This is not the case with the detrital component of benthic diets. If we were able to follow the passage of a piece of 'protodetritus' as it slowly falls through the water column, we would see it lose soluble organic compounds through leaching (about half its original weight being lost in this manner) and others would be taken out as it passed through several guts before finally arriving on the sea bed. By this time, it would probably contain only skeletal or

other refractory substances and be of little direct food value to animals. In and on the sediments, however, it will be colonized by various types of fungus and bacterium, and they will eventually support populations of protists and meiofaunal metazoans. Bacteria occur in littoral and shelf sediments in densities of up to 10^{11} ml^{-1} or 10^9 g^{-1}, although they still only comprise < 2% of the particulate organic carbon in the substratum. The abundance of fungi has rarely been quantified; they are often assumed to be unimportant, but this may merely reflect the extent to which they are grazed down by detritivores (equivalently to the position of filamentous algae on coral reefs, see Section 6.10.1).

Hence a unit of detritus, when it is ingested by, say, a deposit-feeding polychaete, will in fact be a microcosm in its own right, with bacteria, fungal hyphi, ciliates, amoebae, flatworms, nematodes, etc., all associated with the original detrital flake. In the littoral zone, the same particle would also provide a substratum for various cyanobacteria and photosynthetic protists. The question therefore arises: What are the detritus feeders actually assimilating when they ingest this aggregate? The answer cannot be provided with absolute certainty, but the presumption that they are digesting the bacteria, fungi, protists and meiofauna is very strong.

The evidence for this presumption comes from several sources. By comparing the organic content of the deposits consumed and the faeces produced afterwards, it is possible to estimate the efficiency with which the food is assimilated. For many detritus-feeding species the assimilation efficiency is very low (1–10%), which indicates that most of the organic content of the food is indigestible. Yet when in the laboratory these species are fed on diets of pure bacteria, for example, efficiency is greatly increased (to some 70%), and by sampling the ingesta from different regions of the gut it can be shown that the bacterial component decreases on passage, particularly in the region between stomach and intestine. When the relatively bacteria-free faecal material is released to the environment, an increasing microbial metabolism is observed through time until a value corresponding to the steady-state system maintained in the rest of the sediment is achieved (see below). After this point, no further increase is noted until and unless the detrital material is rejuvenated by passage through the gut of another detritus feeder. Most detritus feeders also appear enzymatically ill-equipped to cope with the refractory organic compounds of pure detritus that is anything other than fresh, although some decapod crustaceans, including the burrowing prawn *Upogebia*, do possess an assemblage of gut bacteria capable of breaking down some at least of this material. Some species, however, appear able to select which items from within the matrix to ingest and in these, assimilation efficiencies are much higher—up to nearly 90%. Utilization of meiofauna by the deposit-feeding macrofauna is still under debate (see Kuipers *et al.* 1981; Kennedy 1993), although it appears increasingly likely that the meiofauna are an important dietary component. Circumstantial evidence for a nutritional role can be found in low meiofaunal densities in regions of high detritus-feeder abundance and vice versa, and in the fact that energy budgets calculated for benthic systems will not balance unless the macrofauna are consuming meiofaunal species (we will return to this topic in Section 3.2.3).

Such lines of evidence suggest that consumers of deposits are ingesting the sediment and its associated detrital materials (with various degrees of selectivity), are digesting the living components of the detrital aggregate, and are voiding the relatively inert non-living components back into the environment, where they can be recolonized by bacteria, fungi, protists and meiofauna and the cycle repeated. In the shallowest waters, photosynthetic protists make up sometimes more than 70% of the total particulate carbon and hence it is highly likely that these micro-algae are the component of greatest importance. The production of the deposit and suspension feeders is then available to benthic carnivores (although, of course, the deposit and suspension feeders are themselves carnivorous to a large extent, as are many of the meiofauna) and to benthic-feeding members of the nekton.

3.2.3 Production and biomass

The productivity of suspension feeders will be determined by the rate of supply of their suspended food and, ultimately at least, that of the deposit feeders will be governed by the rate of fall-out of potential food particles. But some shallow-water sediments are so rich in deposited organic compounds that 50% of the respiration of fixed organic matter may be accomplished in the anoxic regions and undecomposed tissues may be removed from circulation by burial —food may therefore appear superabundant in the surface deposits. As pointed out above, however, not all the organic content of marine sediments may immediately be available as food, and it may be largely bacterial production that is really fuelling the system in aphotic regions. What then may determine bacterial productivity? We saw earlier (Section 2.4) that for every unit amount of carbon, bacteria have high demands for nitrogen and phosphorus, and these they obtain from the surrounding water, rather than from the nutrient-poor detritus. Sediments, however, have low permeabilities and accordingly it appears likely that local nutrient shortage could limit bacterial production, not shortage of detrital carbon. It is here that the role of detritus feeders (or consumers of bacteria in general) is important. By consuming the bacteria, the various protists and animals release their incorporated nutrients back into circulation again and permit continued bacterial production: in effect, the consumers maintain the bacteria in growth phases. The faeces of deposit feeders in general are a further important source of nutrients. Movement of consumers within the sediment (and burrow irrigation, etc.) will also cause local turbulence of the interstitial water, resulting in replenished stocks of nutrients for the bacteria. Indeed, it is now known that, contrary to earlier postulation, bacteria do not cover anything like the entire surface of detrital particles; colonies are often small and in a static (no growth) phase. Predation may be, through the nutrient regeneration caused, the single most effective factor in maintaining and stimulating bacterial production. It has even been argued that, although both the meiofauna and macrofauna are dependent on bacteria for food, they do not compete for this resource, as the meiofauna, by their movement and by recycling nutrients, induce a bacterial productivity which could not occur without them, and then they subsist on that production. In a similar manner, in shallow waters irrigation of burrow-systems will create a flux of oxygen and nutrients through the sedimentary environment and can thereby stimulate the productivity of photosynthetic flagellates and diatoms. Some polychaetes (e.g. lugworms, maldanids and *Pectinaria*) have been described as, in effect, 'gardening' such algae by drawing water currents into their burrows. At least one nematode may

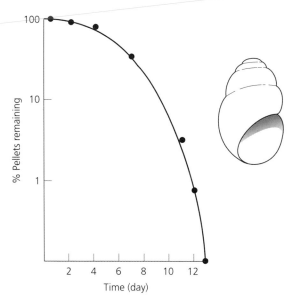

Fig. 3.5 The breakdown rate (into diffuse sediment) of the faecal pellets of a deposit-feeding gastropod, *Hydrobia truncata*. (After Levinton & Lopez 1977.)

also garden microalgae; the latter multiplying in the trails of the nematodes and later being consumed as these meiofaunal organisms retrace their movements.

A further limit on benthic production may arise from the cohesiveness of the faecal pellets produced by some detritus feeders (Fig. 3.5). Although in effect the surface sediment is being recycled continually through animals' guts, some species are disinclined to ingest what can be recognized as freshly produced faecal pellets. (This would clearly be of selective advantage in that such pellets require an 'incubation time' before yielding digestible food (see above) during which they also break down again to diffuse sediment.) Hence abundant populations of deposit feeders may render their own habitat temporarily unattractive, both reducing their own productivity and permitting the bacteria to decline to a steady, nutrient-limited state.

The effects of these limitations in determining general levels of abundance and productivity are poorly known. Unfortunately, early work on the distribution of benthic biomass (little attention then being paid to production) was often quoted in wet or fresh weights per unit surface area, and water and ash content vary markedly from one group of

organisms to another. For the purposes of comparison with data for the deep sea given in Chapter 7, wet weight biomasses on the continental shelf generally lie within the range 150 and 500 g m^{-2}, with a maximum of over 1 kg m^{-2} on the rich Antarctic Shelf. Littoral biomasses are not dissimilar.

Only relatively recently have productivities been measured (as opposed to inferred) and biomasses quoted as ash-free dry weights. Gerlach (1978) provided average figures for the abundance of the various size groups of organisms composing the infauna of the shallow shelf: deposit feeders (120 μg ml^{-1}); other macrofauna (80 μg ml^{-1}); meiofauna (50 μg ml^{-1}); protists (20 μg ml^{-1}); bacteria (100 μg ml^{-1}). He also estimated that up to 105 bacterial duplications per year would be required to support this system. On an areal basis, biomasses of shelf macrofauna recorded to date range from some 50 g (ash-free dry wt) m^{-2} down to less than 4 g m^{-2}, with the highest biomasses in inshore waters. In these same waters, overall annual production by macrofauna is usually only one or two times their average annual biomass, decreasing even further to less than half the average biomass in deeper areas; values of 2–3 g C m^{-2} per year can be regarded as the norm over most of the shelf, although the productivity of the individual macrofaunal species is, of course, variable within fairly wide limits (see Fig. 3.6). The smaller meiofauna and meiofaunal-sized juveniles of the macrofauna probably produce annually more than five times, and sometimes more than 10 times, their standing stock. As one might expect, secondary production in organic muds and muddy sands, and in areas supporting dense beds of suspension feeders, is usually larger than that supported by clean sands, both in absolute terms and as a rate of turnover of biomass. Even so, the production of the benthic macrofauna rarely exceeds 5% of that in the overlying water and the component organisms are often slow growing and long lived.

Because measuring benthic secondary production is an extremely time-consuming process, various workers have tried to find acceptable short-cuts, especially by deriving equations from data in the literature on size and productivity of various organisms. Thus Edgar (1990) advocated use of the following expression to estimate macrobenthic productivity (P, in μg per day) from the mean ash-free dry weights of the animals concerned (B, in μg):

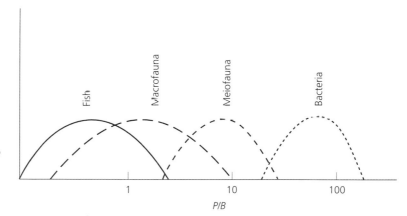

Fig. 3.6 Ratios of annual production to average annual biomass characterizing the populations of various types of benthic species. (Mostly after Valiela 1984.)

$$P = 0.005 \ B^{0.80} \ T^{0.89} \qquad (3.1)$$

where T is water temperature (°C) within a range of 5–30°C. B can be measured or can itself be estimated from the equation

$$\log B_1 = -1.01 + 2.64 \log S \qquad (3.2)$$

where B_1 is the ash-free dry weight (in mg) likely to be retained by a sieve of mesh S (in mm). The database underpinning all such equations is, however, as yet rather small.

3.2.4 Interactions between plankton and benthos

Many of the macrobenthic species of littoral and shelf regions produce larvae which swim and feed in the pelagic system for about 4 weeks and in this manner bypass the losses of potential food inherent in feeding on phytoplanktonic production only after it has sunk to the sea bed. The efficacy of this strategy will vary with latitude, as the pattern of primary production shows latitudinal differences (Section 1.4). Production is continuous over the tropical shelves and here some 80–85% of benthic species have planktotrophic larvae; in temperate regions, phytoplanktonic production is seasonal and less predictable in its duration, and the percentage with planktotrophic larvae falls to near 60%; and in the polar zones, where photosynthesis is possible only for a few weeks, such larval types do not occur. A further 15% of tropical and temperate species, and 5% of those near the poles, produce relatively short-lived larvae which form part of the plankton but do not feed there or only do so to a very minor extent. For the latter, pelagic life serves as a mechanism for short-distance dispersal and, of course, dispersal will also result from the possession of planktotrophic larvae (see Chapter 9). However, if larval life in the plankton is to be interpreted in part as a means of occupying a habitat in which food is relatively abundant, it is legitimate to question why the pelagic larvae should ever leave these areas of plenty and return to the comparatively food-poor benthos. Obviously, it is difficult to conceive of a pelagic cockle or clam, but the answer to this conundrum may be that as many as could manage to achieve permanently pelagic life have done so. Many researchers have suggested that animal life evolved within the benthos and that the individual groups forming the zooplankton (in both fresh water and the sea) have originated by paedomorphosis from the larval stages of benthic species (see De Beer 1958; Gould 1977).

Passage of elements in the food web is never entirely one way, however. We noted earlier (Section 2.4) that pelagic animals may be important in releasing the nutrients required by the phytoplankton and returning them to their dissolved inorganic state in the water. Such a role is also performed by the benthos, as the following example will illustrate. Nixon *et al.* (1976) investigated the regeneration of nitrogen by three, different, shallow-water, benthic systems in Narragansett Bay, Rhode Island, and found that the release of ammonia by benthic animals was responsible for the seasonal pattern of ammonia availability in the pelagic zone. Not all the nitrogen regenerated was in the form of ammonia, however, although nitrite and nitrate release were insignificant. A ratio of oxygen to nitrogen atoms of disturbance from cockles and other bivalves, lug-worms, ragworms, and others is to leave its burrow and attempt to construct a new one elsewhere. While in the water column, however, it is easy prey to small fish. The end result is that *Corophium* is abundant only in the absence of other species or after the numbers of these have been reduced by other agencies (e.g. cold winters, see Section 3.3.5). Other interactions are aggressive. Thus when *Nereis* breaks into the burrow of *Arenicola* during its own burrow constructing activity, it harries the larger worm, up to and including biting it and even nipping off the end of its tail. As with *Corophium*, the lugworm reacts by leaving its burrow and seeking other regions in which to burrow afresh. In a comparable manner but on a larger scale, the physical disturbance caused by large crustaceans, cephalopods, fish such as rays and flatfish, walrus, and organisms up to the size of the grey whale digging into the surface sediment in search of food or for concealment can result in wholesale destruction of burrows and even physical burial of small organisms living in the excavated areas.

Other interactions intuitively seem likely to be widespread. It seems self-evident from the above examples that burrowers by destablizing the sediment would render it unsuitable for species constructing permanent tubes or requiring non-turbid water for feeding, and conversely areas dominated by tube-dwelling species would leave relatively little open sediment available, for example, to more freely burrowing, errant polychaetes. A quarter of a century ago, in one of the pioneer experiments in this field, Woodin (1974) showed that the burrower *Armandia* was more abundant when the tube-dweller *Platynereis* was prevented from monopolizing the available space; and the tubes of the amphipod *Ampelisca* interfere with the freedom of movement and possible ingestion rate of the gastropod *Ilyanassa*, leading to inverse correlation in their numbers (see Fig. 3.11). The burrowing prawn *Callianassa* has been shown to exclude potentially sympatric bivalves, both in South Africa and in the USA, largely by the sediment that it throws into suspension imposing excess feeding costs on the molluscs. We noted in Section 3.1 that *Heteromastus* can bring a layer 20 cm deep to the surface; the annual sediment excavated and transported by *Callianassa* populations can be equivalent to a 75 cm layer! Similarly, the fiddler crab *Uca* can turn over nearly 20% of the surface sediments yearly, and deposit-feeding polychaete and bivalve mollusc individuals rework some 100–400 ml of sediment per annum. It is small wonder that species requiring stable substrata for their lifestyle cannot coexist with them.

The result of all this indiscriminate disturbance and covert competition for sedimentary space is a mosaic of patches in both space and time: in time, as when bioturbating fish create defaunated areas, and in space where high densities of one species may exclude most others (see Hall *et al.* 1994). In a number of cases, the patches are to a degree predictable in that broadly inferior competitors characteristically occur only in marginal habitats—high on the shore, in brackish water, in certain grades of sediment (e.g. Snelgrove & Butman 1994), etc.—lying outside the ranges of the dominants. Regardless of feeding type, space in both the vertical and horizontal planes may be partitioned between a range of species. In these circumstances, where or when the superior competitors are absent, for biogeographical, temporal or chance reasons, the range of patches inhabited by inferiors is often demonstrably broadened. Often, however, patch utilization is not permanent, but may vary from month to month and from year to year dependent on recruitment, which can be dependent not only on the processes considered in this section but on the action of predators.

3.3.3 Predation

The effect of predators on the abundance and composition of the benthic fauna is, as elsewhere (Section 2.2.4), somewhat variable and problematic. Most work has been carried out in the shallowest of regions and has used the technique of enclosing experimental areas of sediment within cages that exclude (or sometimes include) various size categories of potential predators. Three general results of predator exclusion have frequently, though certainly not universally, been observed: (1) the densities of individuals within the cages are up to 500 times larger than in controls; (2) more infaunal species are present in the absence of predators; (3) no marked increase in dominance by any single species appears to correlate with decrease in predation mortality. Table 3.2 displays a typical result of such experimental exclusion of predators. Results (2) and (3) contrast

Plate 1 Although bizarre, these carapace projections help this beautifully adapted mantis shrimp larvae spend its early life in the plankton off Heron Island, southern Great Barrier Reef. (L. Newman & A. Flowers.)

Plate 2 The largest known fish, whale sharks, *Rhincodon typus*, are the gentle giants, feeding on enormous amounts of zooplankton, baitfish and squid from shallow reefs waters. Seasonally they aggregate to feed together off the warm shallow waters off Ningaloo Reef, Western Australia. (L. Newman & A. Flowers.)

Plate 3 Coralline algae are among the major building blocks for coral reefs since they act to cement the reef together. These red calcareous algae are found in exposed areas cementing the reef together as they grow. They are quite variable in form, ranging from thin encrusting to intricately branching structures but they always maintain their unique pink colour (Heron Island, southern Great Barrier Reef). (L. Newman and A. Flowers.)

Plate 4 The solitary ascidian, *Rhopalaea crassa*, is usually found under ledges where it is permanently attached by its abdomen. The transparent test (cellulose-like protective covering) contains a large perforated pharynx with a sieve-like lining for filtering microorganisms and particulate matter from the surrounding water column (Madang, Papua New Guinea). (L. Newman and A. Flowers.)

Plate 5 Commonly known as sea sawdust, the blue-green alga, *Tricodesmium*, is extremely successful in tropical waters. This microscopic plant can fix its own nitrogen and can out compete many other phytoplankton in nutrient poor areas (L. Newman & A. Flowers.)

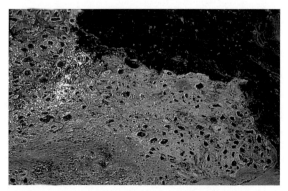

Plate 6 Massive algal blooms of the toxic blue-green algae, *Tricodesmium*, occur periodically at Heron Island, southern Great Barrier Reef. Surface waters can become so discoloured with huge numbers of this phytoplankton that slicks form. Occasionally the breakdown of algal cells can pollute the water with noxious chemicals smothering marine life. (L. Newman & A. Flowers.)

Plate 7 Shallow soft mud with seapen, *Virgularia mirabilis* and the anemone *Sagartiogeton laceratus* (Loch Laxford, Scotland). (R. Holt.)

Plate 8 Tunicate *Ciona intestinalis* and brittlestar *Ophiopholis aculeata* (Haaf Gruney, Shetland). (K. Hiscock.)

Plate 9 A salt-marsh and its drainage creek on the north Norfolk coast. (R.S.K. Barnes.)

Plate 10 Stand of the mangroves *Avicennia*, *Bruguiera* and *Hibiscus* fringing an estuary in Kwa Zulu-Natal, South Africa. (R.S.K. Barnes.)

Plate 11 An underwater view of a sea-grass meadow in Panama. (D.K.A. Barnes.)

Plate 12 A kelp forest in North Wales, showing fronds of *Laminaria digitata* as seen at low spring tide. (R.N. Hughes.)

Table 3.2 Abundance (numbers per 0.1 m^{-2}) of various macrofaunal molluscs, annelids and crustaceans within cages (1 mm mesh) and in the uncaged, control mudflat, after an experimental period of 3 months. (After Reise 1985.)

Species	Uncaged controls	Caged mud
Mollusca		
Littorina littorea	0	3
Hydrobia ulvae	3	10
Mya arenaria	0	88
Cerastoderma edule	8	1283
Abra alba	0	23
Macoma balthica	0	5
Scrobicularia plana	0	3
Spisula subtruncata	0	100
Mytilus edulis	0	13
Annelida		
Tubificoides benedeni	820	3055
Pygospio elegans	18	350
Spio filicornis	3	50
Polydora ligni	0	533
Malacoceros fuliginosus	0	405
Ampharete acutifrons	0	5
Lanice conchilega	0	5
Tharyx marioni	8	5323
Scoloplos armiger	0	3
Capitella capitata	93	140
Heteromastus filiformis	240	223
Pholoe minuta	0	3
Nereis diversicolor	0	8
Nephtys hombergi	3	5
Anaitides mucosa	0	8
Eteone longa	0	118
Crustacea		
Corophium volutator	0	490
All macrofaunal individuals	1193	12343
All macrofaunal species	9	28

markedly with those obtained by equivalent studies on rocky shores (Section 5.1.7). These have often shown that predation, by maintaining the numbers of potentially superior competitors at low levels, prevents dominance by one or a few species and the competitive exclusion that can result, and thereby maintains high species richness. In the absence of predation on rocky shores, the numbers of species declines and dominance by a few species increases, in at least some components of the fauna (these studies have rarely investigated the whole fauna). One might argue from the results of several caging experiments on soft sediments that predators often keep both density and species richness below carrying capacity and that the larger benthic species are therefore not food, but predator limited (Peterson 1979; Holland et al. 1980; Reise 1985; Barnes 1994). When predation is removed, one might expect competition to increase but eventual competitive exclusion in the absence of predators might take a long time—longer than most caging experiments have lasted—and the more immediate effect of interspecific competition might in any event be reduced growth and/or reproduction, not exclusion. There is some evidence in favour of this. On the other hand, the act of installing a cage not only creates a disturbance but also alters the conditions inside it (increasing sedimentation, decreasing rates of water flow, etc.), and it can be argued that the increases in numbers of individuals and of species inside the cages are to some extent caused by colonization of the caged habitats by opportunistic, *r*-selected species absent from nearby control areas. There is evidence supporting this too.

To understand the role of predation in structuring soft-bottom benthic communities, it is probably essential to distinguish (a) between indiscriminate mortality on the young stages of the benthic prey, i.e. on the newly settled larval and post-larval stages in the surface few millimetres of sediment, and the more specialist consumption of the larger, older individuals; (b) between shallow regions supporting considerable productivity by diatoms, etc., and the deeper areas dependent on detritus and bacteria; (c) between predators approaching their prey from the surface (epibenthic predators) and those living with the prey within the sediments.

Reise (1985) has summarized the extensive work carried out in the German Waddenzee which has shown that small epibenthic fish and decapod crustaceans, both in the form of the juveniles of larger species using the area as a nursery ground (flatfish, crabs, etc.) and species that are small as adults (gobies, shrimps, etc.), can exert a profound effect upon the numbers of both meiofaunal and juvenile macrofaunal organisms, especially on open expanses of bare sediment. Here, and in a number of equivalent areas, this indiscriminate consumption of their young stages does seem not uncommonly to prevent the adult annelids, crustaceans and molluscs from attaining the carrying capacity of their habitat, except on the rare occasions when a prey species locally escapes this control and successfully establishes a monospecific patch. The overall effect of these generalist epibenthic consumers of anything small, however, is mitigated by predation within the guild itself (consumption of small crustaceans and fish by larger fish and birds), by stochastic climatic variables (Section 3.3.5), and by the presence in some areas of surface structures, especially sea-grasses but also shell debris, which reduce the foraging efficiency of consumers, and thereby form refuges.

Those prey individuals that escape this period of heavy mortality may be relatively safe from further predation, especially if they construct deep burrow systems. Predators moving over the surface of the sediment can often obtain only part of such individual adult infaunal prey: predators bite off those parts of the larger species that, temporarily or permanently, project near or above the surface. Most prey possess fast escape or withdrawal reactions, but these do not confer immunity. The polychaetes *Arenicola* and *Heteromastus*, for example, may lose the tips of their tails in this manner, fan-worms such as *Sabella* their tentacular crowns, and bivalves their siphon tips. They are not killed by such cropping of their exposed parts and can regenerate the missing portions, but energy has to be diverted away from growth and/or reproduction to accomplish this; the bivalve *Tellina* may be prevented from breeding by the cropping of its siphons by young flatfish, a loss that can account for one-half of its annual biomass. One-third of the mean annual biomass of *Arenicola* may be removed in an equivalent manner. Epifaunal species are, of course, more exposed to such predatory attack and they often bear specific antipredator devices such as thick shells or spicular skeletons.

By no means all experimental manipulations have demonstrated such strong control by epibenthic predators as shown in Table 3.2, however. Indeed, in some cases, paradoxically the densities of some infaunal prey actually decline when they are protected within cages. It is clear that part, at least, of this complexity is caused by trophic interactions within the sediment, and is a consequence of the generalist, omnivorous diets of many of the species. A complex food web in shallow waters that has been investigated on both sides of the Atlantic is: the gastropod *Hydrobia* → anemone *Nematostella* → prawn *Palaemonetes* → higher-order predators (Fig. 3.8). The prawn will consume some *Hydrobia*, although one of the main predators on the gastropod is the burrowing *Nematostella*, which is itself a favoured food of the prawn. It can be observed in this system that *Hydrobia* is abundant only in the presence of the prawn, because of the consequent diminution in anemone numbers, and vice versa. Further, densities of the anemone are high in the presence of fish, such as *Fundulus* in the USA and *Gasterosteus* in the UK, because of their consumption of the prawn. (Numbers of *Hydrobia* are of relevance to many other organisms besides the anemones and prawns above: this genus dominates many soft sediments of the northern Atlantic seaboard, contributing up to 80% of macrofaunal biomass and being the gate through

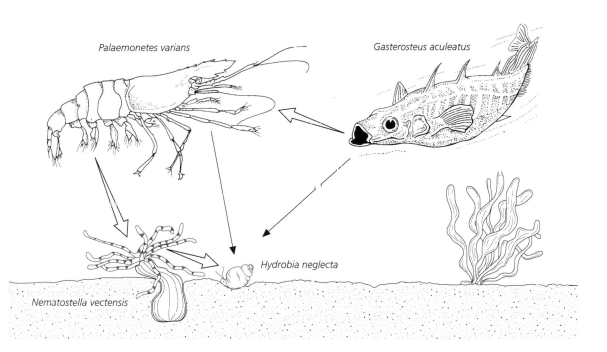

Palaemonetes varians

Gasterosteus aculeatus

Hydrobia neglecta

Nematostella vectensis

Fig. 3.8 A three-level interaction between some predators and prey in a North Atlantic soft-sediment food web. Thick arrows indicate preferred prey species, thin arrows prey that will be taken if no other choices are available. Not to scale. (Diagram courtesy of V. McArthur.)

which 66–90% of the energy flows to predators, including many birds and fish.) Other equivalent complexity is generated by the presence in most sediments of predatory worms (nemertines and polychaetes) which consume other infaunal worms and crustaceans. If then, the epibenthic crabs and fish preferentially take the larger of the available infaunal invertebrates, which often are themselves predatory species, the effect is to enhance the survival of the prey species nearer the base of the food chain. Protect the predatory infauna within a cage and the result can be their enhanced survival but consequent reduction in numbers of, for example, *Hydrobia*. Unfortunately, it is much more difficult experimentally to manipulate infaunal predators within exclosure cages, without causing unacceptable disturbance to the sediment, so less is known of their interrelationships and effects. A final known cause of lesser prey abundance within the protection of exclusion cages concerns the interference or disturbance

covered in Section 3.3.2, although here intraspecifically. Large *Corophium* interfere negatively with smaller individuals, but they are also the preferred prey size range of vertebrate epibenthic predators. Because in the normal course of events large, old individuals are less numerous than small, young ones within any population, the presence of vertebrate predation permits many small *Corophium* to survive whereas in its absence, population density is lower but average size larger. Avoidance of such adverse effects of established adults is the most likely explanation of the dispersal of juvenile (post-larval) macrofauna by 'tidal drifting' that has been widely reported from littoral soft-sediment areas (see, e.g. Jaklin & Günther 1996).

In the absence of equivalent experimental evidence from deeper areas of the continental shelf, it is tempting but unwise to extrapolate from the better known intertidal and sublittoral regions. It does seem, however, that fish and crabs are likely to increase in importance as predators and disturbers, whereas the productivity fuelling the system will be less. But just as in the deep sea (Chapter 7), counter arguments can be mounted as to the likely structuring role of predation. Thus although pelagic consumers of benthic organisms are likely to be more important, so will second-order consumers of those species. Which exerts the dominant control is still open to question.

3.3.4 Competition for resources

It can be argued from theoretical considerations that deposit feeders should be limited by competition for food and suspension feeders by competition for space (e.g. Levinton 1979). Considering the densities of larvae attempting to settle out of the plankton, which can attain more than one million per square metre, it is also clear that were all to survive and grow to adult size neither the available space nor food could possibly support them. If, therefore, other agencies such as disturbance, predation and environmental perturbations (Section 3.3.5) do not control numbers first, there is no question that competition is to be expected. What is not yet known is how often it really does occur in nature.

Consumers are certainly known to be capable of affecting the abundance of their food. Several studies have shown that, for example, gastropods and amphipods deposit feeding in shallow waters can have a dramatic effect on the numbers of diatoms and other photosynthetic microalgae in the sediment. *Hydrobia* could eliminate diatoms from patches of sediment that had originally supported 5×10^5 diatoms per cm^2. It could home in on rich patches by behavioural means, detecting the difference between areas artificially enriched with diatoms and control areas of the natural substratum. Similarly, *Ilyanassa* has been shown to remove more than 10% of the surface chlorophyll per day. When these snails were removed from experimental areas of their respective natural habitats, dramatic increases in algal numbers occurred. Two problems, however, remain. Experimental demonstrations of these effects have often used unnaturally high densities of the consuming species (to 'force' results in short spaces of time), and the consumer affecting the consumed is not the same thing as showing that the consumed affects the abundance of the consumer—which is the crux of competition for food.

There is considerable evidence that intraspecific competition for space does occur amongst suspension-feeding bivalve molluscs (and see Section 3.3.2), as well as within individual deposit-feeding species including various spionid polychaetes, *Corophium* and *Hydrobia* where the resource at stake is most likely to be food. Both the population density and

individual growth rate of several *Hydrobia* species, for example, have been shown to be linearly related to the diatom or chlorophyll content of the ambient sediment. Some suspension-feeding bivalves also compete for food, and display reduced individual growth in inverse proportion to their population density. Several species can feed on both suspended and deposited materials, and in one such species, the bivalve *Macoma*, Ólafsson (1986) found evidence that the limiting resource did indeed vary with feeding method as theorized by Levinton. The evidence for interspecific competition (except in respect of interference competition), however, is less well established and is still too reliant on the circumstantial or on studies of a few sympatric congeneric species, such as *Hydrobia*. This is all the more surprising in that it is of more than academic interest. Presumed competition for food between the slipper limpet *Crepidula* and the oysters *Ostrea* and/or *Crassostrea* has, for instance, led to expensive attempts to eliminate *Crepidula* from oyster beds. The more similar two species are, the more likely are they to compete (on the basis that their resource requirements are likely to overlap more). This is perhaps borne out by demonstration of interspecific competition between different species of *Hydrobia* and *Corophium* (e.g. Jensen & Kristensen 1990).

Other demonstrations of intrageneric competition have proved more contentious, especially where they revolve around character displacement. For many years, the classic 'textbook example' of competitively induced character displacement was that within *Hydrobia* species in the Danish Limfjord. In a series of papers from 1975 onwards, Fenchel showed that (1) where *H. ulvae* and *H. ventrosa* occurred allopatrically within the fjord they were of the same size, (2) where they occurred together, *H. ulvae* was larger than 'normal' so that there was a height ratio of some 1.3 : 1 (a weight ratio of 2 : 1) between it and *H. ventrosa*, and (3) the intensity of intra- and interspecific competition were equivalent. A character displacement in linear dimension of 1.3 : 1 is the equally classic Hutchinsonian ratio reflecting the minimum size difference necessary to allow the exploitation of sufficiently different resources as to permit two species to coexist indefinitely without competitive exclusion. Equality of intra- and interspecific competition intensity is the minimum necessary condition for potential competitive exclusion. The exact nature of the competitive

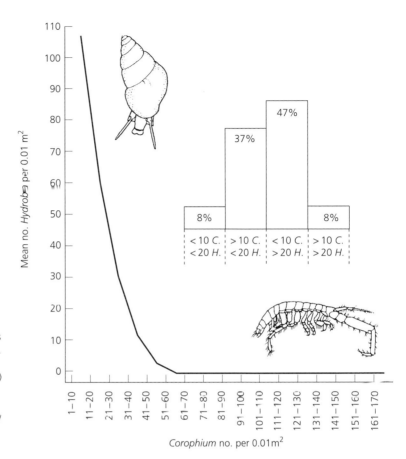

Fig. 3.9 Average numbers of the gastropod *Hydrobia ulvae* found in the same 0.01 m² area as differing densities of the amphipod *Corophium arenarium*. The inset histogram displays the proportion of samples (out of 315 total) with different relative abundances of these two, deposit-feeding species. The numbers of the two are clearly inversely correlated. (Barnes, unpublished data from Norfolk, UK.)

processes was not elucidated, but it was suggested that, when larger, *H. ulvae* ingests larger food particles, so that interspecific competition for food was implicated. Although it seems fairly certain that *Hydrobia* species do compete with each other for food, in the last two decades several lines of evidence have suggested modification of the character displacement aspect of this story. These include: demonstration that size of snail is not necessarily a good indicator of size of food particle ingested; failure to find equivalent size divergence in sympatry at other locations; and the occurrence of an important confounding variable: habitat type. *H. ulvae* is most commonly an intertidal marine–estuarine species, whereas *H. ventrosa* is largely restricted to non-tidal environments. *H. ulvae* nevertheless does occur, albeit rarely, in the habitat type preferred by *H. ventrosa* and when it does, regardless of whether *H. ventrosa* is also present or not, it is larger than when it occurs intertidally, with a lagoonal to intertidal height ratio of some 1.3 : 1. The apparent

character displacement in height may be solely a response to the hydrodynamic environment. In the Limfjord, to some degree the various *Hydrobia* species divide the salinity gradient between them, and the same phenomenon is also shown by the local amphipod *Gammarus*, five species of which have distributions approximating 'beads on a string' along the gradient. In these amphipods, however, the division of resource spectra between the suite of closely related species, whether salinity in the Limfjord or depth, as in the Baltic Sea, proved to be a facultative response to the need for reproductive isolation rather than any effects of competition (see, e.g. Kolding 1986).

We saw above (Section 3.3.2) that distributions are often patchy: high densities of one species correlate with low densities of another and vice versa. This may occur even when there is no evidence of hostile interaction between the two (e.g. Fig. 3.9) or any need for reproductive isolating mechanisms. Interspecific competition is an obvious assumption,

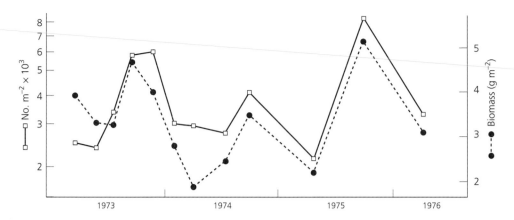

Fig. 3.10 Annual fluctuations in the density and biomass of benthic macrofauna off the Northumberland coast, UK. (After Buchanan *et al.* 1978.)

and faced with failure to obtain experimental evidence of this (as is the case with the *Hydrobia* vs. *Corophium* of Fig. 3.9), recourse to Connell's (1980) 'ghost of competition past' has been made. The argument runs: they may not compete with each other now, but it is only by assuming that once they did, that their present differential use of the resource spectrum makes any ecological or evolutionary sense. We have seen above that there are alternative explanations and although the argument may otherwise be true, the ghost of competition past is nevertheless an unfalsifiable hypothesis of last resort.

The above examples have all been derived from very shallow water. Deeper areas, specifically those without *in situ* benthic microalgal production, would appear to have a bacterial base to the food web. In so far as is known, bacteria are widely but thinly dispersed through shelf sediments, unless in association with large items of organic debris. Perhaps accordingly, except in certain well-defined regions —such as some around Antarctica—deeper shelf populations of consumers are sparser than in the immediately sublittoral fringe. Competition for space appears unlikely, although that for food is possible but as yet uncertain.

3.3.5 Environmental perturbations

Discounting human impingement, several benthic assemblages seem fairly stable in the long term (see, e.g. Jensen 1992), although many fluctuate in the short term, not least because of seasonal cycles of recruitment and mortality (Fig. 3.10). Others, particularly those in shallow waters, are less stable as a result of wind action, receipt of freshwater discharge, temperature extremes, biologically induced anoxia, and other physical processes. The effect of interference competition by *Ampelisca* on *Ilyanassa* was mentioned above (Section 3.3.2). This occurs in the spring as recruitment to the amphipod population progressively establishes dense lawns of tubes. In the autumn, however, storms extend their influence to the sea bed and the massed tubes of *Ampelisca* are 'rolled up like a carpet', freeing the sediment for domination by *Ilyanassa* during the winter months (Fig. 3.11). The consequence is an oscillating pattern of dominance.

An obvious effect of hurricanes and other tropical storms is the precipitation of very large quantities of rain over small intervals of time—over 10 mm of rain per hour for 50 h or more—and this can cause river levels to rise several metres. Mass mortalities of relatively stenohaline marine organisms in the bays receiving the discharge of this fresh water have frequently been reported. A less obvious effect is the transport of large quantities of fine sediment by such river flow and its eventual deposition in the receiving area (equivalently to the transport of organic matter considered in Chapter 4). A 'once-in-a-hundred-years rainstorm' in California in 1978 mobilized vast quantities of silt and clay and caused the deposition of a 10 cm layer of mud at one locality. The bivalves *Protothaca* and *Chione* suffered a 100% death rate; and among the suspension-feeding molluscs generally, those with short siphons, large individuals and dense populations all experienced

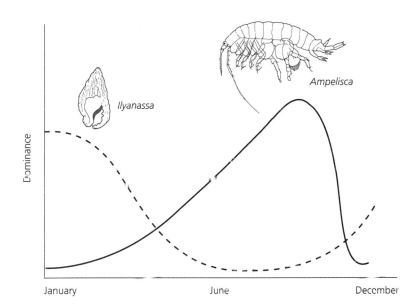

Fig. 3.11 An oscillating system of dominance between the tube-building amphipod *Ampelisca* and the deposit-feeding gastropod *Ilyanassa*. (After Mills 1969.)

substantial mortality. The effect of extremes in another climatic variable, temperature, are most evident in the shallowest of regions. In the European Waddenzee, for example, severe winters with littoral ice cover occur once every 10 years on average, and the reduction in temperature is capable of killing all the cockles. *Cerastoderma* is not the only organism to suffer during these Waddenzee 'ice winters'; amongst other populations dramatically reduced are those of the epibenthic predators considered in Section 3.3.3, and it is notable that recruitment of cockles there and elsewhere is unusually high in the summers following widespread mortality of the consumers of their spat. It is also only immediately after one of these perturbations to the cockle populations that *Corophium* can dominate widespread regions of the substratum (see Section 3.3.2).

Some species, particularly those without dispersive stages in their life history, may take a considerable period of time to recover ground lost during these episodes of mortality. This is well illustrated by a study of the recolonization by benthic animals of Old Tampa Bay in Florida after anoxia consequent on a *Gymnodinium* bloom killed 77% of the infaunal species and 97% of the macrobenthic individuals (Fig. 3.12). The rate of recolonization was roughly proportional to the dispersal faculties of the fauna, after an initial rapid influx of opportunistic polychaetes. One year later, the fauna was similar in composition to that present before the mass mortality, but differences of detail were still present after 2 years. In particular, the polychaete *Onuphis* and the gastropods *Nassarius* and *Prunum*, which together occurred at a density of over 300 m^{-2} before the event, had only achieved a joint total of 3 m^{-2}: none has a free-living planktonic phase.

3.3.6 Conclusions

Not surprisingly, different workers favour different sources of mortality as being the major structuring forces in soft-sediment environments, and all potential sources have their advocates: predation, competition, interference, inadequate recruitment, physical effects and environmental stress, juvenile mortality, and so on. A fundamental question still to be answered is whether populations are generally at the carrying capacity of their habitat or whether other factors keep numbers below this level, and if so how frequently. It has been pointed out that the importance of competition and competition-like interference is mainly argued by those working along the Pacific coast of the USA, whereas studies carried out around the fringes of the Atlantic have stressed processes keeping populations well below carrying capacity. Population densities may well be higher in the Pacific than in the Atlantic and it does appear that shallow-water Pacific communities are generally

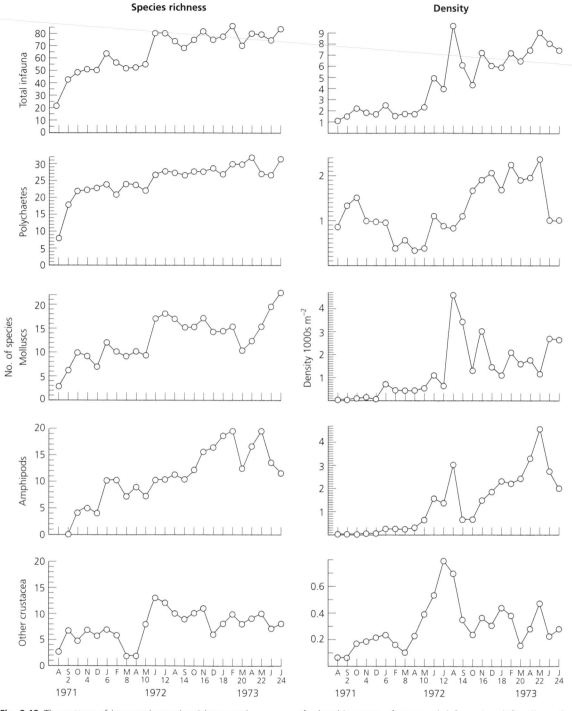

Fig. 3.12 The pattern of increase in species richness and density per unit area of different elements in the macrofauna of a benthic system after natural defaunation. (After Simon & Dauer 1977.)

the richer in species. The dense, single age-class dominated benthic assemblages clearly demarcated from other such assemblages by sharp boundaries described for Pacific coasts of the USA do not seem to have counterparts in European seas, for example; and whereas these Pacific communities would not appear to be highly susceptible to predation, several of those in the Atlantic are dominated by such species. Unfortunately, we do not have nearly enough information to be able to say whether this Atlantic–Pacific dichotomy is a real one or is only an artefact of the very limited number of sites investigated. Neither is there abundant evidence yet available from outside the northern-hemisphere waters of these two oceans.

There is therefore little to be gained by endeavouring to argue the merits of any of the various sources of mortality (including problems of lack of available recruits or colonists as a notional source of mortality) as *the* factor controlling soft-sediment systems. All those considered above are doubtless important at some times in some localities: no study has yet assessed the relative contribution of each to the long-term dynamics of any single system. It has been said that we know more about the surface of the Moon than we do of the sea bed around our shores: we do perhaps know more of the forces structuring sea-bed communities than lunar ones, but not as yet that much.

3.4 Diversity

In Chapter 1 it was pointed out that the global pattern of species richness over the continental shelf was determined almost solely by variation in the epifauna (Section 1.4), the tropical benthos being species rich and diversity declining polewards, and we can now attempt to analyse this pattern (see also Ormond *et al.* 1998). We must start, however, by stressing that differing species richnesses can be consequent on two entirely separate phenomena: (1) the generation and spread of new species in evolutionary time; (2) the persistence of given levels of richness in ecological time. On the one hand, few species may be present because of the absence of barriers to gene flow and therefore because of low rates of speciation (we suggested in Section 1.3 that this might account for the paucity of planktonic species), or because an area in question is isolated

geographically from potential sources of colonists (the Atlantic is isolated from the Pacific, for example, by north–south aligned continental barriers, and the eastern Pacific is isolated from the western Pacific by the huge distances separating shallow-water habitats; see Chapter 10). For the shelf benthos, the geographical configuration most conducive to the generation of species richness would appear to be a series of island shallow-shelves between which gene flow is difficult in the short term but dispersal is possible in the longer term. This can be seen clearly in Fig. 1.22: diversity is high in the extensive Malaysian, Indonesian and Philippino island systems, and declines with both latitude and longitude away from that combined centre.

Alternatively, species richness may be low because of failure of various species to persist at any given location. This is the truly ecological aspect of the study of diversity, and benthic marine ecology has contributed much to general ecological theory in this subject area: the 'predation hypothesis' of Paine (1966), the 'stability/time hypothesis' of Sanders (1968), the 'predation/competition synthesis' of Menge & Sutherland (1976), and the 'disturbance hypothesis' of Connell (1978) were all derived from marine research. In essence, the problem is one of prevention of competitive exclusion. How have so many epifaunal species managed to coexist in the tropics without displacing each other? In part this question is more apparent than real in that small-scale maps such as Fig. 1.22 hide the fact that many species have very restricted distributions, and although large numbers may be present in the same general region most do not coexist but instead replace each other geographically. On a smaller scale, many species may indeed occur on the same beach, for example, but their distributions do not necessarily overlap: some may occur only in the sandier regions and others in pockets of mud. It is generally true that the more subhabitats or spatial heterogeneity a system contains (for example, the presence of rocks, shell debris and sea-grasses on the surface of a mud flat), the more restricted-habitat species it can accommodate—the same general principle as underlies the island-size effect in biogeography. We noted earlier (Section 1.4) that such considerations may in part account for the differential diversities of the infauna and epifauna.

Finally, there remain suites of species which could each occupy the same specific patch of ground, could be in competition with each other, and could exclude each other. For competitive exclusion to be prevented, the potentially competing populations (and especially those of the dominant competitors) must continually or sporadically be kept below the carrying capacity of their habitat by external mortality factors. Such factors may include the indiscriminate predation and disturbance referred to in Section 3.3.3, frequency-dependent predation, environmental fluctuations and disturbances, and so on. Wherever the frequency of population reduction is just sufficient to prevent competitive exclusion, the diversity *status quo* will be maintained; if, however, competitive exclusion is not prevented or if the frequency and magnitude of population decreases become too severe, diversity will decrease as some species are rendered locally extinct (Huston 1979, 1994). (Ecological factors *per se* cannot increase diversity, only maintain or decrease it.)

Are, then, low-diversity areas of the sea bed regions which have been impoverished (former high levels of diversity have declined through ecological time), or does the pattern of species richness only reflect processes operating in evolutionary time? In respect of most regions of the shelf, and on a relatively large scale, the answer would appear to be that they are as rich now as they have ever been, and if they are not richer it is as a result of lack of spatial heterogeneity, lack of speciation, and presence of barriers to long-distance dispersal. On the very small scale of individual patches of sediment, however, predation and disturbance, at least, would appear to be powerful structuring agents reducing local diversity (Section 3.3.3), and such may also, locally, be true of interspecific competition, but expanses of sediments are mosaics of patches and no species appears capable of being displaced from *all* patches so that diversity is maintained in the habitat as a whole. Successful recruitment to any given areas of the substratum may be a lottery, but individuals of one species are as likely to survive through to adulthood as are those of another (see, e.g. Dayton 1984).

3.5 Benthic communities

Early work on the nature of the organisms inhabiting the sea bed identified a series of 'communities' —regular and characteristic associations of macrofaunal species—each named after the one or two species regarded as being most indicative and characteristic of that particular association. In this manner, large areas of the continental shelf were labelled as supporting various named communities; parallel communities were identified in geographically separate areas with similar environments; and different variants or subcommunities were described in local regions. These developments paralleled the phytosociological approach to botanical ecology and, similarly, precise rules were laid down for the naming and characterization of communities, subcommunities, etc. (At the same time, a similar approach was also applied to pelagic ecology and various 'indicator species', indicative of different water masses, were proposed.)

Thus Thorson (1957) diagnosed dozens of 'communities' grouped into seven major types. *Macoma* communities dominated sheltered waters down to 60 m; *Tellina* communities occurred in shallow, sandy bottoms; *Venus* communities in deeper, but still sandy substrata (7–40 m); *Abra* communities typified the organic muds of estuaries and other sheltered regions; *Amphiura* communities were characteristic of soft sediments at greater depth (15–100 m); *Maldane–Ophiura* communities replaced those of *Abra* in the soft muds of the continental shelf down to 300 m; and a group of amphipod-dominated communities (including those of *Ampelisca* mentioned above) were found in localized areas of sheltered, muddy deposits (see Fig. 3.13). *Macoma* communities, for example, included that of *M. calcarea* around the Arctic Ocean, of *M. incongrua* in the northwest Pacific, of *M. nasuta* and *M. secta* in the northeast Pacific, and of *M. balthica* in the boreal northeast Atlantic, whereas the *M. balthica* community included variants dominated by cockles, lugworms, *Mya* or *Scrobicularia*.

Although the names of many of these communities are still used for descriptive purposes—in much the same manner as one might talk of oak woodland or heather moor—their existence as discrete entities has fallen completely from favour. This has resulted from the application of more objective, numerical methods of analysis in benthic ecology, which have disclosed that the subjective 'typical' assemblages of species characterizing the various 'communities' each graded smoothly into the others (see, e.g. Basford *et al.* 1990). Intermediates are the rule rather than the exception, and each benthic species

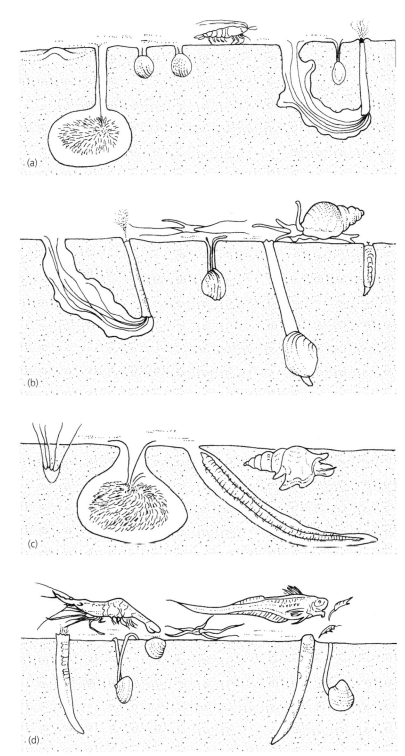

Fig. 3.13 Four benthic shelf 'communities': (a) a *Venus* community; (b) an *Abra* community; (c) an *Amphiura* community; (d) a relatively deep-water community. (After Thorson 1971.)

appears to be distributed in relation to an individual series of requirements, such as substratum type and water depth, rather than as one of many species sharing a common response to any given environmental gradient. In many respects, the whole community story is equivalent to the impression which may be gained of the intertidal zone during marine field-courses. The latter often take students to 'typical' sandy shores, 'typical' mud-flats, 'typical' rocky shores, and so on, and it is easy to gain the impression that the intertidal zone comprises a rather limited series of discrete and distinguishable habitat types. If, however, one carries out a transect along a stretch of coastline, it is evident that habitat types grade into each other and that very few areas are uniform and 'typical'. Sharp boundaries, where they occur, are due not to interactions between different assemblages of species but to major, sudden changes in the environment, as when an area of rock outcrops on an otherwise sandy beach. In like manner, indicator species may characterize certain environmental regimes but they do not necessarily indicate the existence of a predictable suite of associated species.

This is not to suggest that several species do not occur together regularly, and interact with each other; clearly this is often the case. However, each local area of sediment supports an assemblage of species which differs in minor, and sometimes major, respects from all other such local areas, in part for historical reasons, in part as a result of biogeographical processes, in part because of ecological interactions *in situ*, in part as a consequence of minor differences in the environment, and often to a considerable degree as a consequence of stochastic events. Each local area therefore supports its own community at any one time, and a limited number of stereotyped 'communities' do not occur in the sea, except in very broad terms.

4   Salt-marshes, mangrove-swamps and sea-grass meadows

This short chapter will be devoted to the role of the stands of vegetation of terrestrial ancestry which fringe the sea in sandy or muddy areas and function as coastal food factories. Their ecological interest is very much wider than we can cover here; their semi-terrestrial nature places them on the periphery of the main themes of the book and for other details the reader is referred to Pomeroy & Wiegert (1981),

Daiber (1982), Long & Mason (1983), Phillips & Menez (1988), Adam (1990), Jaccarini & Martens (1992) and Stafford-Deitsch (1996).

Sea-grasses are distributed over most of the globe wherever there is shallow water (except in high polar regions), but salt-marsh and mangrove-swamp replace each other geographically: salt-marshes typify cooler and/or drier coasts and mangrove-swamps hot, wet areas (Fig. 4.1). Only in the Gulf of Mexico states of the USA and in southeastern Australia are both vegetation types present along the same

Fig. 4.1 World distribution of salt-marshes and mangrove-swamps. (After Chapman 1977.)

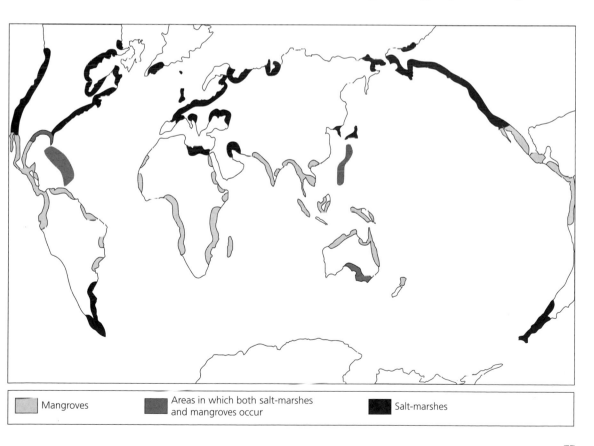

Mangroves | Areas in which both salt-marshes and mangroves occur | Salt-marshes

stretch of coastline, although in dry regions within the tropical mangrove belt a form of salt-marsh can replace the mangrove flora on some parts of the shore where there is minimal shading from the trees. A major difference between salt-marshes and mangrove-swamps on the one hand, and sea-grass meadows on the other, is that the first two are intertidal and rarely extend below mean tide level, whereas the sea-grasses extend well below low water to depths of some 15 m, and 30 m in clear water (a few sites appear to support living sea-grasses at depths of 60 m and even 90 m in one species). All three, however, are dominated by tracheophyte plants which can achieve productivities of up to 8500 g (ash-free dry wt) m^{-2} yearly, and all three floras comprise a suite of phylogenetically unrelated species which have convergently adapted to life in their respective habitats. The terms 'mangrove' and 'sea-grass' therefore have no phylogenetic validity: some 20 tracheophyte families (in 12 orders) have produced mangroves; and six families of monocotyledons are represented by grass-like plants which can survive submerged in sea-water (none, however, is a true grass).

4.1 Salt-marsh and mangrove-swamp

Although salt-marshes (Plate 9, facing p. 64) are dominated by grasses, herbs and dwarf shrubs whereas mangroves (Plate 10, facing p. 64) are large shrubs or trees of up to 45 m in height, they share many characteristics and hence can be considered together. Both facilitate sedimentation by reducing local water velocities and by retaining many of the particles deposited. Development from a muddy or sandy intertidal zone is initiated by germination of the most salt or submersion tolerant of the component species. These begin the process of trapping sediment (if not already begun by the benthic microflora), thereby raising the level of the shore and decreasing the time of tidal submergence. Other less tolerant, but often competitively stronger species can then establish themselves, similarly to be ousted later by yet further species as shore levels continue to rise. Thus a succession of species results with passage of time which may, partly, be reflected by a zonation of species in space as the marsh or swamp progressively colonizes the upper part of a shore. Cycles of erosion and accretion are common although the causes of the different phases are largely unknown. Colonization or recolonization of new areas is effected by vegetative spread or by water-borne seeds or fragments. The physiography of both systems is dominated by a network of creeks which ramify through the densely vegetated areas (Fig. 4.2) and provide the route through which sea-water floods and ebbs. Many species characteristic of shallow, marine sediments in general live within

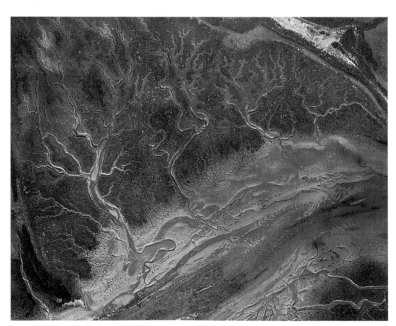

Fig. 4.2 Aerial photograph of a salt-marsh (Norfolk, UK) showing the dendritically branching creek systems. (Cambridge University Collection, copyright reserved.)

these creeks, in pools enclosed by the plant cover, on the marsh or swamp surface, and—in mangroves —as an epifauna or epiflora on the extensive mangrove trunks and aerial roots. These species include single-celled, filamentous and mat-forming algae, deposit-feeding invertebrates and a variety of suspension feeders. The aerial shoots and branches, however, are essentially terrestrial habitats and support an essentially terrestrial fauna. The surface of the sediment therefore forms an interface between the marine and terrestrial ecosystems and there crabs and prosobranchs share the environment with insects, spiders and pulmonate snails.

Both vegetational assemblages are also highly productive: published values generally cluster around 3 kg (ash-free dry weight) m^{-2} yearly. In high latitudes light is the limiting factor to production in winter, whereas elsewhere and at other times inorganic nutrient levels probably set the ultimate limit. Nutrient shortage is only relative, however. As the plants are rooted and fixed whereas sea-water fluxes tidally through the system, nutrients are present in greater abundance than in most other marine habitats. Grazing pressure on the tracheophytes appears insignificant and is confined to the aerial portions of the plants, which are consumed by terrestrial insects (e.g. grasshoppers) and vertebrates (e.g. leaf-eating birds and monkeys), and by a few gastropod molluscs (e.g. *Terebralia*) and grapsid crabs of the genus *Sesarma*, which may consume 10–80% of the annual leaf fall in some mangrove-swamps of the Indian and Pacific Oceans. To some extent, gecarcinid land crabs take their place in the west Atlantic. The larger proportion of the plant production (excluding any associated algae), however, is available for consumption only after conversion to detritus. From our present standpoint, the important question is how much of the productivity of these coastal marshes and swamps is transported to the adjacent sea (see, e.g. Wafar *et al.* 1997).

Early studies (in the 1960s) indicated that some 50% of the net plant production of these areas was exported by tidal flushing and, for example, a *Spartina* marsh in Georgia, USA, was calculated to export 14 kg ha^{-1} of organic matter per tidal flushing during periods of spring tide and 2.5 kg ha^{-1} during neaps. The water ebbing from this salt-marsh contained between two and three times the quantity of organic matter present in the flooding water.

Table 4.1 A general budget for the salt-marshes of the southern and southeastern states of the USA. (After Weigert 1979.)

Net *Spartina* production	1573 g C m^{-2} yearly
Net algal production	180
Total	1653
Metabolized on the marsh	773 g C m^{-2} yearly
Exported by tidal action	880
Total	1653

A series of comparable findings in similar regions suggested an export rate of 0.9–1.3 g C m^{-2} daily for salt-marshes and 0.5–2.4 C m^{-2} daily for mangrove-swamps (with a maximum in the order of 20 tonnes of mangrove litter per hectare per year). If this was generally true, then the marshes and swamps of the coastal fringe would supply some 100 g C per year to each square metre of the continental shelf; a quantity equivalent to the input from the neritic seas, although of a much more refractory nature. Not all marshes behave in this fashion, however (Nixon 1980).

Intertidal marshes may be arranged in a continuum on the basis of their geomorphological and hydrographic setting. At one extreme are those which are open to the sea, are subject to offshore winds and are dominated by patterns of water movement tending to move floating material away from the coast; at the other are those enclosed within land-locked bays, subject to onshore winds and dominated by water movements tending to move flotsam landwards and to strand material along high-level drift lines. The former, open type, including the classic *Spartina* marshes of the coasts of Georgia and South Carolina (Table 4.1), produce an export of salt-marsh detritus, whereas the latter, sheltered type, exemplified by several of those around the southern shores of the North Sea, may actually import fixed organic matter from elsewhere. These two types are the extremes of a continuum, however, and several marshes may be more nearly in balance (Table 4.2). Even this is an over-simplification in that the import–export balance of temperate marshes, at least, may vary within a single year, importing some materials at some times and exporting them at others. This is well illustrated by a salt-marsh on Long

Table 4.2 A nitrogen budget for a Massachusetts salt-marsh. (After Teal 1980.)

	Input (%)	Output (%)	Balance (kg × ha^{-1})
Rain	0.5		3.9
Run-off	22.5		152.2
Nitrogen fixation:			
algal	1.0		6.3
bacterial	9.5		65.2
Tidal exchange:			
NO$_3$ and NH$_4$	21.0	30.0	−78.6
particulate	45.0	49.0	−65.2
Denitrification		18.0	−134.8
Sedimentation		3.0	−24.1
NH$_4$ volatilization		0.5	−0.4
Shellfish harvest		0.5	−0.4
Gull faeces	0.5		0.4
Overall			−75.5

Island Sound, New York (Woodwell *et al.* 1979), in which fluxes of nitrogen, phosphorus, and dissolved and particulate organic carbon included both periods of import and export at different times of the year (Fig. 4.3). Overall, 53 g C m^{-2} yearly were imported, and 1 g N m^{-2} yearly and 0.25 g P m^{-2} yearly were lost. Large fragments of organic debris showed a net loss, but there was a net gain in finely particulate organic carbon. Table 4.3 summarizes annual fluxes in organic carbon and nitrogen in salt-marshes on the Atlantic seaboard of the USA.

Clearly, one cannot generalize the role of these intertidal tracheophyte communities: some do indeed produce a large export of carbon to the coastal sea and thereby partially fuel the food webs of shelf seas, but others are net sinks for organic matter not exporters. They are all sinks in another sense. Grazing on the coarse tracheophytes being insignificant, food webs on the marshes and swamps themselves are in part based on the products of plant decay. These accumulate on the sediment surface and if not remineralized before being buried by sedimentation, become incorporated into the sediment itself. Thus in Woodwell *et al.*'s study, 37% of the net salt-marsh production was buried beneath the surface and K. Paviour Smith suggested that the ratio of animal to plant to dead organic matter biomass which formed a New Zealand salt-marsh was of the order of 1 : 30 : 680.

The fact that large particles of debris may be distributed differentially to finely particulate material (as in the Long Island marsh) introduces the question of the precise nature of the food available on, and as an export from, marshes and swamps. The larger angiosperms produce much litter in the form of dead leaves, etc., and most work has concentrated on this component of the organic input into marine food webs. Such litter decays relatively slowly (Fig. 4.4) and is soon leached of those soluble organic compounds remaining in it (many angiosperms translocate much of the soluble material into their root systems before leaf fall). The litter, together with flakes of particulate matter formed from leached organic compounds, is then colonized by bacteria and fungi in much the same manner as outlined in Section 3.2.2 (see Table 4.4), it requiring some 2 months of decomposition and microbial enrichment in the tropics to raise the nutrient status to levels acceptable to animal consumers. Large particles may often be moved, either to the adjacent sea or onto tidal strand lines, before complete decomposition has occurred. Smaller

	Carbon (g C m^{-2} yearly)		Nitrogen (g C m^{-2} yearly)	
	Dissolved	Particulate	Dissolved	Particulate
Great Sippewissett, MA		−76	−9.8	−6.7
Flax Pond, NY	−8.4	61		
Canary Creek, DE	−38	−62	−0.9	−2.9
Gott's Marsh, MD		−7.3	−2.1	−0.3
Ware Creek, VA	−80	−35	−2.3	0
Carter Creek, MD	−25	−116	−9.2	4.6
Dill Creek, SC		−303		
Barataria Bay, LA	−140	−25		

Table 4.3 Annual fluxes of organic carbon and nitrogen between the salt-marshes and coastal waters on the eastern and southeastern coasts of the USA. Negative values are net exports; positive values are net imports. (From data of Nixon 1980.)

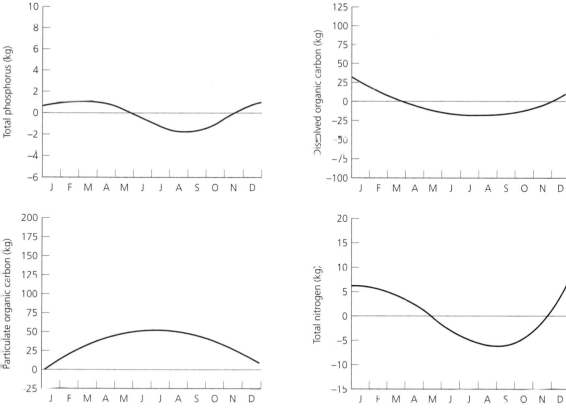

Fig. 4.3 Net tidal exchanges of phosphorus, dissolved and particulate organic carbon, and nitrogen between the Flax Pond salt-marsh and the adjacent Long Island Sound, New York. Negative values are losses to the sea and positive values gains by the marsh. (After Woodwell *et al.* 1979.)

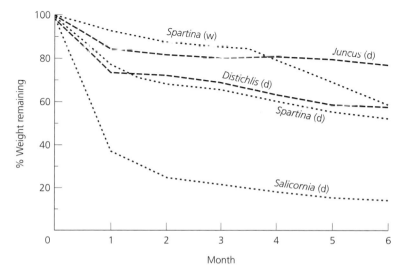

Fig. 4.4 Decay rates of salt-marsh plants in Floridan waters, USA. d, dried; w, fresh materials. (After Wood *et al.* 1969.)

1–14 days	Rapid leaching of sugars and tannins
14–28 days	Depletion of leachable sugars; continued leaching of tannins; initial colonization by bacteria
28–49 days	Depletion of leachable tannins; colonization of outer leaf surface by bacteria, fungi and protists
49–70 days	Erosion of leaf surface; rich microbial community of cellulytic bacteria and fungi; many protist and meiofaunal organisms
> 70 days	Fragmentation of leaves

Table 4.4 Microbial degradation of mangrove leaves in a Florida estuary. (After Cundell *et al.* 1979.)

particles created by mechanical or biological breakdown are available to be consumed by deposit feeders, however, although it seems unlikely that these are the preferred food source on the marshes themselves. The 'detritus-feeding' gastropods and crabs of vegetated intertidal zones probably subsist largely on the abundant (but somewhat neglected) algae which are invariably associated with the larger angiosperms, growing over their surfaces and on the sediment between the plants (see Section 3.3.4). The biomass of all these algae may exceed 25% of that of the salt-marsh tracheophytes, and productivities of the microalgal benthos can reach—and probably exceed—30% of that of their larger associates. Algae make up a much smaller percentage of total mangrove-swamp biomass; in Laguna Joyuda, Puerto Rico, for example, those associated with *Rhizophora* as epiphytes average only 1.25 g (dry weight) m^{-2} (although with a maximum of nearly 5 g per mangrove prop root or 2.5 kg m^{-2}). Nevertheless, they turn over this biomass 4–5 times a year and contribute almost as much as mangrove leaf fall to the system (Rodriguez & Stoner 1990). Recent evidence suggests that littorinid, hydrobiid and nassariid gastropods and a variety of crustaceans graze this microalgal productivity as fast as it is produced. The larger plants may therefore contribute relatively little to *in situ* secondary production, less even than the phytoplankton stranded on the sediment and amongst the vegetation by the tide.

4.2 Sea-grasses

Sea-grasses (Plate 11, facing p. 64) are more thoroughly marine and only a few species extend into the intertidal zone. Their productivity and its fate are therefore entirely marine phenomena and it is less realistic to discuss them in terms of exports and imports. Nevertheless, dead sea-grass leaves can be transported over huge distances. They have been recorded at depths of nearly 8000 m; sea-grass remains were identified in nearly every one of a series of 5300 photographs taken of the sea bed at an average depth of 4000 m; and after hurricanes, mats of leaves up to 50 m across have been reported from the Florida Current (see below). Yet in life they are confined to the fringe of the sea. Long-distance transport of leaves clearly depends on hydrographic conditions, but it also varies with the properties of the leaf itself. Some species (e.g. *Thalassia*) sink after being detached, whereas others (e.g. *Syringodium*) float, at least for some time (Fig. 4.5)—a similar dichotomy occurs

Fig. 4.5 The effects of grazing by parrotfish (*Sparisoma*) on two sea-grasses: *Thalassia* and *Syringodium*. (After Zieman *et al.* 1979.)

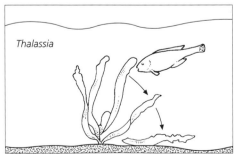
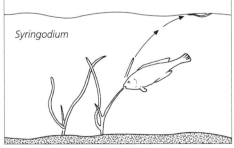

amongst mangrove leaves. Thus in the sea-grass meadows of the Virgin Islands, for example, *Syringodium* leaves account for 92.5% of the loss of sea-grass material (15×10^4 g or $0.2–0.4$ g m^{-2} daily) even though *Thalassia* makes up 94% of the total sea-grass biomass and 90.5% of the productivity.

The term meadows aptly describes the appearance of stands of sea-grasses. Up to 15 000 shoots of the smaller species, with several leaves per shoot, may arise from each square metre of subtidal sand or mud flats, the leaves often being strap-shaped. Biomass and productivity vary latitudinally, being greatest in tropical waters. In temperate areas, mean biomass probably lies close to 500 g dry wt m^{-2} (with a maximum recorded value of over 5 kg m^{-2}) and productivity is of the order of 2 g C m^{-2} daily during the growing season (up to a maximum of less than 10 g C m^{-2} daily). Comparable figures for tropical sea-grasses are: biomass averaging over 800 g dry wt m^{-2} (maximum 8 kg m^{-2}) and productivity averaging some 5 g C m^{-2} daily, with a maximum of nearly 20 g C m^{-2} daily. Similar to the macrophytes considered in Section 4.1, few animals appear directly to use sea-grass production as food; some fish, turtles, sirenians and a few sea-urchins with ruminant-like cultures of cellulose-splitting bacteria in their guts are notable exceptions. Once again, however, most consumers are dependent on the decomposing grasses, and bacteria and fungi are critical in rendering the plant material digestible. Experiments have shown that aged sea-grass detritus is assimilated to a much greater degree than 'artificial detritus' prepared by macerating living sea-grass leaves (thereby permitting access to the cell contents). Perhaps 5% of sea-grass production is consumed directly: a higher value than that of the coarser salt-marsh and mangrove-swamp plants with their more extensive supporting tissues.

Sea-grasses decay slowly (otherwise they would not occur in identifiable form in the deep sea) but nevertheless more quickly than intertidal tracheophyte material (Fig. 4.6). In the tropics, *in situ Thalassia* leaves may lose 10–20% of their initial dry weight per week (varying with the extent to which they are broken down mechanically by water movement) and hence they would be completely decomposed in less than 1 year. Decomposition is slower in higher latitudes and there eel-grass (*Zostera*) leaves may lose only 20% of their original weight in 100 days

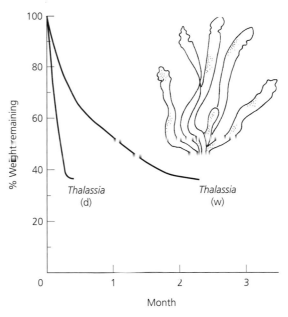

Fig. 4.6 Decay rates of sea-grass *Thalassia* in Floridan waters, USA. d, dried; w, fresh materials. (After Wood *et al*. 1969.) (Compare with Fig. 4.4.)

and much eel-grass material will be remineralized in and below the redox discontinuity layer. Because sea-grasses are normally submerged, the products of their decay can enter the planktonic system to a larger extent than can, for example, leaves of salt-marsh plants. At times of leaf die-back and decomposition, heterotrophic protists abound in the water column over the meadows and take up both the dissolved organic compounds leached from the sea-grasses and the rapidly multiplying bacteria. Sea-grass detritus is also rich in micro-organisms. One gram (dry weight) has been calculated to support, on average, $10^9–10^{10}$ bacteria, $5 \times 10^7–10^8$ heterotrophic flagellates and $10^4–10^5$ ciliates, yielding a total biomass of some 9 mg of monerans and protists to deposit feeders.

Living sea-grass leaves also provide an attachment site for numerous types of epiphytic algae (both macroscopic and unicellular), and other algae occur between the sea-grass shoots and in the surface layers of the sediment. As in the other tracheophyte communities, it is these softer, more digestible algae which support the abundant grazers associated with the meadows. In relatively open stands, the benthic algae may account for 70% of the total primary

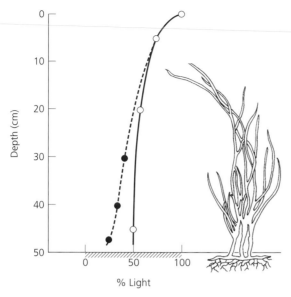

Fig 4.7 Light penetration through shallow waters in the absence of eel-grass, *Zostera* (continuous line), and when eel-grass causes self-shading (dashed line). (After Short 1980.)

production, but thick carpets of leaves reduce the light available to the algal understorey (Fig. 4.7) and in such regions their productivity is less than on open expanses of sediment. Estimates of epiphytic productivity are relatively scarce but biomasses of the same order as those of the leaves to which they are attached are known. An average contribution of over 15% of the total carbon fixation is not unlikely.

Sea-grass meadows are therefore the most productive of shallow, sedimentary environments and not only does their primary production support a rich resident fauna of epifaunal species, for which reason, together with their role as a refuge from epibenthic predation (see Irlandi & Peterson (1991)

and Section 3.3.3), they are frequently used as nursery areas by nektonic species (see Chapter 8). Seagrass production also consistently enters into the food webs of areas far removed from the coastal zone. A question mark does still hang over the importance of exports from all three of the stands of vegetation considered in this chapter, however, and any error is likely to be in the direction of underestimation. Most measurements of loss rates of fixed organic matter are carried out under reasonably clement weather conditions (for understandable reasons). Yet it is likely that freak weather (hurricanes, etc.) can remove more tracheophyte material in an hour than is accomplished during one or more years of normal wind and wave conditions, although for practical reasons it is impossible adequately to sample these irregular storm losses. One such measurement has indicated that during severe storms losses of debris from salt-marshes can be at a rate of over 11 kg ha^{-1} h^{-1}. The effect of fresh-water input during more normal rainfall is also to mobilize organic materials from the sediment surface, increasing the loss rate of carbon by a factor of 5–6, and that of nitrogen by factors of the order of 40. Observations such as these and that after the hurricane in Florida (see above) cannot do other than suggest that estimates that do not take storm losses into account must seriously underemphasize the transport of material from the coast into the open sea. It is probably no coincidence that the harvest of commercial fisheries around the southeastern states of the USA is directly proportional to the area of salt-marsh on the adjacent shoreline, and the same is also true of mangroves. Each hectare of mangrove-swamp may be responsible for supporting an annual production of 400 kg of coastal fish and 75 kg of prawns.

5   Rocky shores and kelp forests

Rocky shores and kelp forests (Plate 12, facing p. 64) are treated sequentially in this chapter because, in cool, temperate regions, the one type of ecosystem is often an extension of the other. Macroscopic algae are the main primary producers, fucoids intertidally and kelps subtidally. Underlying mechanisms of zonation and community organization are basically similar in both ecosystems, so that examples from each can be used in a complementary fashion. The common thread is the predominance of sedentary organisms, a feature also shared by coral reefs that replace kelp forests in suitable tropical environments; indeed, many of the principles of community organization encountered in this chapter will appear again in slightly different forms in Chapter 6.

5.1 Rocky shores

5.1.1 Colonization of the marine–terrestrial gradient

Most students, if asked to describe the biological features of a rocky shore, would mention zonation. This spatial sequence of organisms up and down the shore is universal, involving much the same sort of organisms anywhere in the world. Why should this be so? The shore is the transition between marine and terrestrial environments, but because of water movement associated with tides, waves, and spray, the transition is gradual, creating a habitat in which neither fully marine nor fully terrestrial organisms can prosper. Emersion is stressful to marine organisms, just as immersion is stressful to terrestrial ones. Organisms able to withstand different degrees of stress replace each other along the marine–terrestrial gradient, and this fundamental biological phenomenon applies on any rocky shore from the tropics to the poles. Alternating inundation and exposure to the air are the environmental features of overriding importance determining the kinds of organisms living on rocky shores, and that is why rocky intertidal organisms are rather similar world-wide, despite gross dissimilarities in climate.

One might expect that any transitional gradient between two dissimilar environments would be colonized more or less equally by organisms from either source, yet in rocky intertidal habitats the great majority of organisms are of marine origin. The intertidal zone, however, does not embrace the entire gradient between marine and terrestrial environments. The effects of wave splash and spray extend beyond High Spring Tide Level, and organisms in this spray zone are predominantly of terrestrial origin.

5.1.2 Patterns of zonation

In spite of its ecological importance as the 'other half' of the marine–terrestrial transitional gradient, the spray zone has been studied relatively little, so that most of what is known about life on rocky shores pertains to the intertidal zone. People familiar with temperate shores will be impressed by the dark festoons of fucoid algae covering much of the intertidal rock surface in moderately sheltered localities. The fucoids illustrate very well the phenomenon of zonation (see Fig. 5.5 and Plate 13, facing p. 136) that is characteristic of intertidal sedentary organisms. Mobile animals also tend to be zoned but their limits are not so clearly demarcated as are those of sedentary animals. This is to be expected, as mobile animals can seek refuge in crevices, under weed, or beneath stones at low tide and these protective microhabitats tend to be spread throughout the intertidal zone.

5.1.2.1 Zonal variation between shores

Zonation patterns vary among shores. The order in which species are zoned remains the same, very

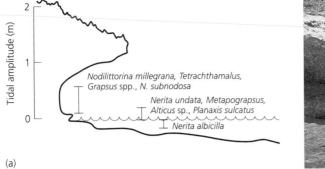

Nodilittorina millegrana, Tetrachthamalus, Grapsus spp., N. subnodosa

Nerita undata, Metapograpsus, Alticus sp., Planaxis sulcatus

Nerita albicilla

(a)

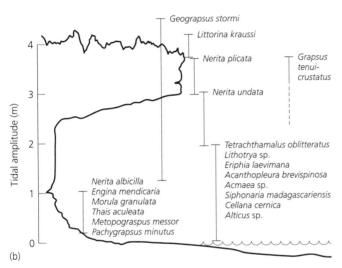

Geograpsus stormi

Littorina kraussi

Nerita plicata

Grapsus tenui- crustatus

Nerita undata

Tetrachthamalus oblitteratus
Lithotrya sp.
Eriphia laevimana
Acanthopleura brevispinosa
Acmaea sp.
Siphonaria madagascariensis
Cellana cernica
Alticus sp.

Nerita albicilla
Engina mendicaria
Morula granulata
Thais aculeata
Metopograpsus messor
Pachygrapsus minutus

(b)

Fig. 5.1 Increasing tidal amplitude vertically extends the intertidal habitat, allowing more species to colonize the shore. (a) Midway along the Red Sea, where there is no appreciable tidal oscillation; (b) Aldabra Atoll, Indian Ocean, where there is a spring-tidal amplitude of 3 m. (After Hughes 1977.)

rarely do species transpose positions down the shore, but the widths of the zones and their degree of overlap vary, and some species disappear altogether and may be replaced by others.

Both physical and biological factors may be responsible for this variation. The biological factors, competition and predation, differ in importance from place to place, often interacting in subtle ways with physical environmental conditions, and are dealt with in Section 5.1.5. Zonal width must ultimately depend on the extent of suitable microclimatic conditions. For sedentary organisms, several physical factors may influence the extent of suitable microclimates:

1 The slope of the shore will determine the amount of suitable rock surface; the flatter the shore the greater the width of substratum lying between a given interval of tidal height.

2 Tidal amplitude will be of great importance. Most localities around Britain have a rise and fall of 3–6 m on spring tides, but the range may be as much as 9–12 m in the Bristol Channel and as little as 0.6 m off southwest Argyll. The greatest tidal range in the world is in the Bay of Fundy, Nova Scotia, where the spring tidal range may exceed 15 m. In central regions of the Red Sea there is virtually no tidal rise and fall, but because wave action wets the rock surface, intertidal organisms occur, but their zones are narrower than at localities experiencing larger tides (Fig. 5.1).

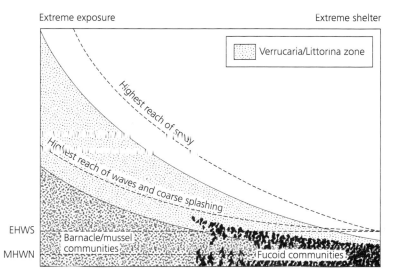

Fig. 5.2 Increasing exposure to wave action elevates zonal boundaries and increases the vertical extent of the splash zone. EHWS, Extreme High Water Spring Tide; MHWN, Mean High Water Neap Tide. (After Lewis 1964.)

3 The average amount of wave action will influence the degree of wetting a substratum receives at a given tidal height; the greater the exposure to wave action the further the zonal boundaries are pushed upshore (Fig. 5.2).

4 The aspect of the shore will affect the amount of drying out at low tide; on shaded slopes surfaces will remain damp for longer than on sunny slopes, so that other things being equal upper zonal limits should be somewhat higher.

These factors will also affect the zonal width of mobile animals in a similar way, but an additional variable for them is the availability of suitable refuges that are used during low tide. *Melarhaphe neritoides*, for example, is sometimes confined to rock crevices above Mean High Tide Level, but on shores with an abundance of empty barnacle shells, which can be used as an alternative to rock crevices, *M. neritoides* may extend down to Low Spring Tide Level.

Superimposed on changing zonal boundaries between shores is the replacement of organisms occupying equivalent tidal heights. In Britain, this amounts to the predominance of fucoid algae on more sheltered shores and their replacement by barnacles and mussels on exposed shores (Fig. 5.2). This is discussed further in Section 5.1.7.4.

5.1.2.2 Geographical patterns of zonation

On a world-wide basis, the characteristics of rocky intertidal faunas and floras are influenced by gross changes in climate in addition to the factors already considered.

In polar regions, ice scrapes the rocks for much of the year and organisms survive only in crevices and hollows, or by migrating to the sublittoral zone. Deep freezing during winter and dilution during the summer thaw create a harsh environment. Very little macroscopic life occurs in the spray zone and is confined to summer flushes of annual algae in temporary pools. The midlittoral zone becomes colonized by annual filamentous and membranous algae that are cropped by a few herbivores including amphipods, littorinids and limpets. A much richer algal flora occupies the sublittoral zone, some species reaching very large sizes, and among this lush vegetation are representatives of most of the major animal groups present in temperate latitudes.

In tropical regions, high light intensities, high temperatures and desiccation at low tide are too severe for most macroscopic algae to survive high in the intertidal zone. In contrast to the festooned rocks of temperate shores, tropical intertidal rocks seem rather desolate at first glance, and it is only below Mean Low Tide Level that life takes on the richness one expects in the tropics (Chapter 6).

The tropical spray zone lacks the tufted lichens characteristic of temperate regions, but is colonized by encrusting lichens and blue–green algae. Littorinid gastropods graze the flora when dampened by spray or by rain. At night, small, terrestrial hermit crabs scavenge over the area and carnivorous

Fig. 5.3 *Grapus tenuicrustatus*, showing spoon-shaped chela used for scraping microalgal film from rock surfaces (Aldabra Atoll, Indian Ocean).

grapsid crabs come up from the intertidal zone to search for prey.

The tropical mid-intertidal zone is colonized by blue–green and filamentous green algae. So close is the growth that often only the greenish colour of the rock surface indicates the presence of plant life. Nevertheless, the productivity of the intertidal algae is evidently very high, because it supports a substantial guild of grazers. Many of these animals are similar to those in other latitudes, for example limpets, chitons, winkle-like gastropods, isopods, and amphipods, but in addition there are grazers not present in cool climates. Among these are numerous herbivorous fishes and grapsid crabs. The latter have spoon-shaped claws with which they scrape algae off the rock surface during low tide (Fig. 5.3). Crabs that forage out of the water, like grapsids and terrestrial hermit crabs, peter out in cold, temperate regions. Presumably at higher latitudes cool air temperatures would lower their metabolic rate too much, causing the crabs to lose their vital agility (see Chapter 10). Further details of British shores have been given by Lewis (1964) and of world-wide shores by Stephenson & Stephenson (1972).

5.1.3 Determinants of zonation

Why are rocky intertidal organisms zoned? There are two levels of approach to this question, one concerning ultimate factors that prevent both the random mingling of species and the monopoly of the intertidal habitat by a few 'super organisms', and

another concerning proximate factors that impose spatial limits to the zones.

5.1.3.1 Ultimate factors

Two ultimate factors are responsible for the general phenomenon of zonation: competition for resources and the restricted potential of any species to perform optimally under different environmental conditions. Competition among species for a common limiting resource is an inevitable process if populations overlap and are not kept at very low densities by grazing, predation or physical factors. Prolonged competition will cause populations of the better competitors to expand and those of the poorer competitors to dwindle to local extinction. Competitive ability depends on various aspects of performance, for example, growth rate, growth form, reproductive rate, aggressiveness and feeding efficiency, and to excel in any such field requires a degree of specialization. Specialization in one field is usually achieved at the expense of poorer performance in another (e.g. fucoids, Section 5.1.4.1; barnacles, Section 5.1.5.1; limpets, Section 5.1.5.2). High performance at one position along an environmental gradient will therefore be accompanied by poorer performance elsewhere. Different species will perform best at different positions, each with the potential to exclude the others from its optimal location. Thus we find that species replace one another, i.e. are zoned, along any pronounced environmental gradient, whether it be the marine–terrestrial transition, or any other kind of transition, such as one of salinity along an estuary, depth on a coral reef (see Fig. 6.7), altitude on a mountainside or aridity at the edge of a desert. Zonation is a particular manifestation of resource-partitioning among potentially competing species. Linear spatial replacements (zones) reflect the strongly linear nature of the environmental gradient, and in the absence of pronounced gradients, resource-partitioning results in more complicated spatial patterns among species (e.g. tropical predatory gastropods, Section 6.7), but the ultimate factors of competition and specialization remain the same.

5.1.3.2 Proximate factors

Proximate factors fixing the limits to zones could,

in principle, be: (a) behavioural, where the organism 'chooses' to settle or remain within its zone of optimal performance; (b) physiological, where the organism is unable to survive in the environmental conditions outside its zone; or (c) ecological, where the organism's potential distribution is curtailed by competition or predation. In practice, remarkably little is known about the proximate factors affecting zonation in many of even the commonest rocky shore organisms. It is so easy to make the unjustifiable assumption that because organisms from higher levels on the shore have greater physiological tolerances to desiccation or extremes of temperature and salinity when subjected to these stresses in the laboratory, then the same factors must limit distribution on the shore. For many years, people tended to be satisfied with demonstrating the almost inevitable correlation between physiological tolerance to stress and zonal height on the shore (e.g. fucoids, below; barnacles, Section 5.1.5.1 limpets, Section 5.1.5.2). Yet as a rule, environmental conditions measured on the shore are far less extreme than the limits of tolerance measured in the laboratory. Not considered was the alternative possibility that the correlation between physiological tolerance to laboratory-induced stress and zonal height on the shore reflects the adjustment of physiological mechanisms so that they work optimally under the appropriate natural environmental conditions. This interpretation is congruent with the concept of ultimate factors determining zonation: optimal physiological performance at a particular tidal level being sharpened on an evolutionary time-scale by interspecific competition, but costing poorer performance at other levels.

Among the relatively few species for which the proximate factors of zonation have been identified, it appears that physiological factors often fix upper zonal limits, especially of high-shore species, but that ecological and behavioural factors tend to fix lower zonal boundaries.

5.1.4 Zonation of macroalgae

Optimal performances at the appropriate heights on the shore, reflecting the ultimate factors of zonation, have been demonstrated for a series of Californian intertidal algae. The capacity of these algae to sustain photosynthesis in air varies according to zonation. Species from the lower shore photosynthesize less well in air than underwater. Species from the mid- to upper-shore reach maximum photosynthesis after some degree of drying, whereupon the photosynthetic rate in air can be as much as six times that underwater at the same illumination and temperature, possibly because removal of surface water increases the rate of diffusion of CO_2 into the tissues. Further drying retards photosynthesis, by uncoupling part of the electron transport system, and the time taken for desiccation to reduce the photosynthetic rate to half its maximum value increases successively in algal species from higher levels on the shore (Fig. 5.4). These differences are due to lower desiccation rates of the higher shore algae rather than an increased tolerance of desiccation itself. Desiccation rates are low enough under normal conditions that these algae are capable of sustaining high rates of photosynthesis throughout tidal emersion, when most of their carbon is probably fixed.

5.1.4.1 Zonation of Scottish algae

Five fucoid algae occupy overlapping but well-defined zones on a sheltered rocky shore on the Isle of Cumbrae, Scotland (Fig. 5.5). As with the Californian algae, the ability to tolerate exposure to desiccating conditions, and then maintain normal growth, increases progressively in species from low to high shore levels. Among naturally situated plants, damage is sometimes caused during the spring and summer when neap tides leave the upper shore continually emersed for several days of calm, warm, dry weather, leading to the death of *Ascophyllum nodosum*, *Fucus spiralis* and *Pelvetia canaliculata* at the upper edges of their zones. Prolonged exposure by neap tides to winter frost and rain, or even hard freezing, has no ill effects. Schonbeck & Norton (1978, 1980) found that *Fucus spiralis* transplanted into the *Pelvetia* zone grew poorly and eventually died even though not crowded by *Pelvetia* itself. Desiccation is evidently an important proximate factor determining the upper zonal boundaries of *Ascophyllum nodosum*, *Fucus spiralis* and *Pelvetia canaliculata*. Drought damage was not observed among *Fucus vesiculosus* or *F. serratus*, and this, together with the less clearly defined upper zonal limits, suggests that ecological factors may curtail the upshore distribution of these algae below the potential level set by physiological factors.

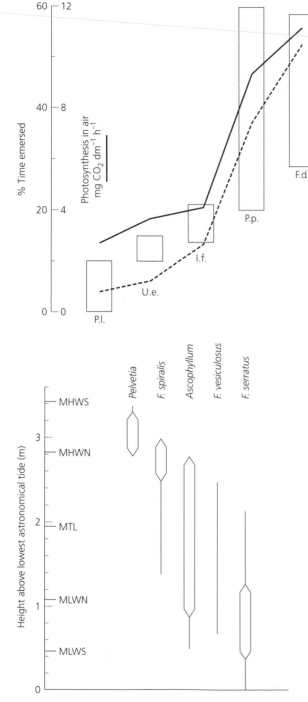

Fig. 5.4 Californian algae from an increasing zonal height on the shore exhibit an increased capacity to maintain photosynthesis during tidal emersion. P.l., *Prionitis lanceolata*; U.e., *Ulva expansa*; I.f., *Iridaea flaccida*; P.p., *Porphyra perforata*; F.d., *Fucus distichus*. (After Johnson *et al.* 1974.)

Fig. 5.5 Zonation of fucoids on the Isle of Cumbrae, Scotland. MHWS, Mean High Water Spring Tide; MHWN, Mean High Water Neap Tide; MTL, Mean Tide Level; MLWN, Mean Low Water Neap Tide; MLWS, Mean Low Water Spring Tide. (After Schonbeck & Norton 1978.)

Pelvetia and *Fucus spiralis* transplanted downshore grew well (Fig. 5.6a), proving that the lower zonal boundaries of these algae are not determined by physical environmental conditions.

Factors limiting the lower zonal boundaries of the other fucoids were not investigated experimentally, but the generally increasing growth rates of species from lower shore levels (Fig. 5.6b) suggest that competitive exclusion may be a common mechanism.

5.1.4.2 Zonation of Australian algae

On sheltered rock platforms in New South Wales, foliose macroalgae such as *Ulva lactuca*, *Colpomenia sinuosa* and *Corallina officinalis* cover most or the substratum at low shore levels, but peter out abruptly in the midshore region. Here the rock is extensively covered by the encrusting red alga *Hildenbrandia prototypus* and supports dense populations of grazing gastropods. Harsh physical factors, such as desiccation and high temperatures because of intense insolation at low tide, or intense grazing by the numerous gastropods, could limit the foliose algae to low shore levels. To determine the relative importances of these factors, Underwood (1980) constructed a series of stainless steel mesh fences and cages on the rock platform. The

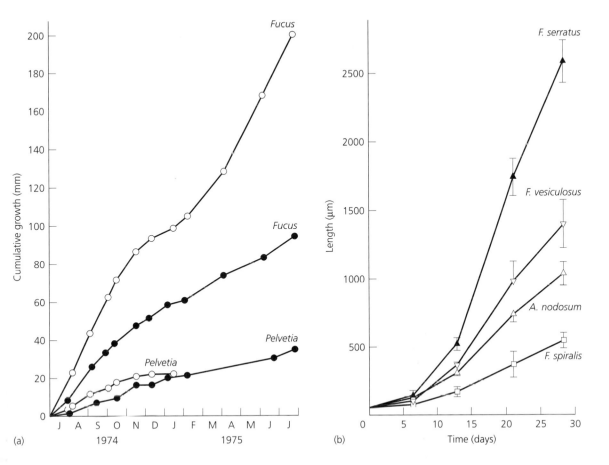

Fig. 5.6 (a) Cumulative linear growth of *Pelvetia canaliculata* and *Fucus spiralis* transplanted into their own zones (●) and onto the middle shore (○). (b) Linear growth of fucoids cultured from zygotes in the laboratory. (After Schonbeck & Norton 1980.)

fences excluded grazers but cast virtually no shade, whereas the cages both excluded grazers and cast sufficient shade to reduce the temperature of the covered areas. Roofs were also constructed like cages without the sides, so as to cast shade without excluding grazers. The experiments demonstrated that the grazing gastropods were removing all spores and sporelings of foliose macroalgae in the midshore region, and that even when grazing was prevented, harsh physical factors killed the algae as they grew beyond the sporeling, 'algal turf' stage.

The encrusting alga *Hildenbrandia* can withstand the physical conditions and although gastropods graze over its surface they merely remove spores of other algae.

5.1.5 Zonation of animals

5.1.5.1 Zonation of sedentary animals

Proximate factors determining the zonal boundaries of intertidal algae often also determine those of intertidal animals, but the zonal distribution of animals may be influenced by two additional features: mobility and food. After settlement and metamorphosis, many sedentary animals such as sponges, hydroids, tunicates, tubicolous polychaetes, vermetid gastropods, oysters and barnacles, are immobile, although barnacles can be pushed along to some extent by pressure from growing neighbours. Other sedentary animals such as sea anemones and mussels are capable of very slow movement and therefore have a limited capacity to adjust their position according to circumstances, whereas non-sedentary animals are potentially capable of moving throughout the intertidal zone within one or a few tidal cycles. Increasing mobility may be expected to reduce the effects of physical environmental stress on zonation, which may be influenced more by the distribution of protective microhabitats or food supply. Of course, the latter are also important resources for sedentary animals: the distribution of protective microhabitats may influence larval settlement, and the daily food supply of sedentary animals, all of which are 'particle' feeders, will decrease with the duration of tidal immersion upshore.

Mussels. Whether time available for feeding ever limits the intertidal distribution of filter-feeders is unknown, but its effect on the growth of mussels is apparent. *Mytilus edulis* from high shore levels may live for many years, but grow very slowly and never reach a large size. When transplanted to low shore levels, the same mussels put on a spurt of growth and achieve large sizes characteristic of low-shore mussels. Large *M. edulis* transplanted from low to high shore levels die, being unable to acquire sufficient food to maintain their large biomass. Desiccation stress, rather than limited opportunities for feeding, probably fixes the upper zonal limit of *M. edulis* on most shores, however. Hundreds of thousands of *M. edulis* at the upper edge of the mussel zone on parts of the Washington coastline have been found dead and gaping during extremely hot summer days. This summer mortality causes the upper edge of the mussel zone to move several centimetres downshore as the dead shells are gradually washed away. During late winter to early spring, mussels move up from the main population to occupy the newly vacated space, a behaviour presumably driven by intraspecific competition. The lower boundary of the *M. edulis* zone on this shore is set by starfish predation. In subtidal regions *M. edulis* survives in refuges among patches of filamentous algae, hydroids, bryozoans, or in kelp holdfasts where starfish cannot reach them.

Barnacles. Two common barnacles on British shores, *Semibalanus balanoides* and *Chthamalus montagui* (previously referred to as *stellatus*), are limited by physical factors in their upward distribution and by biological factors in their downward distribution on the shore Plate 14, facing p. 136). Cyprids of both species settle on suitable substrata throughout most of the intertidal zone down into the sublittoral zone. *Semibalanus* cyprids can cement down their antennules

on the rock surface in about 20 min and are therefore able to settle at high shore levels where tidal immersion is brief. However, the highest settling cyprids are killed by desiccation before the next tidal immersion. Resistance to desiccation increases as the barnacles grow, but even adults may be killed by desiccation during calm spring or summer weather coinciding with neap tides (cf. *Pelvetia*, Section 5.1.4.1). As a result, the upper zonal boundary of adult *Semibalanus* is considerably below the limit of larval settlement. Cyprids of *Chthamalus montagui* apparently can settle out on surface films left by wave surge, enabling them to attach at higher shore levels than *Semibalanus*. Resistance to desiccation at all growth stages is greater in *Chthamalus* so that the adults extend higher on the shore than *Semibalanus*. By continually removing *Semibalanus* from a strip of rock-face on the Isle of Cumbrae, Connell (1972) showed that *Chthamalus* was able to settle and grow down through the *Semibalanus* zone into the sublittoral zone. The lower limit to the natural distribution of *Chthamalus*, therefore, is set by competition with *Semibalanus*. At sites where annual recruitment saturates the substratum, *Chthamalus* lives in a competitive refuge from *Semibalanus* high up on the shore, and is able to do so because of a high temperature tolerance, reduced activity when the tide recedes, a tightly fitting operculum and a nonporous shell. These attributes are gained at the expense of a lower potential growth rate. *Semibalanus* has a looser, more porous shell that is evidently quicker to produce. On the Scottish shore, the lower zonal limit of *Semibalanus* is set by predation, as Connell showed by enclosing the barnacles in stainless steel cages that excluded the predatory dogwhelk, *Nucella lapillus*.

5.1.5.2 *Zonation of mobile animals*

The mobility of non-sedentary animals makes the study of their intertidal distribution less straightforward. Limpets represent a 'half-way stage', sharing some of the hazards of a sedentary life and some of the benefits of mobility. Zonation of limpets has been examined extensively in Britain, South Africa, New Zealand and Australia, but Wolcott's (1973) study of Californian limpets serves well for illustration. Three species live on the bare rocks in the splash zone. *Collisella scabra* occupies mainly horizontal surfaces fully exposed to the sun. It 'homes' to a scar on the rock surface and the shell margins grow to form a precise fit to the scar, thereby reducing desiccation. *Collisella digitalis* occupies primarily vertical or overhanging surfaces shaded somewhat from the sun. *Collisella persona* retires during the day into dark crannies or beneath boulders, emerging to feed only at night. When not feeding during the day, *C. digitalis* and *C. persona* secrete a mucous sheet between the shell margin and the rock surface. This acts as a diffusion barrier and reduces water loss.

All these limpets respond to desiccation by ceasing movement. *C. scabra* returns to its scar and the others clamp down on the rock surface as it dries out, so that just like sedentary animals, they must endure any stressful conditions until the tide returns. Rock-surface temperatures were never observed to exceed the thermal tolerances of the limpets (which increase in species from higher shore levels, suggesting the adjustment of physiological processes for optimal performance under the appropriate conditions (see Section 5.1.3.2)) and desiccation seems to be the main hazard.

Death ensues through the concentration of the body fluids to lethal levels. The desiccation tolerances of the three splash-zone limpets, amounting to about 80% total water loss, are amongst the highest recorded in any animals and involve physiological adaptation to high electrolyte concentrations at the cellular level. Nevertheless, desiccation sometimes takes its toll by concentrating the body fluids to lethal levels. Hundreds of *C. scabra* and *C. digitalis* up to 11 years old were killed in August 1971 after the sea had been exceptionally calm, leaving the rocks dry for over a week where they would normally be wetted by spray.

Lower boundaries to zones and partial segregation within areas of zonal overlap are probably determined by competition. For example, *Collisella digitalis*, owing to its ability to secrete a mucous diffusion barrier that can be made to fit any irregular surface, does not need to forage near to a close-fitting 'home site' as does *C. scabra*. *C. digitalis* is able to outcompete *C. scabra* throughout most of the splash zone, except in the most exposed, sun-baked places where *C. scabra* has a competitive refuge. *C. persona* feeds only at night, avoiding contact with *C. scabra* and *C. digitalis*.

5.1.6 Population ecology

5.1.6.1 *Sedentary organisms*

Space on the substratum is an essential resource for sedentary organisms and may be provided by the rock surface (primary space) or by biological surfaces (secondary space) such as those of algal fronds or mussel shells. Space is non-renewable in the sense that once occupied, no more is forthcoming. This contrasts with renewable resources such as food, which can be replaced as it is eaten by growth of the food organisms. Primary space is fixed in extent, imposing a limit to the number of sedentary individuals that can occupy the rock surface of a given area of shore. Secondary space is not fixed, but may wax and wane according to the population dynamics and individual growth of the organisms providing the surface (see *Fucus serratus*, Section 5.1.7.6).

The availability of primary or secondary space is one of the most important population-controlling factors among sedentary organisms. As space becomes used up, individuals within a population are thrown into competition amongst themselves (intraspecific competition) and some will compete with other species (interspecific competition). The restriction of space available to an individual may be detrimental in two possible ways: physical and biological. Physical constraints may involve reduced area for attachment, deformed growth, undercutting and smothering or crushing due to the encroachment of adjacent competitors (see *Chthamalus montagui*, Section 5.1.5.1). Biological constraints may involve reduced accessibility to light (plants) or water-borne particles or algal films (animals), thereby reducing growth and reproductive output. Important consequences of interspecific competition for space among sedentary organisms are illustrated in Sections 5.1.7 and 5.2.4 dealing with community structure.

Intraspecific competition has an effect that is particularly noticeable among sedentary organisms occupying primary space. If, after dense settlement, individuals within a cohort proceed to grow and monopolize an increasing area of substratum, they will soon occupy all the available space. Room for further growth therefore depends on the death of competitively inferior individuals (e.g. kelp, *Egregia*,

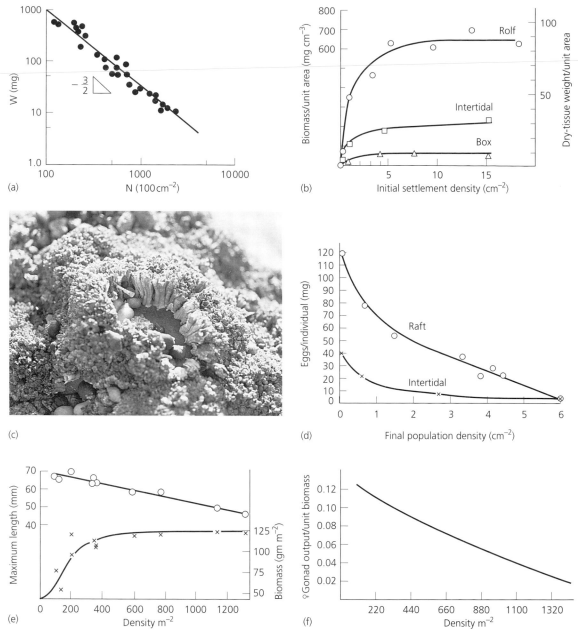

Fig. 5.7 (a) Self-thinning in a cohort of *Semibalanus balanoides*. The data approximately follow a slope of –3/2, derived from the interaction between volume and area. As the cohort ages, it moves from lower right to upper left on the graph (from data of Hughes & Griffiths, 1987). (b) Regulation of biomass in the barnacle *Semibalanus balanoides*. Total biomass per unit area reaches an asymptote when barnacles are grown at increasing densities. The asymptotic biomass increases as the general level of food supply increases. (c) A crowded population of *Semibalanus balanoides* forming hummocks (see 5.1.7.7) and showing the tall, cylindrical shape adopted under severe competition for space. (d) Fecundity is proportional to body size, and so declines in barnacles grown at increasing densities. (e) Regulation of biomass in the limpet *Patella cochlear*. Limpets attain a smaller size when growing at higher densities. (f) Fecundity is proportional to body size and therefore declines among limpets growing at higher densities. (g) Population density scaling, inside and outside two marine preserves (Las Cruces and Montemar). The slopes of the regression lines are not significantly different from the value of –0.75 generally observed for aquatic and terrestrial ecosystems. ((b–d) After Crisp 1962; (e) and (f) after Branch 1975; (g) after Marquet *et al.* 1990.)

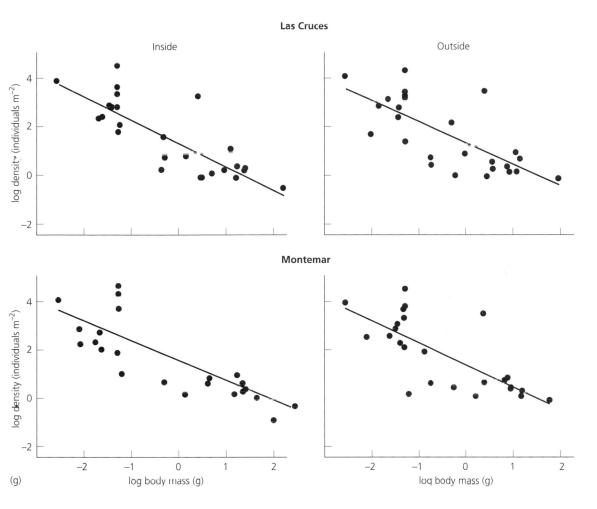

Fig. 5.7 Cont'd

Section 5.2.3), although this can be alleviated by changes in shape in barnacles, or by multilayered packing in mussels. As individuals grow, mean individual mass becomes a power function of population density. So long as cohort biomass is increasing, the power function has an exponent of about −3/2 (Fig. 5.7a), relating to the geometry of packing. But once cohort biomass reaches a limit, set by the environmental productivity, the exponent converges to −1.

Although self-thinning is pervasive among sedentary organisms, the risk of density-dependent mortality is lessened by adjustments of body shape and growth rate. Lacking the ability to move to better conditions, sedentary organisms have highly adaptable growth forms (Chapter 9) that can be modified to suit local conditions. The barnacle *Semibalanus balanoides* develops a squat, volcano-shaped body when not crowded, but becomes taller and thinner as crowding increases (Fig. 5.7c). When grown at increasing densities, the final size of *Semibalanus* decreases, but total biomass remains constant once the 'carrying capacity' of the habitat has been filled (Fig. 5.7b). Correspondence of the total final biomass to the carrying capacity is illustrated by rearing barnacles under conditions with different food supplies: final biomass increases as food supply becomes richer (Fig. 5.7b). At high densities, barnacles become stunted, and as their fecundity is directly proportional to body size, crowded individuals are able to produce fewer eggs (Fig. 5.7d).

A similar effect on growth and fecundity is found in the South African limpet *Patella cochlear*. Although mobile, *P. cochlear* feeds within a territory only a few centimetres in diameter and therefore behaves as a quasi-sedentary animal. The limpet grazes a 'garden' of red algae in a controlled manner, exploiting the 'interest' but not the 'capital' so that a renewable food resource is sustained. As the limpet grows it expands its territory accordingly, but as population density increases, space becomes limited, restricting the expansion of territories of new recruits. Final body size (correlated with territory size and the sustainable yield of algae) declines at high population densities and although total biomass per unit area is regulated (Fig. 5.7e), individual fecundity declines (Fig. 5.7f).

Conservation of biomass in *Semibalanus* and *Patella* is not achieved 'for the good of the populations' but is merely a consequence of competition among individuals coupled with their plasticity of growth. Moreover, the accompanying density-dependent regulation of fecundity (Fig. 5.7d, f) could not normally affect local recruitment, because many larvae are likely to come from populations further afield (see Section 5.1.7.8.).

Interactions between population density and body size occur even at the community level. Among component species, population density scales approximately to the power −0.75 (Fig. 5.7g). After community structure was experimentally altered on Chilean shores, species reassembled and restored the original scaling. This suggests that specific population densities and body sizes become adjusted to overall density by a complex of interacting mechanisms, as yet little understood.

5.1.6.2 Mobile animals

Potentially limiting resources for mobile, intertidal animals include food and protective microhabitats such as crevices or empty barnacle shells (see *M. neritoides*, Section 5.1.2.1). Quantifying the availability of these resources and their effect on consumer populations proves to be much more difficult than with sedentary organisms using space. Problems arise because the animals move about, making it difficult to assess what conditions they experience during their travels, and because in the case of grazers it may not be clear exactly what they are feeding on. Do they assimilate all detritus, microbes, spores, macroalgal germlings, diatoms and endolithic algae that can be scraped off the rock, or are they more selective? Underwood (1978, 1979) circumvented these difficulties by quantifying the effect of experimentally controlling crowding on growth and survivorship in several Australian grazing gastropods. When the snail *Nerita atramentosa* was kept in wire-mesh cages fixed to the natural rock-face, the growth rate of juveniles decreased, but their survivorship remained constant, as density was increased. Adults gradually lost weight when kept at high densities and the survivorship declined as density increased. These effects could only have been due to competition for food, although the food supply itself could not be quantified. Thus, in a crowded population, juvenile *Nerita* survive at the expense of adults because their smaller individual biomass can be maintained on less food. Adult mortality is therefore likely to follow periods of heavy recruitment. As with barnacles and limpets, *Nerita* has planktonic larvae that can recruit onto the shore from diverse origins. There can be no feedback, therefore, between local adult density and future levels of recruitment. The unpredictable fluctuations in numbers of recruits, characteristic of planktonic larvae (Chapter 9), will consequently be reflected in the density and size structure of adult populations (see Section 5.1.7.8.).

Distribution on the shore of slow-moving predators is often governed by the distribution of prey and of places such as crevices, ledges and pools that can be used as refuges from desiccation or wave action. The dogwhelk, *Nucella lapillus*, may wander for several metres among populations of prey, barnacles and mussels, using small-scale topographical features as temporary refuges. Larger-scale features, such as deep crevices, are used more regularly for refuging during winter. Such regularly used crevices may be surrounded by bare zones, where prey have been eliminated by the feeding activities of emerging dogwhelks.

5.1.7 Community structure

Rocky-shore organisms lend themselves particularly well to experimental manipulation. Algae and sedentary animals can be thinned out, and herbivores and carnivores can be excluded or enclosed by fences and cages. Food webs are relatively simple and the limiting resources are often readily apparent. This great potentiality of rocky shores for the experimental investigation of community structure has been heavily exploited, with the result that more is known about the species interactions of rocky-shore communities than of almost any other ecosystem. The experimental manipulation of intertidal organisms dates back at least to the 1930s and 1940s when researchers such as Hatton in France and Jones in the Isle of Man manipulated populations of barnacles, limpets and fucoids. Theoretical ideas on community ecology at that time, however, had not reached the critical stage necessary for these elegant experiments to blossom, and so they were not developed to their full potential. This had to wait for almost another two decades when Paine and Connell took up the reins and developed schools of research that advanced not only our understanding of rocky-shore ecology, but also of the fundamentals of community ecology itself. Paine's (1966) classical study showed how the starfish, *Pisaster ochraceus*, by thinning out populations of the mussel, *Mytilus californianus*, makes space available for colonization by inferior competitors, such as various barnacles, limpets and algae, thereby maintaining a high species diversity within the community. Others, including Dayton, Castilla, Lubchenco and Menge, have extended Paine's approach and applied it to a wide range of intertidal communities. All these studies tell fascinating stories, but just a few have been chosen to illustrate general principles that have emerged.

5.1.7.1 West-coast community structure

Dayton (1971) built upon Paine's original work by examining in detail the effects of physical environmental factors and biological interactions on the upper intertidal mussel–barnacle community of the Washington coastline. The barnacle zone is colonized by three species, *Balanus cariosus*, *B. glandula* and *Chthamalus dalli*, and *Mytilus californianus* occurs in patches among the barnacles on all but the very sheltered shores and the algae *Fucus distichus* and *Gigartina papillata* occur on all but the very exposed shores. Foraging throughout the zone are four species of grazing limpet and three species of predatory dogwhelk. The large starfish

Pisaster ochraceus moves onshore to feed during high tide.

Two important physical factors affecting community structure are wave action and battering by drifting logs (lost by the local logging industry or derived naturally from the heavily forested coastline). Drifting logs, thrown against the rocks by waves, knock off patches of sedentary organisms, making the surface available for recolonization. This effect was quantified by measuring the 'survivorship' of nails embedded into the rock with a rivet gun. Measured in this way, the probability of a patch of rock being struck by a log was well correlated with the percentage cover of mature barnacles on the shore (Fig. 5.8a). Producing a similar effect to log-battering is predation by the starfish *P. ochraceus*. *Pisaster* feeds preferentially on mussels and barnacles, carrying them back down the shore if they are too large to consume during high tide. The starfish everts its stomach over the prey, and in this manner 20–60 barnacles may be removed simultaneously. Among large patches of *Mytilus californianus*, the effects of log-battering and starfish predation are complemented by wave action. After log impact, or a feeding starfish, has removed some mussels, those exposed around the edge of the clearing are twisted and torn off the rock by wave action, so that even a small clearing can be considerably enlarged.

Two potentially important biological interactions were experimentally investigated: interspecific competition among the sedentary organisms and disturbance by the grazers and predators. An ingenious experimental design using cages and fences was made to evaluate the importances of these factors working separately and in combination (Fig. 5.8b). As a result of the experiments, the sedentary organisms could be arranged in a competitive hierarchy. *Mytilus californianus* requires intricate surfaces, provided for example by algae, barnacles, or mussel byssal threads, upon which to settle, but once established it can outcompete all the other sedentary organisms and so is the potential competitive dominant on these shores. Next in the hierarchy are barnacles, which outcompete the algae. Among the barnacles, *Balanus cariosus* is dominant over *B. glandula*, which in turn is dominant over *Chthamalus dalli*. By grazing spores and cyprids or by bulldozing newly settled barnacles off the rock, limpets reduced the recruitment of algae and

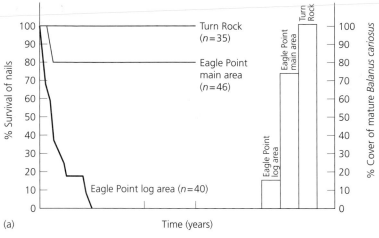

(a)   Time (years)

Stainless steel mesh covers
(approx. 100 cm^2)

Plastic dog dish barriers in *Pisaster*-exclusion
areas (approx. 177 cm^2 —corrected to
100 cm^2 for comparison)

No biological disturbers — Total exclusion cage
Control–*Pisaster*–*Nucella*–*Collisella*

Normal *Collisella* density — Limpet cage
Control–*Pisaster*–*Nucella*

Access to *Nucella* and *Collisella* — Roof
Control–*Pisaster*

Access to all biological disturbers — Control

Access to *Nucella* — O Limpets
Control–*Pisaster*–*Collisella*

Normal *Collisella* density; Access to *Nucella* — Normal limpets
Control–*Pisaster*

→ Tests *Pisaster* effect with *Nucella* and *Collisella* present

→ Tests *Nucella* effect with *Collisella* present and *Pisaster absent*

→ Tests *Collisella* effect with *Nucella* and *Pisaster* absent

→ Tests *Collisella* effect with *Nucella* present and *Pisaster* absent

→ Tests *Nucella* effect with *Collisella* and *Pisaster* absent

(b)

Fig. 5.8 (a) 'Survivorship' of nails embedded in rock compared with the abundance of barnacles at the same localities. (b) Experimental design used to test the effects of competition and biological disturbance (predation and grazing) in the barnacle–mussel association of a west North American shore. (After Dayton 1971.)

barnacles in all natural situations. In the artificial situation where dogwhelks were excluded, limpet disturbance increased the survival of *Chthamalus* by reducing the survival of the competitively superior *Balanus* spp. Dogwhelks depressed the population of all barnacles in natural situations, but when limpets were artificially excluded, dogwhelks increased *Chthamalus* survival by preferentially eating the *Balanus* spp. The effect of limpets and dogwhelks acting in concert was therefore different from either acting alone, illustrating the subtlety that can occur among biological interactions and the need for experiments capable of revealing it. The competitive dominants, *Mytilus californianus* and *Balanus cariosus*, grow too big for dogwhelks to eat if they manage to escape predation for long enough. Therefore, having survived limpet grazing as newly settled spat or dogwhelk predation as young individuals, barnacles and eventually mussels would monopolize all the space in the upper intertidal zone. This is prevented by the repeated clearance of patches by log-battering and starfish predation, and the enlargement of clearings by wave action. The community structure, therefore, is determined by physical and biological disturbances, which alleviate competition by creating an abundance of the potentially limiting resource—space on the rock surface. Because consumers such as *Pisaster ochraceus* fundamentally influence community structure in this way, they have been called keystone or critical species (Paine 1966, 1994). But for consumers to have a 'critical' effect, ecological and environmental circumstances must be just right.

By monitoring survival in experimental patches of mussels clamped to the rock, Menge *et al.* (1994) found that along the Oregon coastline, *Pisaster ochraceus* acted as a critical predator only on exposed headlands, where recruitment and growth of mussels were high. Here, the *Pisaster*—mussel interaction was strong, the starfish clearing patches of rock within the dense covering of mussels. In sheltered coves, where calmer conditions would reduce the turnover of water masses, recruitment and growth of mussels were low. The relatively unproductive mussel populations were less attractive to *Pisaster*, resulting in a weaker predator—prey interaction. On the other hand, these coves provided a suitable habitat for dogwhelks, *Nucella* spp., large populations of which could be supported by the

mussels. Instead of being controlled by the strong effect of a critical predator, the sheltered cove communities were influenced by the individually weaker, but combined 'diffuse' effects of starfish and dogwhelks. At sheltered sites regularly buried by sand, *Pisaster* was scarce and the predator–prey interaction weak. Large patches of bare rock were created by the smothering of sessile organisms during burial. In Oregon, therefore, the role of *Pisaster* as a critical predator depends on the presence of highly productive prey populations. Diffuse (multispecies) predation was associated with less productive prey populations and only a weak predation regime could operate where environmental stress was high.

Changes in predation regime may occur not only between shores, as above, but also within shores. On moderately sheltered shores in New England the dogwhelk, *Nucella lapillus*, acts as a critical predator in the mid-intertidal zone by eliminating patches of barnacles and mussels (Menge 1976). In the low intertidal zone of the same shores, barnacles and mussels are eaten not only by dogwhelks but also by starfish and crabs. A regime of critical predation therefore is replaced by one of diffuse predation at lower tidal levels.

5.1.7.2 East-coast community structure

The relatively species-rich, mosaic structure of the mussel–barnacle zone on the west coast of North America contrasts sharply with the relatively species-poor, mid- to high-intertidal zone of the east coast. The biogeographical difference seems partly due to the harsher environment (more extreme temperature fluctuations) of the east coast, which reduces the variety of species at all trophic levels (see Chapter 10). Using a similar experimental technique to that of Dayton, Menge (1976) identified the major determinants of community structure in the mid- to high intertidal zone of exposed and sheltered rocky shores in New England. In spite of the more austere environment, community structure on the east coast is largely determined by biological interactions, as it is on the west coast.

5.1.7.3 Human exploitation

In many parts of the world, intertidal plants and animals are harvested commercially, often with dramatic influence on community structure. A striking example is the *Concholepas concholepas* fishery in central Chile. *Concholepas* is a carnivorous gastropod that feeds on mussels and barnacles, but because it grows to a relatively large size it also is attractive for human consumption. Shell middens reveal great antiquity of the fishery, which currently accounts for some 40 000 *Concholepas* per annum per kilometre of shoreline. Typically, the rocky shores of central Chile are dominated by mussels, the preferred prey of large *Concholepas*. In areas fenced off to exclude fishermen, populations of *Concholepas* increased and consumed a large proportion of the mussels, clearing space that subsequently was colonized by barnacles and macroalgae. Only by making such exclusion experiments has it become apparent that the 'typical' mussel domination of shores in central Chile is ultimately the result of human exploitation. Both fishermen and *Concholepas* are 'critical' species (Section 5.1.7.1) controlling community structure on these shores.

Where community structure is influenced by sets of 'diffusely' interacting species (Section 5.1.7.1) rather than by a few critical species such as *Concholepas*, human exploitation may have less clear-cut effects. Nevertheless, impact on the exploited community can be severe, raising the question of conservation policy. In New South Wales, for example, anglers take shore fishes in large numbers, using for bait intertidal ascidians, crabs and gastropods collected on site. Kingsford *et al.* (1991) suggested that sanctuaries totally excluding people from the shore could bolster adjacent fisheries and preserve breeding populations of the exploited invertebrates.

5.1.7.4 Grazers and community structure

Physical disturbances and predation can enhance species diversity by repeatedly removing significant quantities of the dominant competitors in a community. Disturbances, however, do not necessarily increase species diversity (see below). If too severe, as with the scraping of intertidal rock surfaces by ice in polar regions (Section 5.1.2.2), disturbance adversely affects all species, keeping diversity low. If disturbance is too mild to significantly affect the competitive dominants, or if it affects the subdominants disproportionately, it will not prevent the dominants from ousting poorer competitors,

but may even accelerate the process, as shown by Lubchenco's (1978) experimental investigation of algal community structure in the mid–low intertidal zone of moderately exposed New England shores. The potential competitive dominants are *Mytilus edulis*, followed by *Semibalanus balanoides*, but these are heavily preyed upon by dogwhelks and starfish, allowing algae to predominate. Among the algae, the competitive dominants are *Chondrus crispus* (Irish moss) in the low shore region and *Fucus vesiculosus* plus *F. distichus* in the mid-shore region. *Chondrus* competitively excludes the fucoids from the low shore, but if kept clear of *Chondrus* the fucoids grow even better here than in their normal zone (cf. *Pelvetia*, Section 5.1.4.1). *Chondrus* cannot withstand desiccation at mid-shore levels, where the fucoids therefore have a competitive refuge. The main herbivores are winkles, primarily *Littorina littorea*, which readily consume algal spores and germlings but avoid *Chondrus*, and fucoids, that have grown beyond the germling stage. Spores that germinate in protective crevices or on smooth surfaces during the winter when winkles have migrated to the sublittoral zone, are therefore able to escape and grow into a size refuge from grazing. Occasionally *Chondrus* is removed by storms, ice-scouring, or people harvesting the plants, and the vacated rock surface is quickly colonized by ephemeral algae. If protected from grazers, the ephemerals delay recolonization by *Chondrus*, but in the natural situation *L. littorea* prefers the ephemeral algae and by preferentially removing them, accelerates regrowth of the *Chondrus* bed.

The effect of *L. littorea* grazing is very different in rock pools high on the shore. Here *Enteromorpha* is competitively dominant over other potential colonists such as *Chondrus* and several other perennial and ephemeral algae. Lubchenco noticed that pools containing few *Littorina* supported an almost pure stand of *Enteromorpha*, whereas others containing many *Littorina* were dominated by *Chondrus*, which the winkles do not eat. When *Littorina* was present at moderate densities in other pools, many perennial and ephemeral algae coexisted because the winkles preferred the competitively superior species, and reduced them to low densities without eradicating them altogether.

Species diversity therefore depends on two things. The first is whether the grazer prefers the

Fig. 5.9 (a) The effect of predation intensity on species richness. (b) The relative importances of predation (continuous lines) and competition (dashed lines) in determining community structure along a gradient of environmental harshness. (After Hughes 1980a.)

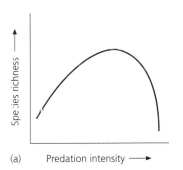

(a)

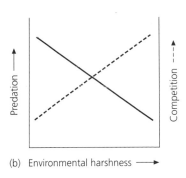

(b)

competitively superior or inferior species. If competitively inferior species are preferred, grazing always enhances competitive exclusion, leading to spatial monopoly by the competitively superior species, such as *Chondrus crispus*, in the low intertidal zone. The second is the intensity of grazing. If competitively superior species are preferred, grazing of moderate intensity increases algal species diversity by keeping the competitively superior species at low densities. Too little grazing is insufficient to prevent the competitive dominant from ousting other species, whereas very heavy grazing prevents all algal germlings from growing to maturity (Figs 5.9 and 12.4).

In the British Isles there is a general tendency for fucoid domination on sheltered shores to give way to barnacle/mussel domination on exposed shores (Fig. 5.2). This change in community structure is correlated with increasing densities of limpets, *Patella vulgata*, as shores become more exposed to wave action. Limpet exclusion experiments (reviewed by Hawkins *et al.* 1992) on shores of inter mediate exposure have demonstrated the powerful effect of limpet-grazing pressure on the distribution of fucoids. A strip of shore initially cleared of limpets became completely covered by fucoids within 6 months, most plants being attached to barnacles that previosly occupied much of the rock surface. Limpets from bordering areas began to aggregate beneath the *Fucus* canopy, whose shade also favoured the recruitment of juvenile limpets. The resulting grazing pressure greatly reduced any further algal recruitment. After 3–5 years the algal canopy began to thin out as plants senesced or were lost when their barnacle substratum eventually sloughed away from the rock. As rock surface became available, the experimental strip was recolonized by barnacles. Meanwhile, on either side of the strip,

fucoid coverage was enhanced by the reduced grazing pressure following the emigration of limpets.

Even on the most exposed British shores, the establishment of fucoids is prevented by limpet grazing rather than by wave action. Application of dispersants after the *Torrey Canyon* oil spill killed limpets over wide areas of shore in West Cornwall and was followed by a dense growth of fucoids (reviewed by Hawkins *et al.* 1992). On sheltered shores, limpets are relatively scarce and the low grazing pressure allows fucoids to establish dense populations.

5.1.7.5 Mesoherbivores

Invertebrates less than about 2.5 cm in length have been termed 'mesoherbivores' (Brawley 1992). They include small crabs, shrimp, amphipods, isopods, small gastropods, polychaetes and dipteran larvae. Often reaching densities of thousands per square metre, mesoherbivores may strongly affect community structure by feeding preferentially on ephemeral algae. Removal of mesoherbivores from experimental areas allows early successional algal species to colonize and persist. These blanketing species hinder the establishment of macroalgae, whose spores are unable to penetrate to the substratum. Mesoherbivores may promote the survivorship of established macroalgae by reducing epiphyte loading, which otherwise would increase drag and susceptibility to dislodgement of the host algae during storms. In some cases, mesoherbivores feed on the reproductive tissues of macroalgae, but paradoxically the process may enhance recruitment. The amphipod, *Hyale media*, feeds on the reproductive fronds of *Iridaea laminarioides*, promoting the release of spores. Moreover, although many spores are destroyed, some survive digestion and become deposited onto the substratum by defecation. Encapsulated in nutrient-rich faeces, the newly germinating sporelings are able grow quickly and establish themselves (Santelices 1992).

Certain mesoherbivores, however, graze destructively on macroalgal fronds. On certain Swedish shores, for example, isopods of the genus *Idotea* cause breakage and frondal loss of *Ascophyllum nodosum* (Åberg 1992). The destructive potential of such mesoherbivores is very great, as seen in the Baltic Sea during the 1980s. Recruitment of *Idotea baltica* was greatly enhanced by a surge in productivity of green filamentous algae, caused by a natural eutrophication event. On growing to a larger size, the isopods switched from filamentous algae to a diet of *Fucus vesiculosus*, completely destroying the macroalgal beds over extensive areas (Salemaa 1987).

5.1.7.6 Effects of renewal of secondary space and of non-hierarchical competition

There are two reasons why disturbances, whether they be physical or biological, are so important in maintaining species diversity on rocky shores. First, the potentially limiting resource, space on the rock surface, is non-renewable (see Section 5.1.4). Secondly, the sedentary organisms form an unchanging, linear, competitive hierarchy in which those higher up can always out-compete those below them. Left undisturbed, such a system of competitors would always lead to monopoly by the competitive dominant. Coexistence of competitors would be possible, however, if either or both of the two conditions were changed. This appears to be the case with the community of epiphytic animals living on low-intertidal fucoids.

Fucus serratus grows prolifically on moderately sheltered British shores and supports a rich epifaunal community of sponges, hydroids, bryozoans, serpulid polychaetes and tunicates. Seed & O'Connor (1981) have examined the population dynamics and competitive interactions within a subset of the epifaunal community. A crucial feature of the system is the summer growth of the plant, which replenishes the substratum available for colonization by up to 75% per annum (violating the first condition above). *Flustrellidra hispida* and *Alcyonidium hirsutum* are two of the bryozoan competitors, among which

Flustrellidra is usually dominant. *Flustrellidra* larvae are released in spring and summer, settling selectively on the distal growing regions of the *Fucus serratus* fronds, so producing a broad band of developing colonies. *Fucus* continues to grow more slowly during the winter, providing new substratum that is selectively colonized by *Alcyonidium hirsutum* larvae released in midwinter. Continued annual cycles of *Fucus* growth and bryozoan settlement result in alternating broad and narrow bands of *Flustrellidra* and *Alcyonidium*, respectively, across the distal segments of the fronds, coexistence being made possible by the annual replenishment of unused frond surface. Two other members of the bryozoan subset, *Electra pilosa* and *Membranipora membranacea*, are competitively inferior to *Flustrellidra* and *Alcyonidium* and survive in refuges unused by the two latter. *Electra* and *Membranipora* settle gradually throughout the year on any part of the *Fucus* frond, making opportunistic use of unoccupied patches.

A second important feature of the *Fucus* epifaunal community is its competitive structure, which does not form an unchanging, linear hierarchy, but involves a certain amount of unpredictability because competitive interaction can be reversed (Fig. 5.10), violating the second condition above. Although in probabilistic terms, *Flustrellidra hispida* is the potential competitive dominant, the stochastic element of the competitive interactions would prevent its monopoly of the substratum even if the latter were not replenished by plant growth each year.

5.1.7.7 Cyclical regeneration of primary space by intrinsic mechanisms

Sometimes the regeneration of bare substratum on the shore is not entirely due to external forces such as predation, storm damage or log impact, but may result from the growth activities of the competitive dominants themselves. Mussels, for example, tend to form multilayered populations as recruits attach themselves to larger residents. Founder members gradually die and sediment accumulates among the shells, as a result of which younger, actively growing mussels form 'hummocks' attached indirectly to the rock base via the layer of dead shells and sediment. As the hummocks increase in size they become mechanically less stable, eventually being washed away by waves. A mosaic of patches in

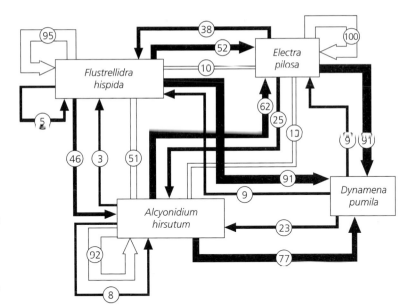

Fig. 5.10 Flow diagram of competitive interactions among three bryozoan species and a hydroid that colonize the fronds of *Fucus serratus*. Arrows point towards competitive subordinates. Encircled numbers are the percentage probabilities of the respective interactions taking place. (After Seed & O'Connor 1981.)

various stages of maturity, from newly bared rock surface to pronounced hummocks, can therefore be maintained in a cyclical fashion. Hummocking cycles also characterize certain populations of barnacles (Fig. 5.7c).

A somewhat similar mechanism allows the small annual kelp, *Postelsia palmaeformis* (see Fig. 5.15b) to succeed in the barnacle–mussel zone of very exposed shores on the Washington coastline. Mussels and then barnacles are the competitive dominants on these shores (Section 5.1.7.1), but *Postelsia* patches are maintained by settlement and growth of the alga on top of mussels and barnacles. Increased drag of the growing kelp eventually causes both *Postelsia* and the animals to which it is attached to be torn away by waves. The vacated rock surface can then be directly colonized by more *Postelsia*.

Rocky shore community structure is seen to be influenced by physiological factors limiting the potential range of an organism and by disturbances and competition that limit distribution within this range. The result is the familiar patterns of zonation along gradients of tidal level, exposure to wave action and salinity. These patterns, however, may not be as stable as would appear at first sight (Connell 1985). Intertidal organisms can have maximum lifespans of several decades, so that complete population turnover will be missed by the short-term studies providing the bulk of available data. Where long-

term studies have been made, interesting changes come to light. For example, *Semibalanus balanoides*, towards the limit of its geographical range in southern England, became locally extinct in the 1950s owing to slight warming of the climate. The reverse trend led to its re-establishment by 1975.

Community interactions, especially the successional replacement of species recolonizing cleared substratum, may depend heavily on the vagaries of recruitment by planktonic spores and larvae and on the life histories of the organisms. Those with rapid growth but short lifespan, e.g. *Ulva*, will predominate at first, later giving way to slower growing but longer-lived species, e.g. fucoids. This is simply a consequence of the correlation between short lifespan, flimsiness and rapid colonization and between long lifespan, robustness and slow colonization. It does not necessarily involve the sequence of interactions once thought to be essential in facilitating the transition from pioneer to climax stages of succession.

5.1.7.8 Supply-side ecology

Factors such as competition and disturbance, which may control zonation and species diversity in more permanently saturated communities (Sections 5.1.7.1 and 5.1.7.4), may have relatively little effect in places where recruitment is highly variable. Instead, community structure may strongly reflect the history of recruitment of dominant species, i.e. the 'supply-side ecology'.

The amount of recruitment onto a shore from water-borne propagules may depend on the reproductive output of source populations, on hydrographical variables at all spatial scales and on properties of the substratum that induce settlement. Large-scale factors driving recruitment may have profound effects on community structure (see review by Underwood & Denley 1984). In the region of Anglesey, North Wales, tidal currents encourage the inshore retention of larvae. Each year, free space on the shore is saturated by larval settlement of the barnacle, *Semibalanus balanoides*. On the Isle of Man, out in the Irish Sea, currents frequently carry larvae away from the shore and barnacle recruitment is sporadic (Hawkins & Hartnoll 1983). Recruitment also is sporadic along the coast of southern England, where *S. balanoides* is close to its southern geographical limit. Here, climatological conditions controlling survivorship within source populations may be a critical factor causing annual variation in recruitment (Southward *et al.* 1995, see also Section 5.1.7.7).

Variation in larval recruitment, and the effect this has on community structure, have been particularly well studied for Californian populations of the barnacle, *Balanus glandula*, occupying the high intertidal zone. In 1988, oceanographic measurements revealed alternating periods of onshore and offshore transport of surface water associated with the waxing and waning of equatorial winds that drive coastal upwelling. Recruitment was enhanced by the onshore advection and severely depressed by offshore transport during upwelling (Farrell *et al.* 1991). Previous observations had shown that settlement could vary among sites by almost two orders of magnitude and that this was correlated with the concentrations of larvae carried by passing water masses (Gaines *et al.* 1985). Larval concentration may depend not only on input to the water mass by a source population, but also on subsequent history. Water masses bringing *B. glandula* larvae onto the shore at the Hopkins Marine Station, Monterey Bay, pass through kelp forest that shelters large populations of planktivores, particularly juvenile rockfish, *Sebastes* spp. By feeding upon the barnacle larvae these predators can reduce recruitment to 1/50 of the level on adjacent areas where offshore kelp is absent (Gaines & Roughgarden 1987).

5.1.7.9 Patch dynamics

Although zonation can be the most striking pattern seen in the distribution of sessile organisms on rocky shores (Sections 5.1.4 and 5.1.5), closer scrutiny usually will reveal patchiness within zones. Spatial characteristics of the patches can vary enormously from relatively small, isolated areas of high population density to extensive carpets perforated by holes and channels of unoccupied habitat. Patches of diameter measured in centimetres may be nested within higher-order patches measured in metres. Shores along a stretch of coastline usually are occupied by the same sets of species whose local populations may be linked through dispersal, even when separated by considerable stretches of unsuitable habitat. Local patchiness therefore is nested within population distribution on a scale of kilometres. Different physical and biological factors are likely to operate on the different spatial scales, and knowledge of larger-scale effects may be necessary for understanding local pattern.

Patch mosaics within zones reflect different phases of space occupation by sessile organisms and their consumers. In low-diversity habitats such as the upper littoral zone of moderately exposed temperate shores, the mosaic may simply represent dense barnacle coverage and bare rock. In the lower intertidal zone the mosaic may represent different stages in succession among competing space occupiers and community diversity depends on maintenance of the mosaic by agents of disturbance (Section 5.1.7.1). Once a patch of bare rock has been created by processes such as predation, grazing or sloughing, recolonization takes place either by immigration from the edges or by the settlement of water-borne propagules. Small patches, centimetres in diameter, may encourage settlement by modification of water flow close to the substratum. Settlement of barnacle larvae, for example, tends to be greater within small empty patches than on wider expanses of bare rock. However, as small or narrow patches have relatively high perimeter/area ratios, they also favour rapid peripheral colonization by individuals able to move over the substratum. As patches become larger and wider, colonization by immigration from the edges will tend to become less important than the settlement of water-borne propagules.

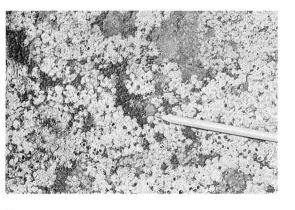

(a)

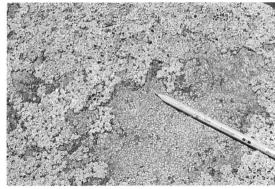

(b)

(c)

Fig. 5.11 Recolonization of empty patches on a rocky shore in central Chile. (a) Small patches colonized primarily by mussels, *Perumytilus purpuratus*, which recruit more successfully against the edges of barnacles. (b) Large patch colonized primarily by barnacles, able to recruit successfully on bare rock. Photos by courtesy of Sergio Navarrete. (c) Number of newly recruited mussels and barnacles in relation to patch area 130 days after clearing. The number 35.5 is the critical patch area (cm²) where the recruitment rate of mussels equals that of barnacles. MA is the minimum patch area for successful barnacle recruitment.

In their classical study of patch dynamics of the mussel-dominated community on exposed shores of the Washington coastline, Paine & Levin (1981) found that the longevity of bare patches could be accurately predicted from patch size, according to the principles outlined above. The bare patches were created by the removal of mussels, *Mytilus californianus*, during winter storms, patch size reflecting the intensity of disturbance. Very small patches disappeared almost immediately as bordering mussels leaned over, closing the gap. Medium-sized patches (< 3 m^2) were filled in at a rate of about 0.05 cm day^{-1} by the immigration of peripheral mussels. Central space of larger patches was colonized by planktonic larvae about 26 months after patch formation and was gradually filled in at a rate of some 2% per month by growth of the newly recruited mussels.

The relative effects of peripheral and central recruitment, however, may also depend on the behaviour of settling larvae. By observing experimentally cleared patches in the mid-intertidal zone of central Chile, Navarrete & Castilla (1990) showed that the spat of mussels, *Perumytilus purpuratus*, recruited only onto the shells of barnacles bordering each patch, whereas barnacle larvae settled anywhere within the patch. Owing to the high perimeter/area ratio of patches smaller than about 35.5 cm^2 (Fig. 5.11a), recruitment of mussels exceeded that of barnacles. As patch size increased above this threshold, recruitment of barnacles progressively exceeded that of mussels (Fig. 5.11b, c). An extensive covering of barnacles disrupted by many small patches therefore would encourage mussel recruitment more effectively than the presence of fewer, larger patches.

Higher-order effects, involving other members of the rocky shore community, are likely to accompany the patch dynamics of any dominant space occupier. On certain British shores, *Fucus* sporelings are more likely to become established among barnacles than on bare rock. The rough surface may facilitate settlement and it provides some protection from grazing limpets. Patches of *Fucus* therefore tend to become established among patches of barnacles, but are unable to do so on intervening, heavily grazed areas. The dynamical nature of this process is ensured by the eventual sloughing of the underlying barnacles when algal fronds, having grown to a critical size, generate excessive forces under the drag of storm-driven waves (Hawkins *et al.* 1992).

5.1.7.10 Energy flow

Although relatively little is known of energy flow through rocky shore communities, recent work suggests that a comparative approach may reveal interesting patterns along major environmental gradients. Hawkins *et al.* (1992) noted that the general increase in algal biomass and decrease in barnacle coverage at lower tidal levels should be accompanied by a downshore trend of increasing net production. This trend will be continued into the highly productive kelp forests, wherever they occur (Section 5.2.5). Surplus production of kelp forests, exported by currents and wave action, may be an important source of energy input to consumer populations at higher tidal levels. On sheltered boulder shores of South Africa, the limpet, *Patella granatina*, grazes epilithic microalgae, but also feeds on kelp fragments drifting onshore with the tide. By experimentally excluding drifting kelp, Bustamante *et al.* (1995) showed that without this energy source *P. granatina* suffers a marked reduction in population biomass.

A second major trend, at least on British shores, is likely to be a shift from net production and export of energy on sheltered shores to net consumption on exposed shores (Fig. 5.12). This trend reflects the decreasing coverage of macroalgae and increasing biomass of consumers with increasing exposure to wave action (Section 5.1.7.4).

Linkage between near-shore oceanographic conditions and littoral community structure has only recently begun to be explored, but may be significant particularly in upwelling regions. Two sites along the Oregon coastline differ in the relative abundance of filter feeders and macroalgae, the former prevailing at the site where there is greater phytoplankton production (Menge *et al.* 1997).

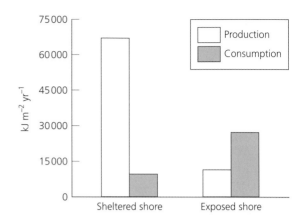

Fig. 5.12 Estimates of community production and consumption for a sheltered and an exposed rocky shore in the United Kingdom. Producers were algal film and fucoids; consumers were limpets (*Patella*), periwinkles (*Littorina*), dogwelks (*Nucella*) and barnacles (*Semibalanus*). (From Hawkins *et al.* 1992.)

5.2 Kelp forests

5.2.1 Introduction

Fucoid algae (Fucales) which perhaps form the most prominent feature on sheltered to moderately exposed, cool temperate, rocky shores throughout the world, are replaced at around Low Spring Tide Level (sublittoral fringe) by another group of large brown algae, the kelps (mostly Laminariales), which extend as far down into the sublittoral zone as light intensities and availability of suitable hard substrata will allow. In cool, clear waters, kelp may flourish to depths of 20–40 m and extend 5–10 km offshore on a gentle gradient.

Kelp forests (see Plate 12, facing p. 64) are just as impressive as coral reefs (Chapter 6) in two respects: some maintain productivities of 1500–3000 g C m^{-2} yearly, comparable with the most productive aquatic or terrestrial ecosystems, and the plants themselves create an intricate three-dimensional topography (Fig. 5.13) that attracts numerous invertebrates and fish. The geographical distribution of kelp forests (Fig. 5.14) is very roughly complementary to that of coral reefs (Fig. 6.3), kelps being limited in general to latitudes between the subpolar regions and the 20°C summer isotherms, and corals to latitudes within the 20°C winter isotherms. Extension of kelps into subtropical latitudes is facilitated by cold currents, usually associated with upwellings (Section 1.4) such as the California Current, Peru Current and Benguela Current that flow along the coast towards the equator.

Kelps alternate between an asexual sporophyte and a sexual gametophyte during their life cycle (Fig. 5.15a). It is only the sporophytes that grow into the large plants that form the kelp forest. The gametophytes are tiny filamentous plants, the ecology of which is very poorly known. The sporophyte basically consists of a holdfast, stipe and blade or lamina (Fig. 5.15b). Sometimes the whole sporophyte is ephemeral or annual, but in other cases the lamina or lamina plus stipe is deciduous and the holdfast perennial. Most of the prodigious productivity is due to growth of the blade, which occurs where the blade joins the stipe. As the blade grows it is worn away at the tip, behaving like a conveyor belt of algal tissue. Many variations exist on the basic morphological theme, ranging from simple stipes, branching stipes, simple blades, subdivided blades to the presence or absence of one or more gas-filled bladders to buoy up the blades. This diversity of form reflects

Fig. 5.13 Kelp impart an important three-dimensional structure to the habitat. Vista of Californian *Macrocystis* forest. (After North 1971.)

Fig. 5.14 World distribution of major kelp forests (L, *Laminaria*; M, *Macrocystis*; E, *Ecklonia*). The 20°C summer isotherms are shown. (After Mann 1973.)

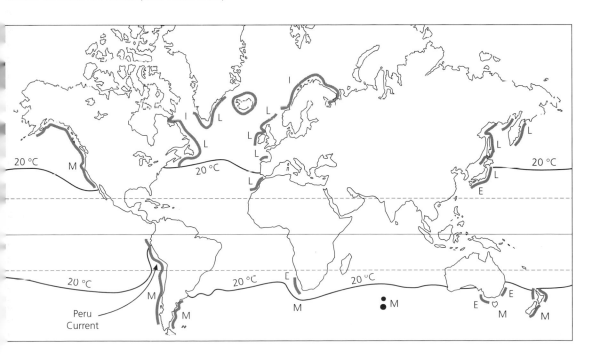

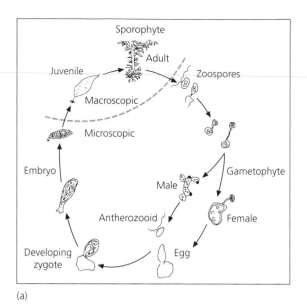

(a)

Mature sporophyte

Young
sporophyte
Length
about 20 cm

Egregia laevigata
California
Length 1–3 m

Macrocystis pyrifera
California
Length sometimes
over 50 m

Postelsia palmaeformis
West North America
Length 10–30 cm

Ecklonia maxima
South Africa
Length 2–5 m

L. hyperborea
Europe
Length 2–5 m

L. saccharina
Europe
Length 1–3 m

(b)

Fig. 5.15 (a) Kelp life cycle. (After North 1971.) (b) Kelp
morphologies. (After Mann 1973.)

the range of habitats occupied by kelps. The palm kelp, *Postelsia palmaeformis*, whose crown of multiple blades on a short, rubbery stipe fit it for life on wave-beaten, intertidal rocks, was mentioned in Section 5.1.7.7. Shallow, sublittoral kelps, such as *Laminaria digitata* and *L. saccharina*, common on European coastlines, tend to have short, flexible stipes and strap-like blades that are kept up in the water column by flotation and turbulence, but are able to bend limply without damage when partially emersed on low spring tides. *Laminaria hyperborea*, a dominant competitor in deeper water, has longer, stiffer stipes that lift the blades into well-illuminated water. In spite of the long stipes, *L. hyperborea* is completely submerged at high tide, and this is true of all kelps in the North Atlantic. Kelp forests in this region are therefore visible from the shore only on low spring tides, when the distal regions of the stipes and pendulant blades appear just above the water surface. In the South Atlantic, Pacific and Indian Oceans, joining the strap-like laminarians is a second category of kelps typified by greatly elongated stipes bearing gas filled bladders in the distal region. Such a kelp is like a moored buoy, the long, flexible stipe tethering the crown of blades to the sea bed, but at the same time enabling them to remain at the surface even at high tide. Buoyed-up kelps often form dense beds parallel to the shore and visible at any state of the tide, notable examples of which are the *Macrocystis* 'giant kelp' beds off California and the *Ecklonia* beds round the Cape Province of South Africa.

Patterns and processes of community organization in kelp forests have basic features in common with those of the intertidal, rocky-shore communities described in Section 5.1.7. Zonation is a prominent spatial pattern and the regeneration of space on the substratum by physical and biological disturbances is an important process in both types of community.

5.2.2 Zonation

5.2.2.1 Vertical distribution

In European kelp beds, *Laminaria digitata* dominates the sublittoral fringe but is replaced by *L. saccharina* and *L. hyperborea* in deeper water in suitable localities. The upper limit of *L. digitata* is determined by an inability to survive more than the minimal tidal emersion occurring on low spring tides. Within its zone, however, *L. digitata* achieves dominance by virtue of a high growth rate persisting from late spring to late autumn (Fig. 5.16). *L. saccharina* and *L. hyperborea* are outcompeted by *L. digitata* in the sublittoral fringe because growth slows down (*L. saccharina*) or ceases altogether (*L. hyperborea*) in late summer. These different growth patterns persist in plants grown experimentally at the same depth, and are due not so much to differences in photosynthetic rate but to the way in which the photosynthates are used. *L. digitata* channels most of its photosynthates into growth right through to the autumn, so that at the beginning of the dark season, from October onwards, stored energy reserves are relatively low. In deeper water, light levels fall below the compensation point sufficiently often during winter that stored energy reserves are necessary to sustain kelp plants until spring. Hence, *L. saccharina* and *L. hyperborea* begin to build up energy reserves in late summer, but at the expense of retarding or ceasing growth. *L. saccharina* therefore has a competitive refuge in deeper water from *L. digitata*, whose low energy reserves prevent it from perennating at these depths. In still deeper water, *L. saccharina* is outcompeted by *L. hyperborea*, which achieves dominance by its long, stiff stipe and overtopping growth form. The lower limit of *L. hyperborea* is often set by grazing by the sea-urchin *Echinus esculentus*.

5.2.3 Population ecology

In spite of their size, kelps seem to have population turnover rates comparable with those of intertidal fucoids. There is a range from competitively superior, perennial species to opportunistic annuals that repeatedly invade frequently disturbed areas. Superimposed on this is the great demographic flexibility of single species, enabling many of them to flourish over a wide range of environmental conditions. Although the giant kelp *Macrocystis pyrifera* behaves as a perennial, living for several years in deeper water, its life expectancy in shallow water on exposed coasts may be less than a year because of removal by storms.

At least under the crowded conditions of kelp beds in the lower intertidal and sublittoral fringe,

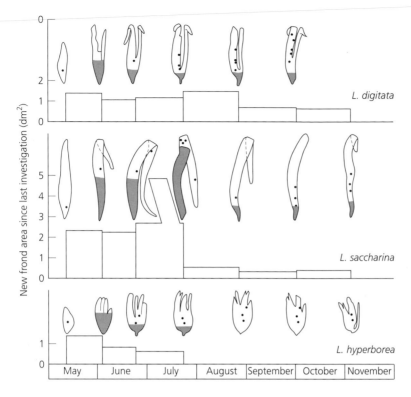

Fig. 5.16 Seasonal growth in three European laminarians. Holes punched in the blades move towards the tip as new tissue (black) is formed by the meristem at the base of the blade. (After Lüning 1979.)

populations of kelp undergo the same processes of competition, self-thinning and regulation of biomass per unit area as seen in higher, intertidal, sedentary organisms (Section 5.1.6). *Egregia laevigata* (Fig. 5.15b) populates the lower intertidal and shallow sublittoral zones of Californian shores. The upper distributional limit of *Egregia* on these shores is set by sun burn and sand scour, but there is extensive recruitment on exposed rock surfaces even where mature kelp never survive. The intertidal plants are annual but *Egregia* may be perennial sublittorally. A large intertidal recruitment occurs in the spring and although this is unaffected by grazers or the presence of other sedentary organisms, *Egregia* plants remaining from the previous year-class reduce recruitment by the abrading action of their stipes during the rising and falling tides. Among new recruits, survival is density dependent because of self-thinning. At high densities, holdfasts become enmeshed and some recruits are attached to the holdfasts of others rather than to the rock surface. The resulting mechanical instability causes whole clumps of recruits to detach from the sub-

stratum. By the time the fronds of the recruits begin to grow rapidly, mortality becomes density independent, but there may still be large differences in the density of recruits in different parts of the population. Biomass per unit area, however, becomes fairly even throughout the population, firstly because plants grow faster at lower densities and secondly because they develop a more branched growth form at lower densities. As a result, plant size is inversely proportional to population density.

Long-term studies by Dayton *et al.* (1992) on another Californian kelp, *Macrocystis pyrifera*, showed that populations of this competitive dominant from deeper water have a remarkable capacity for recovery after severe disturbance. The unusually severe El Niño–Southern Oscillation event of 1982–84 depressed kelp productivity by bringing very warm, nutrient-depleted water onshore and in 1988 the most severe storm for two centuries caused much destruction. Although these large-scale episodic events had dramatic impact, their effects were temporary. Within a few years, kelp populations had resumed their previous characteristics, shape by local conditions. Differences among populations reflected the influence on recruitment and survivorship of gradients in long-shore currents, temperature, light, wave energy, floc, planktonic propagules and herbivory.

5.2.4 Community structure

5.2.4.1 Grazing by sea-urchins

Often, the local dominance of a particular kelp species can be related to the degree of environmental disturbance to which it is adapted, and the overall species diversity of a kelp forest is maintained by disturbances on various scales of intensity and frequency. One kind of disturbance however—grazing by sea-urchins—is of particular importance in kelp communities from diverse parts of the world. Sea-urchins are typical members of the grazer guild in kelp forests, different species replacing each other according to geographical locality. Sometimes urchins prefer the competitively dominant kelp species and promote algal species diversity when grazing is of moderate intensity (cf. Fig. 5.9). However, whatever their dietary preferences, sea-urchins in high population densities are capable of eliminating kelps and, indeed, all frondose algae, leaving a sparse-looking substratum colonized only by encrusting coralline algae, diatoms, and a close turf of filamentous green algae. For various reasons, urchin populations sometimes build up to epidemic proportions and the devastation they cause to kelp forests is reminiscent of the effect of *Acanthaster* outbreaks on coral reefs (Section 6.5). Before 1968, lush forests of *Laminaria digitata* and *L. longicruris* (= *saccharina*) dominated the rocky sea bed in a 140 km^2 bay on the outer Nova Scotian coast. The kelps extended from the infralittoral fringe to a depth of 20 m and were exceedingly productive (see Section 5.2.5). The sea-urchin population of about 37 urchins m^{-2} appeared to be in equilibrium with the kelp, feeding on fragments breaking off the plants. Wave-induced swaying movement of the kelp prevented the urchins from climbing up the plants and causing damage. However, in certain places where urchins were particularly numerous, some were able to climb kelp plants, weighing them down into the reach of the other urchins. In this manner, patches tens of metres in diameter were cleared

within the kelp forest. Six such holes were discovered in 1968, but during the next 6 years they grew larger and coalesced until 70% of the kelp along a 15 km length of shore was destroyed. Complete devastation was caused as 'wavefronts' of urchins advanced on the kelp forest at rates of up to 1.7 m per month. Eventually, the entire kelp forest as such was wiped out, leaving isolated patches of kelp at the sublittoral fringe above the normal range of the urchins. A cause of the urchin outbreak has not been identified conclusively, but a correlation was suggested between the urchin outbreak and decimation of the lobster population by overfishing. Lobsters feed readily on urchins and at normal densities, lobsters would be able to keep a pre-epidemic urchin population under check. Overfishing may have reduced the lobster predation pressure so much that the urchin population 'escaped' and grew into an unchecked epidemic. Restoration of the original predator–prey population densities is hindered by two phenomena. First, elimination of the kelp forest removed an important nursery ground for the lobsters and secondly, it would require unnaturally high densities of lobsters to bring an epidemic of urchins back to pre-existing densities.

Wide temporal fluctuations of urchin population density, however, may be typical of these animals just as they are of many benthic invertebrates with planktonic larvae (Chapter 9). Vagaries of the weather and hydrography affecting reproduction, larval survival and dispersal can cause bumper recruitment years to alternate unpredictably with long periods of almost zero settlement. In the Strait of Georgia off British Columbia, *Strongylocentrotus droebachiensis* undergoes natural periodic outbreaks that affect the kelp community, but not so drastically as did the epidemic in Nova Scotia. An increase in urchin density was triggered by a favourable plankton bloom in 1969 following an unusually long period of cool water temperatures, which extend the period of adult fertility and enhance larval survival of this cold-water species. No recruitment occurred in later years, but the urchin cohort advanced in 'wave-fronts' along the coast, removing all foliose macrophytes in their path. The macroalgal community recovered its former status within 4–6 years. The previous urchin epidemic occurred about 20 years earlier, when sea temperatures had also been exceptionally low.

In the case of the Nova Scotian populations, the urchins were finally hit by a disease, occurring in the early 1980s and sweeping epidemically along the coast until few were left. Kelp has begun to regenerate and it seems as if the community may oscillate in the long term between dense kelp forest and sea-urchin barrens. Contrary to earlier interpretation, the over-exploitation of lobsters may not have been a crucial factor. At infrequent intervals the balance may be tipped in favour of sea-urchins by hydrographic conditions promoting unusually successful recruitment. The urchins exceed a threshold population density and destroy the kelp. Persistently high population density exposes the urchins to the risk of epidemic disease, which eventually wipes most of them out, allowing the kelp forest to regenerate.

On more exposed parts of the Californian coast another urchin, *Strongylocentrotus fransiscanus*, is able to survive in shallow water cooled and aerated by wave action. The grazing pressure prevents most kelps from establishing themselves inshore, except for the annual kelp *Nereocystis leutkeana*, which is less preferred by urchins. Several hundred years ago, the inshore distribution of urchins would have been limited by the predatory sea otter, *Enhydra lutris*. After local extermination of the sea otter, urchins invaded the shallows and eradicated most of the kelps. To boost crops of the commercially exploited giant kelp, calcium oxide is now used in some areas to poison urchins, in place of sea otter predation.

5.2.4.2 Role of the sea otter

Originally, the sea otter was a 'keystone' species of nearshore communities from the northern Japanese archipelago, through the Aleutian Islands, down the Pacific coast of North America to Baja California, but was almost extinguished from the whole of this geographical range by fur traders. Sea otters have now been re-established in many areas, where their effect on the shallow benthic communities has become very evident.

At Amchitka Island in the Aleutians for example, sea otters now occur at a density of $20–30 \text{ km}^{-2}$ and consume about $35\,000 \text{ kg km}^{-2}$ yearly of prey, mainly urchins, molluscs, crabs and fish. Sea otters do not forage effectively below 18–20 m and at these depths *Stronglylocentrotus droebachiensis* becomes very abundant and eradicates nearly all frondose algae. Here, the substratum is covered by encrusting coralline algae and close-growing green algae. Around Attu Island, where sea otters are absent, a similar sparse algal flora extends to the sublittoral fringe, but at Amchitka Island sea otters remove nearly all urchins from the shallow water and also reduce the densities of sedentary animals such as mussels and barnacles. Freed from heavy grazing pressure and from intense competition with barnacles and mussels, a very dense kelp flora is able to flourish within the foraging depth range of the sea otter. So productive is this lush algal vegetation ($1275–2840 \text{ g C m}^{-2}$ yearly) that it once supported Stellar's sea-cow, *Hydrodamalis gigas*, a giant (*c.* 10 t) sirenian, long ago exterminated by hunters. It seems likely that the sea otter was indirectly responsible, by checking populations of invertebrate grazers and competitors, for the high productivity of large algae necessary to maintain the sea-cow populations (Fig. 5.17).

5.2.4.3 Competition in an Aleutian kelp community

The community dynamics of the species-rich Amchitka kelp forests have been elucidated by Dayton (1975), using some of the experimental techniques previously applied to intertidal communities (Section 5.1.7). The algal community has four separate canopies (Fig. 5.18). *Alaria fistulosa* has long, floating blades that form a canopy on the surface and is most dense at depths less than about 5 m. The second canopy level is formed by *Laminaria groenlandica*, *L. dentigera*, *L. yezoensis* and *L. longipes* and occurs from the sublittoral fringe to a depth of about 20 m. The third canopy is formed by *Agarum cribrosum*, whose broad fronds supported by a short stipe lie prostrate on the sea bed. The fourth canopy is a turf of numerous species of red algae together with occasional green algae. In shallow (< 10 m) water, the canopies tend to occupy non-overlapping patches, suggesting the effect of interspecific competition. The experimental design used to test this supposition is summarized in Fig. 5.18. From the experiments it became clear that in spite of forming a surface (primary) canopy, *Alaria fistulosa* is not a competitive dominant, but an opportunistic

Fig. 5.17 Stellar's sea-cow, *Hydrodamalis gigas*, used to browse on North Pacific kelps. The rich supply of kelp necessary to support the sea-cow may have been maintained by the sea otter, *Enhydra lutris*, which feeds on sea-urchins and molluscs that will eradicate kelps if left unchecked. (After Scheffer 1973.)

species, colonizing shallow areas cleared of the dominant *Laminaria* by storms. Among the laminarians, however, *L. longipes* has a rhizoidal growth pattern enabling it to regenerate very quickly after its canopy has been torn away and so this species tends to retain its local monopoly of the substratum. In shallow water, *Laminaria* spp. overshadow and outcompete *Agarum cribrosum*, but in deeper water where sea-urchins are moderately abundant, *Agarum* becomes more successful because it is distasteful to the urchins and is able to colonize areas where urchins have removed *Laminaria*. The *Agarum* (tertiary) canopy prevents the recruitment of *Alaria*, which in deeper water can survive only by attaching high on *Laminaria* stipes. The secondary and tertiary canopies suppress the growth of the red algal

turf, which opportunistically invades newly cleared patches along with *Alaria*.

5.2.5 Productivity

Kelps often live in an environment ideal for plant growth. Photosynthesis is not limited by desiccation or excessive insolation. Wave action keeps the blades in constant motion, providing maximum exposure to sunlight and enhancing the uptake of nutrients. The overall supply of nutrients is kept replenished by wind-induced mixing and upwelling of deeper water. Kelp forests in Nova Scotia maintain a productivity of about $1750\,\text{g C m}^{-2}$ yearly compared with a productivity of $640-840\,\text{g C m}^{-2}$ in the *Fucus*- and *Ascophyllum*-dominated intertidal zone. What happens to this prolific production of organic matter? In certain circumstances a good proportion of it may be consumed by herbivores. Until exterminated in the eighteenth century, the giant Stellar's sea-cow browsed the lush kelp forests of the North Pacific coastlines (Section 5.2.4.2) and in California the giant kelp *Macrocystis pyrifera*

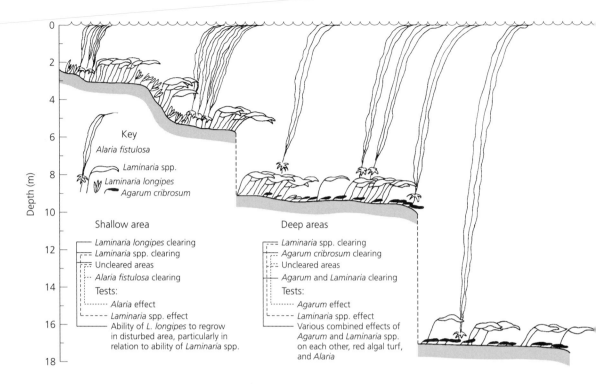

Fig. 5.18 Canopies at three different depths in an Aleutian kelp forest. The experimental design used to investigate the effects of competition among algae is also depicted. (After Dayton 1975.)

provides a commercial harvest of 10 000–20 000 t (dry wt) yearly. In most circumstances, however, it appears that the majority of kelp production enters the community food web via the detritus food chain, just as does primary production in mature terrestrial ecosystems. Some of the detritus is derived as fragmented particles, but a large proportion is derived from the release of dissolved organic matter that is flocculated by bacteria.

5.2.5.1 South African kelp forest

South African kelp forests contain two dominant species, the buoyed-up *Ecklonia maxima* predominating in turbulent inshore areas, and the subsurface *Laminaria pallida* continuing into deeper water. Figure 5.19a summarizes the community energetics of a kelp forest on the Atlantic coast of Cape Province. Wind and waves supplement the energy input from the Sun. Southeasterly winds cause the

upwelling of nutrients and waves keep water flowing over the kelp blades, thereby maintaining concentration gradients facilitating the supply of nutrients and removal of wastes at the kelp surfaces. Kelp production amounts to about 25 900 kJ m⁻ yearly, about 15% of which is represented as plant uprooted by storms. Uprooted *Ecklonia* floats and is washed out to sea or cast onto the shore, where it is either collected for commercial use or is consumed on sandy beaches by teeming populations of the amphipod *Talorchestia capensis*, or on rock shores by the isopod *Ligia dilatata*. Faeces of these crustaceans, together with any remaining kelp fragments, enter the detritus food chain either inter tidally or as particles suspended by wave action and consumed sublittorally by filter feeders. Uprooted *Laminaria* sinks to the bottom, remaining within the kelp forest where it is consumed mainly by the sea-urchin *Parechinus angulosus*. Waves continually erode the tips of kelp blades as in the Nova Scotia kelp forests, releasing small particles and dissolve organic matter that are utilized by bacteria and converted into detritus particles, which are in turn consumed by filter-feeding and deposit-feeding benthic animals.

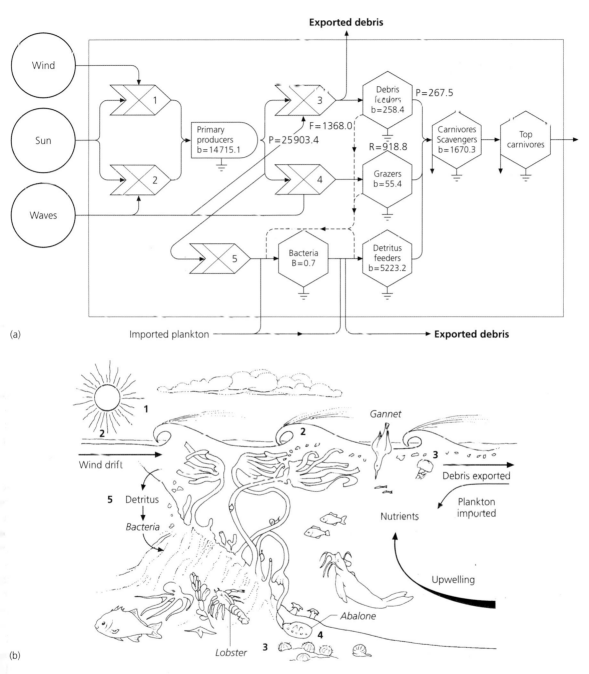

Fig. 5.19 (a) Energy circuit diagram for a South African kelp community. b, Standing crop; B, standing crop in water column under 1 m² sea surface; P, production; R, respiration; F, faeces. (All units are kJ m⁻².) The numbered work-gates are: (1) southwest winds causing upwellings; (2) waves moving water near the kelp surfaces; (3) waves uprooting kelp; (4) waves keeping the kelp in motion; preventing grazers climbing them but also bending them in reach of the abalone; (5) waves eroding the tips of the kelp blades. (From Velimirov et al. 1977.) (b) The effect of Sun, wind and waves in driving a kelp bed ecosystem. The numbers refer to the work-gates depicted in (a). Waves erode the kelp blade-tips; mussels and holothurians filter detritus from the water. Sea-urchins feed on broken pieces of kelp, and abalones trap kelp blades underfoot to graze them. (After Field et al. 1977.)

Energy subsidies from winds even affect the bacterial 'machinery'. Bacterial biomass in kelp forest sea-water is minimal during winter, but southeasterly winds during summer generate upwellings with high densities of bacterial aggregates in the water. In addition to fragmentation, the kelps produce mucilage at a rate of about 16 700 kJ m^{-2} yearly, supporting an annual dry biomass of bacteria of some 43 g m^{-2}, which in turn supports an annual dry biomass of flagellates and ciliates of about 43 g m^{-2}. The bacteria and protozoans are probably consumed directly by the filter- and deposit-feeding benthic invertebrates. In addition to its important effects on primary production and on the detritus food chain, wave action also affects the grazers. By keeping kelp plants in perpetual motion, wave action prevents grazers such as urchins and the large trochid snail *Turbo cidaris* from climbing up the plant and damaging the canopy. On the other hand, some plants are bent over so much by waves that the blades sweep over the sea bed, where the abalone *Haliotis midae* can trap them underfoot. The main carnivore is the rock lobster *Jasus lalandii*, which feeds heavily on mussels, but also takes a variety of other invertebrates, and is in turn eaten by dogfish, seals, octopus and cormorants. These trophic relationships are summarized in Fig. 5.19a and b, from which it is clear that the detritus food chain accounts for the great majority of kelp primary production. The South African study emphasizes particularly well the importance of energy subsidies in the form of wind and waves, which allow kelps to make maximum use of the input of solar energy and sustain productivities rivalling those of coral reefs, tropical rain forests or even the most intensive agricultural ecosystem.

6   Coral reefs

6.1 The coral organism

Coral reefs are a naturalist's paradise (Plate 15, facing p. 136). Diving among them is like entering another world: clear, warm waters allow one to drift comfortably over vistas of intriguing growth forms, among which swarm a kaleidoscope of brightly coloured fish (Plate 16, facing p. 136). Even sound may augment the visual impact, for on diving very close to the substratum, auditory signals of multitudinous pistol-shrimps hiding within crevices can often be heard like the crackle of a breakfast cereal. The vast quantities of limestone produced by corals, their high productivity in otherwise unproductive seas, the richness of species and subtlety of biological interrelationships engendered by them, put coral reefs at the centre of attention of numerous scientists from geologists and biochemists to biologists.

What is the structure of the organism that has such a profound impact on nature? Corals are closely related to sea anemones, and may be visualized as colonial anemones that secrete limestone foundations providing structural support and protection (Fig. 6.1). Each polyp sits in a cup-like depression, the calyx, which has radiating fins projecting from the base. In times of danger or physical stress, the polyp contracts so that the tissue is crammed between and over the fins within the calyx. At other times the polyp extends to varying degrees, depending on time of day and on species. Polyps are connected to their neighbours in the colony by a thin ·

Fig. 6.1 (a) Coral polyp, showing arrangement of tentacular crown, pharynx, coelenteron ('stomach'), mesenteries, mesenterial filaments and coenosarc (b) Calyx secreted by the polyp and upon which the polyp rests, showing arrangement of vertical septa that project into furrows in the wall between the mesenteries of the polyp.

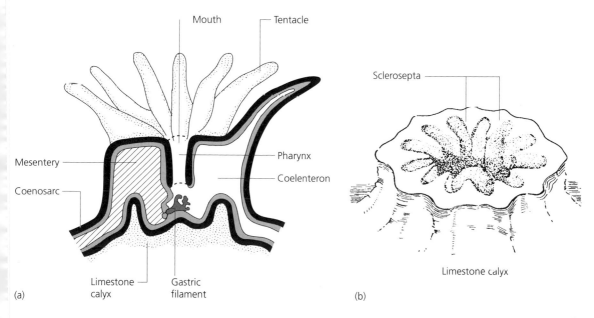

be formed of dense colonies of vermetid gastropods, whose upwardly growing tubular shells are cemented together by the calcareous algae. In certain places, especially near the edges of drainage channels, there may be carpeting clones of zooanthids (colonial 'sea anemones') whose brown or green tissues feel soft and slippery underfoot. Sea-urchins of the genera *Echinometra* and *Echinostrephus* often occur in dense populations, each urchin excavating a depression for itself in the calcareous substratum.

In situations of moderately severe wave action, windward reef crests tend to be dominated by one or two coral species, notably stoutly branching corals such as *Acropora palmata* in the Atlantic, *A. cuneata* in the west Pacific, *Pocillopora* spp. in the east Pacific, or by robustly branching 'stinging corals', *Millepora* spp., on reefs in various parts of the world. The close-growing, robust colonies form ramparts able to withstand the heavy seas. Both algal- and coral- dominated reef crests tend to be dissected by a 'spur-and-groove' system normal to the reef edge and extending some distance down the reef slope (Fig. 6.7a, b). The alternating spurs and grooves are several metres wide and may be up to 300 m long depending on the slope of the reef. They appear to be formed by erosion reinforced by the prolific seaward growth of corals on the grooves. The relative importances of differential erosion and coral growth remain unknown, but development of the spur-and-groove system is evidently self-reinforcing. Incoming waves cause tremendous surges back and forth along the grooves, churning up sediment and rolling boulders so that coral growth is prevented. Less turbulent water on the spurs allows corals to grow, reducing turbulence further. Sometimes corals from neighbouring spurs arch over the intervening grooves, forming canyons or even tunnels leading to blow holes on the reef crest. The effect of the spur-and-groove system is to dissipate the tremendous force of the unabating waves, estimated to be about half a million horsepower (373 megawatts) at Bikini Atoll, so stabilizing the reef structure. So severe and persistent is the water motion in the first few metres down the reef slope, that diving or sampling of any kind is virtually impossible most of the time, and for this reason the uppermost part of the reef slope has been termed the innominate zone (Fig. 6.6).

At greater depths the force of water movements is less, but so also is the light intensity. The less firmly cemented reef structure and the abundance of sponges, molluscs and worms that excavate and weaken coral skeletons at these depths increase the risk of dislodgement by collapse and slumping of the substratum. Coral growth-forms therefore tend to be more foliaceous or plate-like, with large areas for the interception of light from above and with hydrodynamic properties tending to prevent them from resting wrong-side-up after dislodgement or from rolling down the reef slope into lethal depths. Another, not necessarily mutually exclusive, explanation for the change from massive to foliaceous growth forms at greater depths is that diminished light intensities restrict the photosynthesis of zooxanthellae, and hence the rate of calcium carbonate deposition. As polyps can feed on plankton, which does not decrease in abundance with depth, the growth of coral tissue outstrips skeletal growth, resulting in flattened, plate-like colonies. Contrary to this idea, however, the foliaceous coral *Agaricia agaricites* produces calcium carbonate more slowly in relatively shallow water.

Sponges, alcyonarians (seawhips and gorgonians) and non-hermatypic corals become increasingly important down the reef slope and gradually replace hermatypic corals below 30–70 m depth, depending on water clarity. Reduced water movements allow more silting and the substratum below the talus slope usually consists of fine sediments.

On leeward reefs, zonation along the reef flat is essentially similar to that on windward reefs, but the boulder zone, moat, algal ridge and spur-and-groove system are absent (Fig. 6.7c). Coral associations on the outer reef flat merge gradually with those on the upper reef slope. Diverse forms are present, but there is an abundance of branching, quick-growing forms. Corals tend to be aggregated into patches or large mounds, between which are pockets of sand and rubble (Fig. 6.7c). The topographically diverse habitat supports a rich invertebrate and fish fauna, notable members of which are the black sea-urchins (*Diadema* spp.) whose needle-like spines readily pierce unwary feet.

Changes in coral morphology with increasing depths down the reef slope are similar to those on the windward reefs, and in both cases are due more to species replacements in the Indo-Pacific and more to phenotypic plasticity in the Caribbean. On the slopes of some Pacific reefs, for example, species of

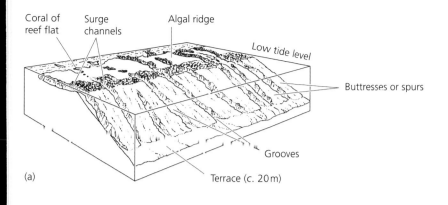

(a)

(b)

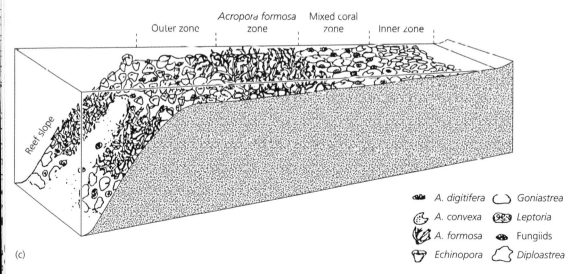

A. digitifera   Goniastrea
A. convexa   Leptoria
A. formosa   Fungiids
Echinopora   Diploastrea

(c)

Fig. 6.7 (a) Spur-and-groove formation on a windward Pacific reef. (After Yonge 1963.) (b) Buttresses (equivalent to spurs) of a Caribbean reef. (After Goreau & Goreau 1973.) (c) Coral zonation on a lagoon (sheltered) reef of a Pacific atoll. (After Spencer-Davies et al. 1971.)

Species	Gastropods	Enteropneusts	Nereids	Eunicea	Terebellids	Other polychaetes
flavidus		4			64	32
lividus		61		12	14	13
pennaceus	100					
abbreviatus				100		
ebraeus			15	82		3
sponsalis			46	50		4
rattus			23	77		
imperialis				27		73

Table 6.1 Major foods (percentages) of eight species of cone shells (*Conus*) on subtidal reefs in Hawaii. (After Kohn 1959.)

partial knowledge for two well-studied groups of animals: predatory gastropods and reef fish.

Predatory gastropods are particularly diverse on coral reefs, either living among the corals or confined to boulders, rubble or sediments on the reef flat. Why so many families, such as the Conidae, Muricidae, Terebridae and others, are better represented in the tropics than in temperate regions is partly a question of biogeography, to be considered in Chapter 10. But how can so many species of rather similar predators coexist on coral reefs?

Extensive work by Kohn on the genus *Conus* (Table 6.1) and other studies of the Muricidae and Terebridae have shown that each species tends to exploit a unique combination of available types of prey and microhabitats. Because of this 'niche specialization', interspecific competition is reduced and competitive exclusions, so common among corals competing for space and light, are probably rare. Unfortunately, it would be very difficult to test this inference by direct observation because of problems of sampling and of quantifying resource usage or of disentangling the effects of different population-controlling factors such as predation, competition and physical hazards.

Kohn (1979), however, made use of a 'natural' experiment by noting that *Conus miliaris* at Easter Island has a much broader diet than when it coexists with other congeners elsewhere in its geographical range. This supports the idea that on an evolutionary time scale, predatory gastropods respond to interspecific competition by niche contraction, thereby allowing coexistence. A similar 'niche diversification' basis to coexistence would not be possible among the great number of coral species inhabiting a reef, because their essential resources

of space and light could not be partitioned in so many ways as can the diverse prey and micro habitats used by gastropods. A certain amount of resource-partitioning is, of course, possible among corals with regard to autotrophy vs. heterotrophy (Section 6.9.1), diurnal vs. nocturnal feeding, depth and exposure to wave action.

The species composition of certain reef fish assemblages requires yet another explanation. Damselfish (Pomacentridae) are small reef fish that use coral heads, or similar blocks of reef material as a home base. Each coral head provides sufficient space for only a certain number of damselfish which are territorial and repel intruders. By a long series of carefully replicated experiments and controls, in which the occupants of coral heads were manipulated, Sale (1980) has shown the local diversity of pomacentrids to be a function of the number of species available in the general area and of local population densities, with the result that the species composition at any site is unpredictable, depending on the history of chance colonization. There is no competitive hierarchy, as there is in corals, and no niche diversification, as there is in predatory gastropods. However, because of the limits to feasibility, Sale concentrated entirely upon a very small scale habitat and a single guild of fish.

Longer-term studies of natural and artificial reefs in the Caribbean (Ogden & Ebersole 1981) suggest that although at the small scale of individual coral heads fish community structure may be a function of chance, at a larger scale encompassing the whole reef, fish community structure is a function of the orderly processes of competition and niche diversification. Fish recruit onto the reef as larvae which may have undergone considerable dispersal

Establishment of the larvae and ensuing juvenile stages depends considerably on interactions with residents and it is especially during this phase that community structure may be biologically, rather than stochastically determined. Communities of decapod crustaceans inhabiting coral heads in the Pacific seem also to be structured by chance events, and it would be interesting to see if evidence for more orderly processes would emerge from studies on a larger spatial scale.

6.8 Coevolution in coral-reef communities

Disturbances and chance have been shown to be fundamental in the structuring of coral-reef communities, and although these factors may explain how species can coexist on the reef, they do not explain why there are so many species in the first place. Probably there are several partial answers to this question, some of which will be considered in Chapter 10, but an important contributing factor is the overall predictability of the coral-reef environment. Despite tectonic earth movements, eustatic changes in sea-level and changes in world climate,

Fig. 6.13 (a) Cleaner wrasse, *Labroides dimidiatus*, attending to the red snapper, *Lutianus sebae*. The mimic, *Aspidonotus taeniatus*, takes a bite out of the 'host's' tail. (b) In addition to mimicking colour, *Aspidonotus* also mimics the undulating movement of *Labroides*. (After Wickler 1968.)

coral reefs rather like those existing today have persisted in the tropics for over 50 million years. Because corals require warm, clear waters throughout the year and are killed by relatively minor deviations from optimal conditions, the environment on reefs must have remained virtually constant throughout all those millions of years. This general stability over geological time has allowed ample opportunities for the coevolution of very specialized, finely tuned relationships between organisms and in this respect the old view of organismal diversity on coral reefs is correct.

The subtlety of coevolved biological interactions is epitomized in the bizarre symbiotic relationships to be found among coral reef inhabitants. Classical examples are the 'cleaner fish', usually small wrasses, butterfly-fish, or gobies, which feed on the ectoparasites attached to the outer surface, gills and buccal lining of larger fish (Fig. 6.13a). Parasite-laden fish queue up at certain 'cleaning stations', perhaps a conspicuous coral head, where they posture so as to allow the small cleaner fish access to the gills and mouth. The bold colour patterns and swerving, jerky movements of the cleaners (Fig. 6.13b) act as recognition stimuli to the hosts, thus ensuring that the cleaners are not eaten. A final twist to this coevolutionary tale is that cleaner-mimics occur (Fig. 6.13), which use the cover of similar recognition stimuli to gain close access to the 'host', whereupon they take a bite of flesh rather than remove a parasite.

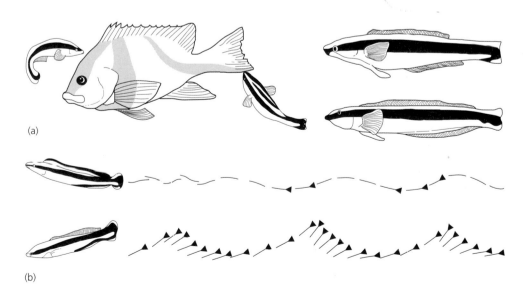

(a)

(b)

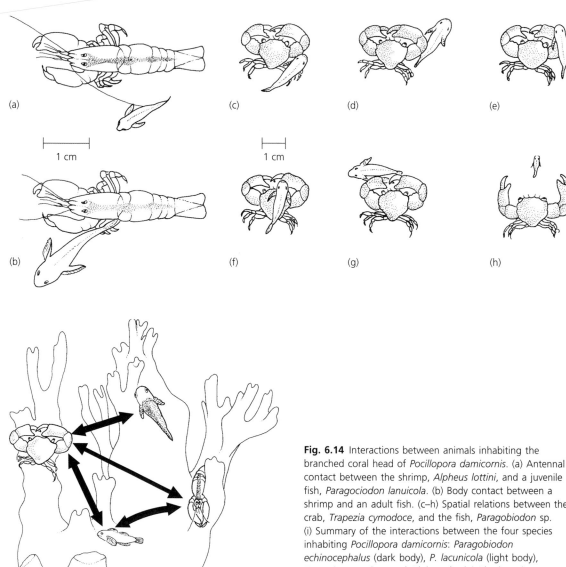

Fig. 6.14 Interactions between animals inhabiting the branched coral head of *Pocillopora damicornis*. (a) Antennal contact between the shrimp, *Alpheus lottini*, and a juvenile fish, *Paragociodon lanuicola*. (b) Body contact between a shrimp and an adult fish. (c–h) Spatial relations between the crab, *Trapezia cymodoce*, and the fish, *Paragobiodon* sp. (i) Summary of the interactions between the four species inhabiting *Pocillopora damicornis*: *Paragobiodon echinocephalus* (dark body), *P. lacunicola* (light body), *Trapezia cymodoce* and *Alpheus lottini*. The broad arrows indicate communication, incorporating the shivering behaviour of fish; the thin arrow represents the cleaning of the crab by the shrimp. (After Lassig 1977.)

The intricate interdependence of organisms on coral reefs is not confined to symbioses. On the Great Barrier Reef, for example, the branching coral *Pocillopora damicornis* harbours up to 16 species of crustaceans and fish that use the coral for food and shelter. Close study of the most conspicuous residents, two fish, a crab and a shrimp, showed that although they have the potential to harass, capture or exclude one another from the coral, a system of signals among the residents facilitates coexistence. The pistol-shrimp, *Alpheus lottini*, uses its legs to scratch the hairy hind surfaces of the claws of the small xanthid crab, *Trapezia cymodoce*. The crab performs similar movements itself to transfer food particles from the hairy margins of the claws to the mouth. By simulating this movement, the shrimp

Plate 13 Fucoid-dominated shore, sheltered from wave action, illustrating down-shore zonation from *Fucus spiralis*, through *Ascophyllum nodosum* to *Fucus serratus* (Anglesey, North Wales) (R.N. Hughes.)

Plate 14 Barnacle-dominated shore, exposed to wave action, illustrating down-shore zonation from the spray zone (orange lichens), *Verrucaria/Littorina* zone (black patches) and barnacle zone. The thinly populated upper edge of the barnacle zone is colonised by *Cthamalus montagui*, whereas the remaining densely populated area is colonised by *Semibalanus balanoides* (Anglesey, North Wales). (R.N. Hughes.)

Plate 15 All available space on coral reefs is covered with a profusion of life. Soft siphonogorgia corals predominate ledges off Madang, Papua New Guinea and grow out into the water currents to catch planktonic prey. (L. Newman & A. Flowers.)

Plate 16 Hanging around together above a shallow coral bommie during the day these Red Throated Sweetlips, *Lethrinus chrysostomus*, feed separately at night on benthic invertebrates and small fish. Sweetlips, are aptly named for their enlarged lips and small mouths (Heron Island, southern Great Barrier Reef). (L. Newman & A. Flowers.)

Plate 17 The large and conspicuous Crown-of-thorns sea star, *Acanthaster planci*, feeds exclusively on hard corals, digesting the polyps by everting its stomach. Unexplained cyclic population explosions can occur throughout the Indo-Pacific where these echinoderms are encountered in plague proportions, eating all live coral in their path. The long spines on the upper surface contain toxins which can inflict a serious wound. (L. Newman & A. Flowers.)

Plate 18 Giant, tubucolous, vestimentiferan worms, *Riftia pachptila*, found abundantly around deep-sea fumaroles. The worms lack guts, but have symbiotic, chemosynthetic sulphur-oxidising bacteria on which the depend for their nutrition. (R.R. Hessler.)

Plate 19 A juvenile dwarf sole (*Parachirus xenicus*) which reaches 8 cm in length and occurs in Indo-Pacific waters. It is not exploited commercially. (A. Connell.)

Plate 20 Blocking out the sun this huge school of Chevron barracuda, *Sphyraena putnamiae* congregate for mating off the warm waters of Madang, Papua New Guinea. (L. Newman & A. Flowers.)

Plate 21 Hussars, *Lutjanus amabilis*, find safety in numbers, their movement and colour patterns are used to confuse predators. They are usually found in large schools in the same location during the day where they rest under overhangs. At night, these active predators range over large areas preying on smaller fish and crabs (Heron Island reef, southern Great Barrier Reef). (L. Newman & A. Flowers.)

Plate 22 Deck-load of hake caught by a deep-sea trawler in the waters of the Beguela upwelling system. (J.G. Field.)

'appeases' the crab. The two fish undergo shivering movements when in close contact with the shrimp or crab (Fig. 6.14), thereby preventing aggression from the crustaceans. Experiments showed that non-resident fish fled from approaching crabs, which became extremely aggressive and snapped at the fish. Newly colonizing fish have to out-manoeuvre the crustaceans among the coral branches until the signal system is learned. Once established, the signal system saves time and energy that would otherwise be spent evading the aggressive responses of the other residents. Each member of the established 'team' benefits by the presence of the others because the effort of defending the coral head against predators or competitors is shared. All inhabitants feed on coral mucus or tissue, and the repulsion of competitors may keep the number of residents at a level that is not detrimental to the food supply.

6.9 The trophic status of corals

Equally impressive as the species diversity of coral reefs is their high biological productivity, amounting to as much as 12 000 g C m^{-2} per year, which is maintained in seas that are otherwise relatively unproductive (20–40 g C m^{-2} per year) because of nutrient depletion (Section 1.2). The key to this high productivity probably is twofold. First, the continual renewal of nutrients by water masses passing over the reef may be sufficient to sustain high productivity, even though concentrations of nutrients remain low. Secondly, the symbiotic machinery of corals and zooxanthellae facilitates a tight recycling of nutrients within the coral tissue. There is, however, much more to the story than that. Corals feed to varying extents on plankton, and perhaps also on bacteria that they trap in mucus or even on organic molecules taken up from solution across the body wall. These food sources contribute nutrients and energy to the total reef ecosystem, yet their relative importance in the nutrition of corals themselves remains in debate. A brief review of the known feeding biology of corals is therefore in order.

6.9.1 Autotrophs or heterotrophs?

In quick-growing, branching corals such as *Pocillopora damicornis*, the coral tissue forms a relatively thin layer over the skeleton, and the proportions of animal and plant protein in the tissue are about equal; in other words, these corals are about half animal and half plant, whereas the proportion of plant material is less in the thicker tissues of massive corals. Evidently the branching corals function more like pure autotrophs and some of the massive corals more like pure heterotrophs. This divergent nutritional trend is reflected in both skeletal and polypary morphology. On passing through water, light is extensively scattered by suspended particles and molecules in solution. In shallow water the overall light intensity is so high that in addition to the vertical light from above, there is a considerable horizontal component of scattered light. The multi-layered growth form of branching corals (see Fig. 6.2) increases the surface area of light-intercepting tissue both horizontally and vertically, enabling these corals to make maximum use of incident and scattered light. In addition to these skeletal modifications, the polyps of 'autotrophic' corals tend to be small, thereby exposing the maximum area of zooxanthellae, which reside mainly in the endoderm, to the light.

The more 'heterotrophic' corals typically have a monolayered, spheroidal skeletal structure (see Figs 6.2 and 6.8). Normal plankton densities are such that after passing over a layer of filtering polyps and mucus traps, most of the potential food particles would be removed, so that the coral would gain very little from multiple layers of polyps. The tissues are thicker and the polyps larger in the 'heterotrophic' corals in accordance with their planktivorous habit. A spheroidal form maximizes the surface area of plankton-intercepting tissue; indeed, as C.M. Yonge pointed out, corals have the highest prey-capturing surface to living-tissue volume of any animal. Encrusting and other growth forms are probably adaptations to non-trophic environmental factors.

Of course, light is important to all hermatypic corals, because without it they are unable to secrete large quantities of calcium carbonate, and probably all corals ingest some plankton; it is only the relative proportions in which these energy sources are used that differs among corals. Estimating these proportions is not easy and data are rather uncertain, but the proportion of energy ultimately derived from photosynthesis ranges from over 95% in the 'autotrophic' corals down to somewhat over 50% in the more extreme 'heterotrophic' species. To what

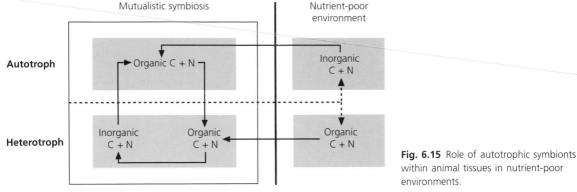

Fig. 6.15 Role of autotrophic symbionts within animal tissues in nutrient-poor environments.

extent corals acquire photosynthates by digesting the zooxanthellae or by taking up substances leaking across the algal cell walls is debatable, but the importance of the second mechanism is indicated by the ability of coral-tissue homogenates to stimulate zooxanthellae to release glycerol and small amounts of other organic molecules. Incubation of *Pocillopora damicornis* with $^{14}CO_2$ on the reef has shown that up to 50% of the ^{14}C fixed by the zooxanthellae during 24 h ends up in the coral tissue, mostly as lipid and protein. Evidently, the glycerol released by the algae is used by the animal in lipid synthesis.

Phosphorus and nitrogen may be limiting nutrients for algae in tropical seas, but zooxanthellae can obtain them from the metabolic wastes of their hosts, and owing to this recycling, the phosphorus and nitrogen turnover by corals is much slower per unit body weight than in other organisms (Fig. 6.15). Inorganic nutrients, however, must ultimately be taken up from sea-water by the corals. Two routes are possible: via consumed food particles or by active transport across the body wall.

6.9.2 Capture of food particles

Planktonic organisms are captured mainly by a mucociliary mechanism that also serves as a cleansing device, keeping the coral surface free of silt. Drifting particles that touch any part of the living surface of the coral become stuck in mucus and transported by ciliary currents either directly to the mouth or along rejection tracts if unsuitable as food. In some corals the food-laden mucus accumulates at the tips of the tentacles, which bend over and transfer the particles to the mouth, or bend outwards

and reject them (Fig. 6.16). A correlation has been found between the amount of zooplankton material in the stomach and polyp diameter. Certain large-polyped corals are capable of using their long tentacles to sense microquantities of amino acids leaking from passing zooplankton and to capture the zooplankton passing within 1 cm of the coral head. Some corals use extracoelenteric digestion to cope with food particles too large to be handled either by the ciliary tracts or tentacles. Mesenterial filaments creep by ciliary action out through the mouth or through temporary openings in the body wall (see Fig. 6.11) onto the food material and proceed to digest it.

Bacteria rapidly colonize the mucus produced by corals as they do the mucus produced by any aquatic animal, but it appears that the amount of bacterial organic matter consumed by coral polyps may be very small under natural conditions.

6.9.3 Possible uptake of dissolved organic matter

Sea-water contains dissolved organic substances, ranging from free amino acids to macromolecules, which are a potential resource of those heterotrophs able to take them up against a concentration gradient (see Section 3.2.1). Some corals, notably those employing extracoelenteric digestion by the extrusion of mesenterial filaments, have an ectoderm covered with microvilli and possessing a high alkaline phosphatase activity. Such corals, therefore, seem well equipped, with an enormous absorptive surface and an enzyme system usually involved in membrane transport, for the active uptake of organic molecules. These features, together with

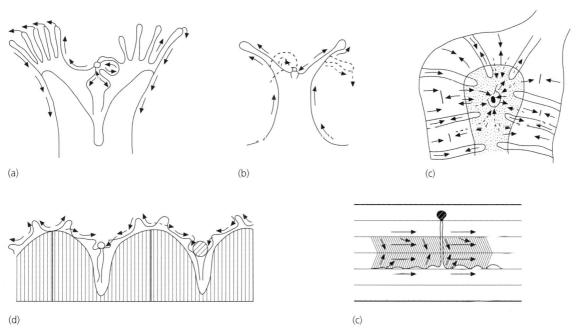

Fig. 6.16 Methods of feeding in hermatypic corals.
(a) *Euphyllia*, with large polyps with all cilia beating away
from the mouth. (b) *Pocillopora*, with small polyps and cilia
carrying particles up to the column. (c) *Merulina*, a brain coral
with short tentacles and reversal of cilia (the resultant current
is indicated by the broken arrow). (d) *Coeloseris*, with all
particles carried by cilia over the mouths for ingestion if
edible and removal by water currents if not. (e) *Pachyseris* has
mouths in rows within parallel grooves and no tentacles, but
food is collected by mesenterial filaments extruded through
the mouth. (After Yonge 1963.)

the observed uptake of ^{14}C-labelled material, leave
little doubt that corals can remove some substances
from solution, but the extent of this remains to be
established. By noting the characteristic proportion
of ^{13}C in the tissues of terrestrial vegetation of central Queensland, Risk *et al.* (1994) could trace the
uptake of terrigenous organic matter by corals right
across the reef to the edge of the continental shelf,
some 110 km offshore. As much as one-third of the
diet of inshore corals may be derived from terrestrial sources, although whether this is directly as
dissolved organic matter or indirectly through the
ingestion of bacteria, remains unresolved.

6.10 Productivity of coral reefs

Measuring the productivity of an entire coral reef
is fraught with difficulties owing to the diversity of
habitats from reef flat to reef slope and the diversity of organisms present in each of them. Added to
these difficulties are uncertainties about the trophic
status of many organisms, not least the corals themselves, and about the magnitudes of organic inputs
to the ecosystem via oceanic plankton, bacteria and
dissolved organic matter.

6.10.1 Primary production

Early attempts to estimate the overall primary productivity of reefs were based on day- and night-
time comparisons of the change in oxygen tension
that occurs in water as it flows over the reef. Such
estimates indicated a high primary productivity of
1500–12 000 g C m^{-2} per year, which is two orders
of magnitude greater than that of the surrounding
seas. From more recent studies, a hazy picture of
the relative contributions of benthic algae and symbiotic zooxanthellae to total primary production is
beginning to emerge.

Benthic algae include the non-calcareous, frondose 'macroalgae' of inner reef flats, the belts of
Sargassum present near the boulder zone on some

reefs, upright calcareous algae such as the super-abundant *Halimeda*, encrusting calcareous algae such as *Porolithon*, filamentous algae growing either on surfaces or penetrating calcareous substrata such as coral skeletons, blue–green algae that grow as films on reef flats, and finally the zooxanthellae, present not only in corals but also in alcyonarians (soft corals), zooanthids (colonial anemones), giant clams (*Tridacna*), certain ascidians, and other animals.

After taking into account the proportions of total surface area covered by living *Acropora palmata*, macroalgae and filamentous turf-forming algae, together with the total surface area of substratum contained within a 1 m² quadrat, the net primary production of these autotrophic categories on a Caribbean reef has been estimated at 630 g C m⁻² per year for the coral, 1170 g C m⁻² per year for macroalgae, and 700 g C m⁻² per year for the filamentous algae. The intrinsic net primary productivity of the zooxanthellae in *Acropora palmata* was about 10% that of the macroalgae and about 60% that of the filamentous algae.

Filamentous green algae are quick to colonize uncovered limestone surfaces, such as dead coral skeleton. They remain inconspicuous, however, because they are constantly grazed down by herbivores, such as guilds of browsing fish (Section 6.5) and gastropods, so that their standing biomass is negligible. As the average herbivore biomass of some 150 g (dry wt) m⁻² may be turned over at a rate of 12 times a year, the filamentous algae must be producing at least 3000–4000 g C m⁻² a year, about five times that estimated directly. This discrepancy however, could easily be accounted for by uncertainties of the total area of filamentous green algae present on a reef. The encrusting calcareous alga *Porolithon* that dominates the algal ridges of Pacific windward reefs (Section 6.2), has an estimated net primary productivity of about 180 g C m⁻² per year. On the Great Barrier Reef epilithic algae are the major energy source for grazers.

The overall picture of primary production is imprecise. However, it appears that zooxanthellae in corals are about equally productive compared with the various kinds of benthic algae, but that because of their relative abundance compared with other autotrophs, zooxanthellae probably account for less than half the total net primary production in the coral reef ecosystem.

Adjacent to many reefs are large sea-grass beds covering sandy areas on the reef flat and producing around 7000 g C m⁻² per year (Chapter 4). Only about 5% of this production is consumed by herbivores, the rest being degraded into detritus, some of which, together with the associated bacterial flora, will enter food chains on the adjacent reef. As the roots of sea-grasses fix nitrogen, they may be an important source of this otherwise scarce nutrient.

Atmospheric nitrogen is also fixed by blue–green algae such as *Calothrix crustacea*, which occurs on intertidal reef flats in the Pacific as a thin, monospecific film (it also occurs in other reef habitats in different growth forms). *Calothrix* can fix nitrogen at the rate of 1.8 kg ha⁻¹ per day—two to five times the rate achieved by fields of lucerne or alfalfa. Fixed nitrogen enters the food web through at least three routes: (1) the blue–green algae are consumed by herbivores, especially by certain fish with low assimilation efficiencies, so that the water over the reef gains nitrogen via the fish faeces; (2) in areas subject to strong wave action, large pieces of the *Calothrix* film are dislodged and washed over the reef, where they will be available to consumers; (3) *Calothrix* releases about 50% of its fixed nitrogen into solution, from which it may be taken up by other autotrophs.

6.10.2 Nutrient input from oceanic plankton

The total autotrophic productivity of coral reefs is certainly very high, but does it balance the consumption rate of the heterotrophs? Some flow-respirometry studies show that it does, others that it does not. Nutrient cycling within the coral-reef ecosystem is highly efficient—for instance, at least 75% of the phosphorus pool of a reef is recycled within a day—but some losses are inevitable. To what extent is the reef ecosystem dependent on an external source of nutrient?

Plankton, as a source of energy and nutrients, is undoubtedly important on all reefs. Measurements of plankton depletion in water masses passing over a Caribbean reef have revealed a 91% reduction in the diatom crop and a 60% reduction in zooplankton, amounting to a total gain of about 65 g C m⁻² yearly, equivalent to 4–13% of the net community metabolism and also 10% of the primary production.

6.10.3 Resident plankton

The figures above refer to net, incoming plankton to the reef, whereas up to 85% of the zooplankton ingested by corals may come from the reef itself. Resident zooplankton migrates vertically from under the coral heads at dusk and although eaten by the corals, this demersal plankton owes its existence to the shelter provided by the corals and associated reef structures, and represents an internal component of the coral-reef ecosystem distinct from the net input of plankton from beyond the reef.

The flow of energy between corals and resident zooplankton may be two-way. Corals produce about 50 mg of organic mucus m^{-3} per day in the Red Sea. Mucus contains energy-rich wax esters and so has a high total energy content of 22 J mg^{-1} (ash-free dry weight). The copepod *Acartia negligens* can assimilate at least 50% of the organic content of the mucus. Moreover, soon after its secretion as a clear, viscous liquid, mucus becomes denatured and transformed into polymeric strands and webs that tend to become suspended as flocs or incorporated into the detritus. The mucus soon gains a high bacterial content and there is little doubt that mucus and bacteria represent significant sources of energy in coral-reef food chains. Bacterial biomass has been estimated at about 60 mg C m^{-3} over the Great Barrier Reef compared with 20 mg C m^{-3} in the open water off the reef.

Carbon fixed by photosynthesis on the reef is lost partly by offshore transportation but most is accounted for by the metabolic activity of intermediate consumers. The production to respiration ratio ($P : R$) therefore is close to unity and there is insufficient fixed carbon left to support high, sustainable yields of large carnivores at the end of the food chain. Evidence suggests that natural coral-reef communities are predator controlled rather than nutrient limited, that is, they are driven from 'top down' rather than from 'bottom up'. Reversal of this pattern, however, may be caused by eutrophication. Particulate organic loading greatly increases the biomass of small invertebrates within sediments, among weed and among the interstices of the reef framework. High nutrient concentrations promote the growth of benthic algae to levels that can choke the community. Organic pollution therefore drives the system from 'bottom up' (Grigg *et al.* 1984).

6.11 Conclusions

Despite all the uncertainties about the relative contributions to the coral-reef ecosystems of different kinds of autotroph, bacteria, dissolved organic matter, and internal vs. external inputs, it is clear that the phenomenally high total productivity is in large measure due to the combination of a tremendous surface area of photosynthetic tissue (either in the form of zooxanthellae or benthic algae and higher plants), optimal light and temperature conditions for photosynthesis, and the tight recycling of nutrients in an otherwise nutrient-poor environment. The efficient recycling of nutrients occurs both at the level of the coral–zooxanthella symbiosis and at the general level of the overall food web. Many consumers are present in reefs and although in absolute terms primary production is high, relative to the number of consumers, food can be regarded as scarce. Hence food is consumed rapidly and utilization is efficient. A high proportion of the environmental pool of nutrients is therefore maintained within living tissues, so reducing opportunities for the loss of nutrients out of the system. Any such losses are compensated by the slow accrual of nutrients from water masses passing over the reef and by the nitrogen-fixing activities of blue–green algal associations on the reef or rhizomes of adjacent seagrasses. Summaries of recent work on coral reefs have been given by Stoddart & Yonge (1971) and Jones & Endean (1973).

That the largest part of the ocean only merits the smallest chapter in this book is not a reflection of its relative importance but of our lack of knowledge. Nevertheless, we know very much more of deep-sea ecology now than we did 30 years ago (see Rex 1981; Jones 1985; Grassle 1986; Wilson & Hessler 1987; Gage & Tyler 1991; Hargrave 1991; Childress & Fisher 1992; Rice & Lambshead 1994), and we are learning more every year, with several of the new discoveries being reported by the popular media. There is currently, for example, much speculation that life on Earth may have originated in conditions equivalent to those of the deep-sea hydrothermal vents (see below), and that vents on Jupiter's moon Europa may support living organisms. In the early days of marine biology many people predicted that the abyssal depths would be found to be lifeless, whereas others thought (hoped?) that they would be populated by relics from bygone eras. We now know that it is neither so extreme nor a haven for competitively inferior lineages. Indeed, except for the absence of photosynthesis, the deep sea is not basically different from the regions described in Chapters 2 and 3, merely less productive and more sparsely inhabited as a consequence of the lower input of food materials. In this respect, it is equivalent to cave ecosystems on land.

7.1 Pelagic zones

From the middle of the surface layer to about 2000 m, planktonic biomass declines almost exponentially and below that depth it declines more slowly down to values of only a fraction of a milligram of living organisms per cubic metre of water (Fig. 7.1). Variation in planktonic abundance at a given depth correlates well with the productivity of the overlying surface waters, re-emphasizing that it is the rain of detrital particles and faecal material

which supports the abyssal plankton. Most species are therefore omnivores or carnivores and these either seize macroscopic items of food individually, or filter the microplanktonic consumers of bacteria. M.E. Vinogradov argued for many years that living food is introduced into the deep sea through an overlapping series of vertical migrations but, although this is an attractive idea, there is very little evidence

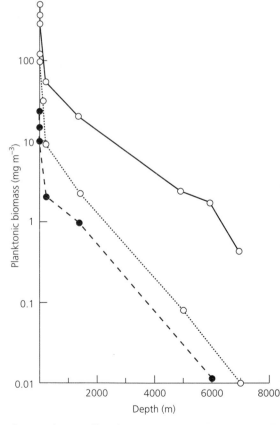

Fig. 7.1 Three profiles of zooplanktonic biomass in the Pacific Ocean. (From data of Menzies *et al.* 1973.)

in support of it. Well-documented cases of vertical movements are all restricted to the upper 2000 m. The presence of algal debris in the guts of deep-sea species has been used in support of such a 'ladder of migration' but these algae—especially the enigmatic 10–15 μm long 'olive-green cells'—are themselves inhabitants of the abyssal depths. Most probably the olive-green algae are the resting stages formed by diatom species as a reaction to sinking out of the photic zone: they have been recorded at all depths down to 5000 m. Populations of coccolithophores and cyanobacteria of up to 350 000 individuals per litre are also known from depths of 1000–3000 m in the Atlantic and Mediterranean, although it is unlikely that they are active photosynthetically. Further, it has become apparent that dead phytoplankton cells from the photic zone can even sediment out on the abyssal plain, especially when, for some reason, they clump together and sink rapidly. Seasonal fluctuations in surface production are then reflected by a distinct seasonal input of phytodetritus into the deep sea; itself reflected in seasonal breeding of some of the consuming species.

7.2 Benthic areas

Like the abyssal plankton, the biomass of the benthos declines markedly with depth, especially if the depth gradient coincides with increasing distance away from the coast (Table 7.1; Fig. 7.2). Overall, the decrease is exponential and is approximated by the expression (Rowe 1983)

$$\log_{10} \text{ biomass (g m}^{-2}) = 1.25 - 0.0004 \text{ depth (in km).} \qquad (7.1)$$

Table 7.1 Approximate average values of biomass (g (wet wt) m^{-2}) at different depths in the ocean.

Depth range (m)	Biomass
0–200	200
500–1000	< 40
1000–1500	< 25
1500–2500	< 20
2500–4000	< 5
4000–5000	< 2
5000–7000	< 0.3
7000–9000	< 0.03
> 9000	< 0.01

Most of the ocean bed supports less than 0.5 g (wet wt) m^{-2} of living organisms (Fig. 7.3) and most macrofaunal populations exist at densities of only 0.5–1.0 individuals m^{-2}. Because the deep-sea benthos is dependent on detritus and the associated bacteria, however, anomalously high biomass can occur where large quantities of detritus reach regions adjacent to a coast. Quantities of plant fragments and waterlogged wood have been found at 7000 m in the Banda Sea, for example, and in conjunction with this munificence were superabundant organisms, 10–12 g (wet wt) m^{-2}, a very high biomass for that depth.

Hydrothermal vents in the deep sea bed (Fig. 7.4) may also support an anomalously large biomass; up to 3000 times the background level has been recorded. Although first seen only in 1977, many such vents have now been discovered, particularly along the central rift valleys of the East Pacific and

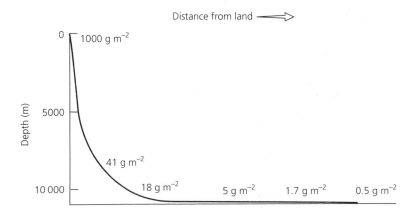

Fig. 7.2 Variation in wet-weight biomass of the benthos of the Pacific Ocean in relation to the depth and to the distance from the nearest continental coast. (After Zenkevitch & Birstein 1956.)

Distance from land ⟹

0 — 1000 g m^{-2}

Depth (m)

5000

41 g m^{-2}

18 g m^{-2} 5 g m^{-2} 1.7 g m^{-2} 0.5 g m^{-2}

10 000

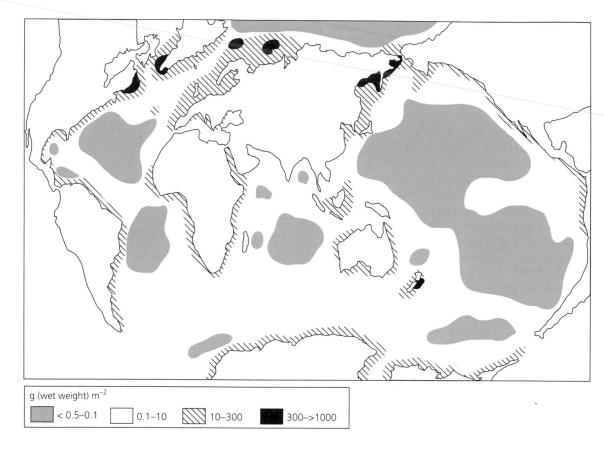

g (wet weight) m^{-2}

| | < 0.5–0.1 | | 0.1–10 | | 10–300 | | 300–>1000 |

Fig. 7.3 The distribution of benthic biomass (wet weight) in the world ocean. (After T. Wolff 1977.)

associated rises at depths of 2000–3000 m. From them issues water, rich in bacterial floc, at up to 350°C although at the prevailing pressure water does not boil until some 450°C. Wet weights of over 70 kg m^{-2} are known around such vents, mainly composed of huge vestimentiferans (Plate 18, facing p. 136) in densities of 175 m^{-2}, and of the large, fast-growing bivalve molluscs *Calyptogena* and *Bathymodiolus* that can grow at rates in excess of 4 cm per year (cf. *Tindaria* below). The vestimentiferans, in contrast, grow very slowly: annual growth rates average some 8 mm and mature worms of 2 m length are probably more than 150 years old. The fauna is dominated by animals with chemoautotrophic bacterial symbionts that oxidize the various reduced compounds which issue from the fumaroles—especially sulphides, sulphur and thiosulphates, but also methane, hydrogen, nitrogen compounds and manganous ions (see Tunnicliffe 1991). The vestimentiferan *Riftia*, for example, possesses 10^9 bacteria g^{-1} in its 'trophosome', a special tissue forming half of its body mass. The other species either feed on free bacteria or are predatory.

Each of these specialized communities, dependent on geothermal sources of energy and on chemosynthesis, is a small island surrounding each vent— some 60 m across at most—and groups of vents may be separated by 100 km or so. Further, the lifespan of any individual vent community may be only one or a few decades before it is destroyed by erupting lava or by change in fumarole location. This raises the question of how the vent fauna manages to persist. It is possible that vent species use large carcasses as 'stepping stones' between fumaroles: the average distance between whale corpses across one area of deep ocean has been estimated to be only 9 km. Transport by rotating plumes of vented water is another possibility. An ephemeral and precarious existence as it appears to be, vent assemblages

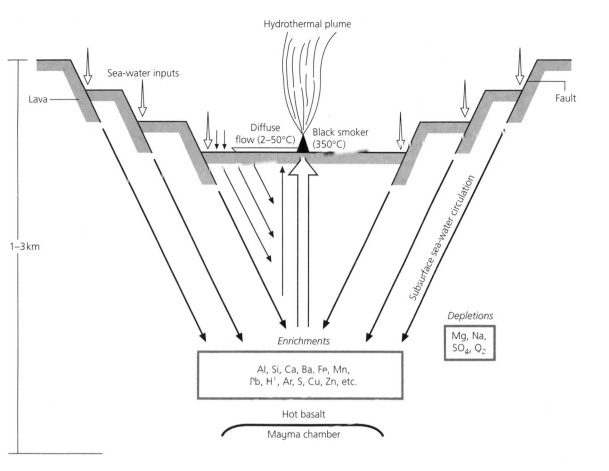

Fig. 7.4 Sea-water can penetrate into the oceanic crust through the cracks and faults occurring at centres of sea-floor spreading, there to react with hot basalt and eventually, chemically modified, to issue out again at hydrothermal vents. Typical chemical modifications of vented sea-water are indicated. (After Van Dover 1990.)

have a long pedigree in that they are known from Devonian and Cretaceous rocks. Not all are in fact geothermal and deep sea: off California, thermal vents occur in 20 m of water; and off Florida and elsewhere sulphur-rich or hydrocarbon seeps are of cold water.

These rich areas, however, are exceptions. Over most of the ocean bed the food input is very much less. Even at the relatively shallow depth of 2000 m, a mere 5.7 mg C m^{-2} day^{-1} was recorded by one study as sedimenting out of the water column, largely in the form of faecal pellets (660 pellets m^{-2} day^{-1}) and 'marine snow'. The export from water column to ocean bed is probably usually within the range of 1–100 mg C m^{-2} day^{-1} or some 1–3% of surface productivity, of which animal carcasses may supply 10% (Fig. 7.5). Surprisingly large numbers of bacteria inhabit deep-sea sediments, however: average, 1×10^6 bacteria g^{-1} sediment between 4000 and 10 000 m (up to a maximum of 84×10^6 cm^{-3}), together with 20 000 protists cm^{-3} and 1.7–17 meiofaunal animals cm^{-3} (meiofaunal biomass is often 3–4 orders of magnitude greater than that of the macrofauna). One might think that these densities would support considerable benthic productivity, but it is very dangerous to use density or biomass as an indication of productivity, and in this case any impression of high potential productivity would be an illusion.

The metabolic rate of abyssal bacteria is up to 100 times slower than that characterizing equivalent bacterial densities maintained in the dark, at the same temperature but at atmospheric pressure. This

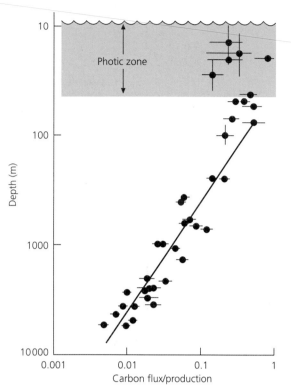

Fig. 7.5 The flux of carbon (per unit production) into the deep sea. (After Suess 1980, with permission.)

hatch open. When the water-filled *Alvin* was recovered almost 1 year later, the lunch was found to be in perfect condition! This stimulated a series of experiments in which organic materials were maintained in parallel: (a) at deep-sea temperatures in the laboratory but at atmospheric pressure and (b) at 5300 m depth in the deep sea itself. These confirmed the very low rates of bacterial metabolism in the abyss. Bacterial productivities in the sea bed probably fall within the range $0.2 \, g \, C \, m^{-3} \, day^{-1}$ (at 1000 m) to $0.002 \, g \, C \, m^{-3} \, day^{-1}$ (at 5500 m). Consuming species must therefore be even less productive, relatively slow growing and long-lived. The small bivalve *Tindaria* (8.5 mm in largest dimension) does not reproduce until it is at least 50 years old and can live to over 100 years. It therefore seems unlikely that the productivity of the deep-sea consumers exceeds $0.1 \, g \, C \, m^{-2}$ a year, except around fumaroles and on detrital bonanzas, and it must frequently be an order of magnitude less. Uptake of oxygen by deep-sea sediments is only of the order of $0.1 \, ml \, m^{-2} \, h^{-1}$ and release of ammonia is less than $1 \, \mu M \, m^{-2} \, h^{-1}$.

Although unproductive, the deep-sea benthos is diverse, especially on the continental slopes between depths of 2000 m and 3000 m. Even at 4700 m, single trawl hauls have yielded 196 macrofaunal species, but numbers of species per haul decline to 20 or less below 8000 m (Fig. 7.6). In part, the paucity of the fauna below 8000 m is an artefact of the sampling technique (trawls operate well only on level ground and cannot sample adequately the sides of the hadal trenches), but a number of animal groups do seem to have an apparent depth limit in the range 6200–8300 m. These include hydrozoans,

was discovered by accident. The Marine Biological Laboratory at Woods Hole, Massachusetts, operates a research submersible, the *Alvin*. In 1968, this was being prepared for a dive when it sank to a depth of 1500 m. The crew escaped but left their packed lunch (bologna sandwiches with mayonnaise, bouillon and apples) on the table and they left the escape

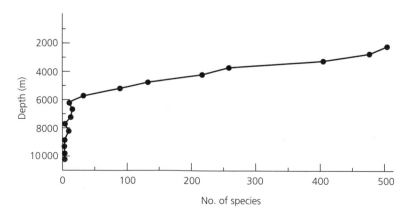

Fig. 7.6 The number of species of abyssal macrobenthos as a function of depth. (After Vinogradova 1962.)

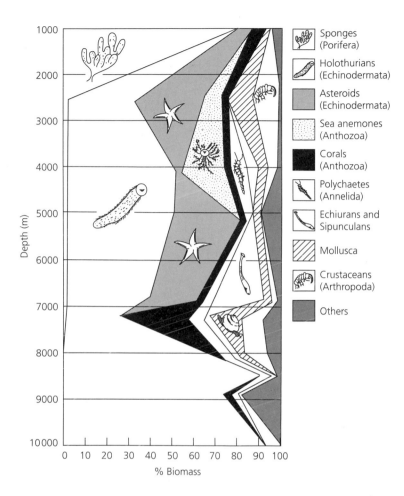

Fig. 7.7 The proportion of the benthic biomass at different depths made up by different animal groups. (After Friedrich 1965, with permission.)

Legend:
- Sponges (Porifera)
- Holothurians (Echinodermata)
- Asteroids (Echinodermata)
- Sea anemones (Anthozoa)
- Corals (Anthozoa)
- Polychaetes (Annelida)
- Echiurans and Sipunculans
- Mollusca
- Crustaceans (Arthropoda)
- Others

flatworms, barnacles, cumacean, tanaid and mysid crustaceans, pycnogonans, gastropods, echinoid, ophiuroid and asteroid echinoderms, and fish, whereas decapod crustaceans only extend to some 5000 m. Many, though not all, of these absentees from the deepest regions are predators, and the groups which dominate the fauna below 8300 m are ooze-consumers such as the holothurians, which make up 90% of the faunal biomass in the deeper trenches (Fig. 7.7).

The meiofauna and the smaller members of the macrofauna (sometimes separated as the 'minifauna', macrofaunal species larger than 10 mm then forming the 'megafauna') are probably feeders on the phytodetritus, which principally arrives as surprisingly large clumps of dead phytoplankton cells, and on the bacteria and protists. The larger macrofauna, however, are distributed between three or four feeding types: deposit (ooze) feeders, suspension feeders, and carnivores–scavengers. It is not possible, however, to assign species to these categories on the basis of the food materials normally taken by the groups to which they belong. Some deep-sea bryozoans and ascidians, for example, appear to feed on the deposits (whereas their shallower-water relatives are suspension feeders); members of otherwise suspension-feeding groups, such as the bivalves and the ascidian tunicates, may even function as macrophagous predators in the deep sea (e.g. the septibranch molluscs and octacnemid and aspiraculatan ascidians); and the giant, foraminiferan-like xenophyophorians of the abyss (some of which can achieve a diameter of 25 cm) may catch living members of the macrofauna. (In some areas, the pseudopodia of foraminiferans may carpet half of the available sediment surface.)

Fig. 7.8 Epifaunal holothurians (and brittlestars) at a depth of 1060 m off California. (Official photograph US Navy.)

The relative importance of these feeding types is somewhat controversial. Russian work has emphasized the importance of suspension feeders, especially in areas of the inert red clay (< 0.25% organic content), the organic poverty of which would be unlikely to support many deposit feeders. The Russian data, however, were gathered by the use of trawls, which might be expected to show a bias in favour of such suspension feeders as project above the sediment surface (sponges, crinoids, serpulid polychaetes, etc.). American work with box-corers, which take an *in situ* sample, has emphasized the importance of deposit-feeders, even in food-poor regions, but correspondingly this technique is biased against the larger suspension feeders. The two contrasting views are by no means mutually exclusive and it is probable that deposit feeders are the dominant component of the fauna in the more organically rich sediments (> 0.25% organic content) whereas suspension feeders, removing the sedimenting material from the water before it reaches the bottom, are most abundant on the hard substrata of the extensive mid-oceanic ridges, in areas of relatively rapid water flow (sediments may be ripple-marked even at 7000 m), and in the relatively barren clays of the abyssal plain. The mobile and wide-ranging scavengers and carnivores are (not surprisingly considering their trophic position) less abundant than either of these two detritus-feeding types, but when bait in the form of dead fish is lowered onto the sea bed, large numbers of ophiuroids, crustaceans (including giant amphipods 0.33 m long) and fish accumulate around it in a few hours. Such organisms have been termed 'croppers' and it is supposed that they roam over the deep-sea floor cropping any food organisms that they can find. Mobile deposit feeders (Fig. 7.8) may also act as croppers in that they will ingest other deposit feeders smaller than themselves. The large amphipods mentioned above are not the only abyssal giants; gigantism is as widespread amongst the opportunist scavengers/predators as is the contrasting miniaturization seen in the deposit feeders: the reasons for both states are conjectural.

Croppers are stressed by one faction in a current debate on the factors responsible for the relatively high diversities of some deep-sea faunas (see Gage (1996) and Chapter 10). One school of thought regards the feeding of indiscriminant predators as maintaining the populations of all the benthic prey species below the carrying capacity of their habitat (a directly comparable hypothesis to that stressing indiscriminant predation as the major structuring process in areas of soft littoral and shelf sediment; see Section 3.3.3). Thus croppers prevent competition for food amongst the prey, competitive exclusion, and dominance by a few competitively superior species. A second school stresses the long-term implications of competition for food in the stable, presumed food-limited deep sea. Under this opposing hypothesis, competition for food has resulted in character divergence (Section 3.3.4), small niches, and the coexistence of many species as a result of specialization on different parts of the total resource spectrum. Much of the basic information that could resolve this debate has not yet been gathered. We know nothing, for example, of the extent to which the abyssal detritus feeders show dietary specialization: detritus feeders in general are often regarded as ingesting material unselectively, but 'unexpected selectivity has been demonstrated in several intertidal species, and we are still ignorant of precisely what even the common littoral detritus feeders actually digest from their food in nature (see Section 3.2.2). Neither do we know to what extent the abyssal benthos is heavily predated. Several aspects of their biology (e.g. slow rates of reproduction, populations dominated by old individuals, etc.) do not suggest that they have evolved strategies to cope with intense predation pressure; and indeed we have seen above that specialist predators are one of the first feeding categories to disappear from the fauna with increasing depth.

In any event, the two schools of thought are really considering the two different aspects of diversity that we distinguished in Section 3.4. One is postulating long-term mechanisms of diversity generation, whereas the other is addressing the question of diversity maintenance at the present time. Over the deep sea as a whole, trenches are clearly centres of the generation of new (endemic) species (Table 7.2), but trench faunas have probably been derived from those of the abyssal plain rather than vice versa, and the pattern of speciation on the abyssal plain is still problematic (Section 10.3.4). In Huston's argument (Section 3.4), predation is but one of the factors involved in the maintenance of diversity: to obtain a complete picture, therefore, we need data on the rates of both potential population growth and population reduction. Population growth rates must be very low in the deep sea and this means that only slight degrees of disturbance will serve to oppose competitive exclusion and maintain relatively high diversity, i.e. there is no need to postulate heavy cropping rates. Correspondingly, potential recolonization rates are generally low. Trays containing defaunated sediment when placed at depths of 1300–4150 m had still not achieved ambient population densities and species richness after an interval of 5 years in one series of experiments (although trays that were screened to exclude croppers did show enhanced densities of some opportunist species). Other probable disturbing factors are bioturbation, interference (Section 3.3.2) and turbidity currents.

Table 7.2 Percentages of species present below 6000 m depth that are endemic to the hadal region. (After data of Belyaev 1966.)

Group	No. hadal species	% endemic
Foraminifera	128	43
Porifera	26	88
Cnidaria	17	76
Polychaeta	42	52
Echiura	8	62
Sipuncula	4	0
Crustacea		
barnacles	3	33
cumaceans	9	100
tanaids	19	79
isopods	68	74
amphipods	18	83
Mollusca		
aplacophorans	3	0
gastropods	16	87
bivalves	39	85
Echinodermata		
crinoids	11	91
holothurians	28	68
starfish	14	57
brittlestars	6	67
Pogonophora	26	85
Vertebrata	4	75

Clearly, much more remains to be learned of the deep sea, including which organisms make up its biota. In the last few years, several entirely new groups have been discovered (the vestimentiferan pogonophorans, and the concentricycloid echinoderms), and the faunal richness of some of the deep trenches is only now beginning to be appreciated. Some trenches support 100 macrofaunal species per square metre. Photographs taken of the sea bed are also disclosing lifestyles, patterns of distribution and abundance, and types of organism unknown from core and trawl samples. One recently published photograph is claimed to show a large arachnid—a group not otherwise thought of as being marine—and potentially a new class of hemichordates, the 'lophenteropneusts', is currently known only from photographs.

The nekton refers to those marine organisms which swim actively and control their movements by swimming, as opposed to planktonic organisms whose movements are governed by passively drifting in currents. All marine food webs, whether pelagic, benthic or fringing, terminate in the nekton: they comprise the upper portions of marine pyramids of consumption and production. Rather than treat nektonic organisms piecemeal in each of the previous six chapters, it makes greater ecological sense to accord these top consumers a section to themselves. Man, of course, sits on the pinnacle of the pyramid and consumes or otherwise utilizes fish, turtle, seal and whale products. In this chapter, the terms fish and fisheries are used in the broadest sense to include all exploited marine life, e.g. fish, mammals, reptiles, crustaceans, molluscs, echinoderms and seaweeds.

During the decade from 1955 to 1965 the world's landings from marine fisheries doubled from approximately 30×10^6 to 60×10^6 t. Since the 1980s the growth has levelled off and the world catch has now stabilized at about 92×10^6 t per annum (FAO fisheries statistics 1995). Another 27×10^6 t of fish is estimated to be caught in nets and discarded as unwanted bycatch (Pauly & Christensen 1995). Most of this is unreported to fishing authorities. The discards redirect energy within the marine ecosystem, having profound effects on populations of scavengers (Section 12.1.3). In addition, about a further 21×10^6 t is caught in inland waters, so that the total world fish catch (including the discard) is about 140×10^6 t.

Figure 8.1 shows a graph of the catch and the numbers of overexploited and underexploited species. The levelling-off of marine catches, coupled with the decline of many stocks (e.g. the northern cod stock off Newfoundland, Canada, in the late 1980s) suggests that most stocks are being fished at or above the maximum sustainable level. Other well-known examples of overfished stocks include:

- Antarctic blue whales, which were nearly fished to extinction at the turn of the century;
- the Californian sardine fishery, which crashed in the late 1940s;
- the Peruvian anchoveta fishery, which crashed in the 1970s but has recently recovered (see Section 8.7.5);
- the white abalone fisheries on the west coast of the USA, where white abalone is now listed as an endangered species (see Chapter 12).

Clearly, fish stocks need conservation and management if they are to be exploited on a sustainable basis. This chapter provides an introduction to

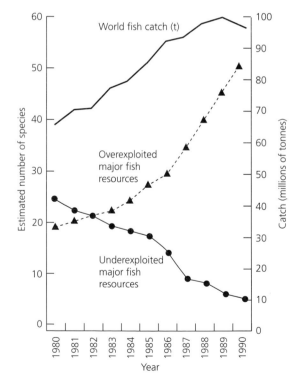

Fig. 8.1 Extent of overfishing—over half the world's fish catch is overexploited. (From Cushing 1996.)

fisheries biology: the science of the conservation and management of sustainable fish exploitation.

There are two complementary branches of fisheries biology, some of which is crucially linked to oceanography:

- life history and migration studies: identification of fish stocks and where they migrate to spawn, grow and feed;
- population dynamics: the rates of reproduction, growth and death of fish populations.

8.1 Life history and movements of the nekton

One of the features differentiating the nekton from the plankton is their ability to move in a horizontal plane relative to their inhabited water mass. Many species (and almost all of those of commercial interest) use this ability to maintain themselves in foodrich surroundings and to select habitat types that are optimal for the different stages in their life history. This is especially important where primary (and hence zooplanktonic) production is seasonal, i.e. outside the tropics.

In general, it is the young stages of most organisms that suffer the highest rates of mortality (see, e.g. Table 8.1) and in large measure this results from their relatively small size (Plate 19, facing p. 136). Small organisms can be consumed by many more predatory species than can large ones, and hence it is selectively advantageous to grow through this critical stage as quickly as possible, to achieve a size beyond the catching range of the abundant, small, sympatric predators. This can only be achieved in areas of relative food abundance (see also Section 3.2.4).

Extensive meadows of sea-grasses or areas with abundant populations of benthic algae provide rich feeding areas for neritic/shelf species. Thus, semi-enclosed bays, lagoons and estuaries are often used as nursery areas by nektonic crustaceans, fish and even by the grey whale, *Eschrichtius gibbosus* (see Barnes 1980a, 1984b), and the littoral zone as a whole often supports high densities of larval and young nekton feeding on the abundant plankton and benthos. Such regions may not provide suitable food for the larger, adult individuals, and, accordingly, after growth to a certain threshold size, nektonic species may migrate offshore into deeper waters.

Purely as a consequence of their size, the youngest stages of most nekton are planktonic and hence they are moved passively by surface currents. This constitutes a hazard with respect to successful arrival at the nursery grounds. If the reproductive adults spawned over the nursery ground itself, the larval stages might be carried out of many such areas by water flow and, therefore, not only be unable to utilize their high productivity but also be carried into potentially unsuitable regions. Hence the adults must spawn at a site that will permit the drifting larvae to gain access to the nursery area.

Figure 8.2 shows the idealized triangle of fish migration, characteristic of many fish stocks, and illustrates how fish life histories have evolved to

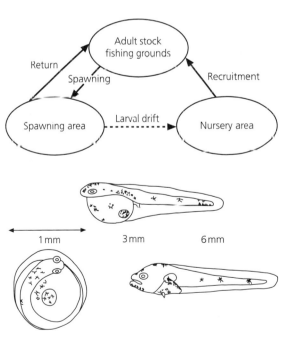

Fig. 8.2 Idealized triangle of fish migration, showing typical egg, and young and old larval stages. (After Harden Jones 1968.)

Table 8.1 Mortality rates for plaice, *Pleuronectes platessa*, of different age classes.

Age group	Mortality percentage per month
0–3 weeks	96
1–2 months	80
4–8 months	40
9–12 months	10
5–15 years	1

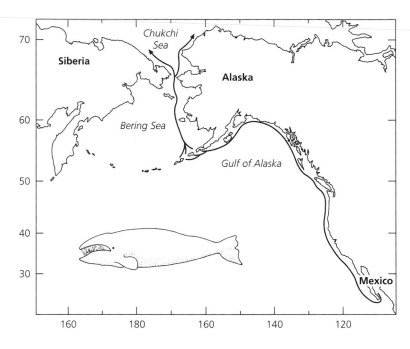

Fig. 8.3 The probable migration route of the grey whale, *Eschrichtius gibbosus*, which feeds in the Chukchi Sea and calves in the bays and lagoons of the Gulf of California. (After Pike 1962.)

take advantage of ocean currents and provide a measure of genetic isolation. Fish are usually captured as adults on their feeding grounds. They migrate to spawning grounds, from which the eggs or larvae drift with the prevailing currents to nursery areas where the larvae find suitable food and conditions to develop into juvenile recruits. The process of recruitment can be defined as the young fish becoming available to fishing gear, either because they are big enough to be worth catching, or because they have migrated into an area accessible to fishing vessels. Recruitment differs greatly from one fishery to another, depending upon the type of fish and upon the gear used in the fishery. Recruitment often also involves a change of habitat and migration from one habitat to another to find suitable conditions for feeding and growth of young adult fish.

In some deep-water species, the triangular form may be manifested in the vertical plane—the eggs rise to the surface waters (or the adults migrate to spawn nearer the surface), larval life is spent in the photic zone, and the young adults migrate down to the aphotic depths. In some lagoons with minimal exchange of water with the adjacent sea, adult nekton may be able to spawn in the nursery area; and the same is true for those species which produce large young (e.g. the nektonic mammals), the migra-

tion circuit here being linear (Fig. 8.3). The basic triangle of migration, however, is shown by a wide variety of nektonic species; for example, by shrimps, squid and fish (Figs 8.4 and 8.5).

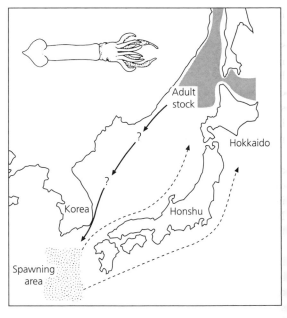

Fig. 8.4 Movements of the squid, *Todarodes pacificus*, in the Sea of Japan. (After Harden Jones 1980.)

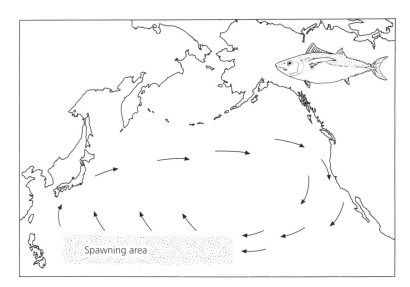

Spawning area

Fig. 8.5 Movements of the albacore, *Thunnus alalunga*, in the North Pacific Ocean. (After Harden Jones 1980.)

Large distances may be covered. The grey whale travels a distance of 18 000 km in each circuit and even the 30-cm-long herring may undertake a round trip of 3000 km. Although capable of swimming against currents, the nekton rarely undertake migrations in the face of opposing current flow; rather, they use existing current patterns to minimize the energetic cost of travel. By regulating their depth in the water, they can make use of water masses flowing in different directions at different depths. Thus hake in the North Pacific may migrate southwards to spawn in the California Current, but return to the north in the deeper counter current (Fig. 8.6). A number of fish species have also been shown to use tidal currents to aid their migrations—by resting on the sea bed when water flow is in the wrong direction but swimming up into mid-water when currents are moving in the required direction of travel.

The timing of the different phases of migration and the routes used appear stable over long periods of time (facts known and exploited by the fishing industry); the mean date of spawning in the Arcto-Norwegian cod stock, for example, has varied by less than 10 days over the last 70 years.

Other species in seasonally productive waters make less spectacular migrations to keep within areas of high productivity. Species exploiting the phytoplankton outburst of the Arctic summer, for example, move southwards during the autumn, spawn in lower latitudes and then follow the spring algal bloom as it spreads northwards. Nevertheless, the majority of nektonic species, especially those living within the shallow tropical regions of relatively constant productivity, do not migrate to any appreciable extent. They may show inshore movements to spawn (Plate 20, facing p. 136), but their powers of locomotion are used to maintain station against water flow or to seek out concentrations or individual items of prey. Many species school (Plate 21, facing p. 136), which may aid locomotion, the finding of prey, and anti-predator defence, and some use sound production as a means of communication and navigation.

8.1.1 Fishing methods

Fishers have developed many different methods to catch fish in the sea. These range from the simple traps of artisanal fishers to large sophisticated factory trawlers (Plate 22, facing p. 136). The three most important large-scale fishing techniques are illustrated. More details have been given by King (1995). Otter trawls (Fig. 8.7) have been used since the early 1900s to drag along the sea floor for demersal (i.e. bottom) fish. These include fish that migrate vertically during the night but move close to the bottom by day. Common fish caught by otter trawls include Alaskan pollock, cod, haddock; Chilean, Argentinean and southern African hakes and kingklip; orange roughy, and flatfish such as plaice, soles, halibut and dabs. Otter trawls can only be used on soft bottoms unless fitted with special rollers to drag over small stones and rocks. Trawls are used from depths of only about 20 m to

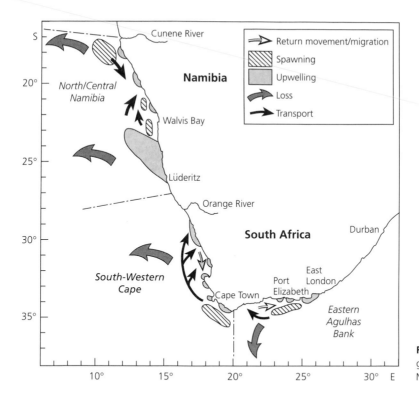

Fig. 8.10 Map showing main spawning grounds of two sardine stocks off Namibia and South Africa.

from Mexico in the south to northern California and have at least three different stocks. In the southeast Atlantic, the South African anchovy stock is distinct from that off Namibia, separated by a very cold area of upwelling off Lüderitz in southern Namibia (Fig. 8.10).

If each stock has different population characteristics (such as mortality rate), it is necessary to know whether there is a single stock, the extent of its movement, and its life history for effective management. It is therefore also important to be able to distinguish stocks from one another. For this, there are many techniques, all aimed at assessing the population genetic status of the fish. Some examples include the following.

• Counts of meristic characters such as the number of dorsal fin rays on bony fish, or the number of vertebrae measured by X-ray photography. In the case of eels, the American stock has a mean of 108 vertebrae, with lower and upper limits of 101 and 111, respectively. On the other hand, European eels of the same species have an average of 116 vertebrae, ranging from 110 to 120.

• Blood serum tests involving red-cell antibody–antigen reactions have been used to separate three sardine stocks off California, and variations of the haemoglobin I allele (measured using electrophoresis) have been used to distinguish cod stocks off the east coast of Canada.

When genetic techniques show differences in the genetic status of stocks, it is clear that they must be separate stocks. However, sometimes stocks spawn in different places and migrate on different routes, but there is sufficient genetic interchange to make them indistinguishable by genetic techniques. In these circumstances, mark–recapture tagging methods provide the best indicator of whether stocks are separate or not.

8.2 Estimation of fish abundance

It is never possible to census all the fish in a stock, so indirect methods have to be used to estimate stock sizes for management. The commonest method is based on catch statistics obtained from the fishery itself.

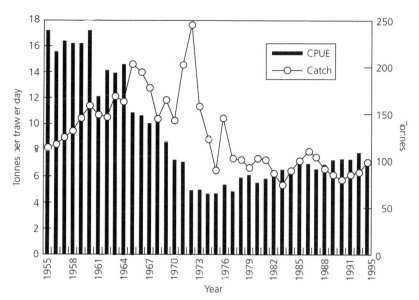

Fig. 8.11 Hake catches—graph of catch per unit effort plotted against time for Cape hakes on the west coast of South Africa from 1955 to 1995.

8.2.1 Catch per unit effort (CPUE)

This is based on the fact that if the fish population is dense, a fishery is likely to catch more fish per hour or day of fishing, or per unit of fuel used, than if fish are sparse. Thus, fishers are asked to keep records of how much time they spend fishing, or how much fuel they use, and how much fish they catch. Catch per unit effort, also known as the catch rate, is then calculated as

CPUE = catch obtained/effort used (8.1)
(e.g. tonnes of fish caught/trawler days).

The record of Cape hake catches off the west coast of South Africa is given in Fig. 8.11. It can be seen that fishing effort peaked in 1972, when foreign fleets were fishing these waters intensively, but when South Africa introduced a 200 mile exclusive fishing zone in 1976, fishing effort dropped off and the fleet has been carefully regulated since then. The record shows that the catch rate declined steadily with heavy fishing from 17.3 t of fish per trawler per day in 1955 until the minimum CPUE of about 4.6 t was reached in 1974–75, indicating that the density of fish was at its lowest then, only about one-third of its former value. The CPUE then picked up from 4.6 to about 7.2, suggesting that the density of fish increased between 1975 and 1993.

There are four main problems with CPUE.
1 It is only a relative index of abundance and does not measure the actual biomass of fish.
2 The catch may vary according to the availability of fish, even if the stock does not change in size. Several factors may affect the availability of fish (see below).
3 The real fishing effort often changes with time, because of improvements in technology, although the measured effort may appear to remain constant. Thus, trawlers may become more powerful, nets larger, aids such as satellite navigation allow pinpointing of fishing grounds, all leading to increased efficiency, yet the number of trawler days may remain constant or decrease, giving the impression that the catch rate has improved when in fact it may not have.
4 False reporting of catches may occur, giving rise to unreliable data.

8.2.1.1 Factors affecting availability

Within a particular fishery, a number of factors may influence the catch rate (CPUE), such as bad weather, which may prevent boats setting out, or may reduce their efficiency by causing them to slow down or by making the nets fish less effectively. The main factors that influence availability are as follows.

- The vertical distribution of fish in the water column: if fish such as sardines, anchovies or herring are too deep they may not be available to surface gear such as purse seine nets; if trawled fish such as cod, hake or haddock are off the bottom they are not available to bottom trawl nets.
- The horizontal distribution because of migration and fluctuating environmental conditions: fish that migrate far from the fishing ports or usual grounds are less accessible than those found close by.
- Gear selectivity: e.g. hooks are designed to catch fish with mouths of a particular size—they are biased towards catching that size of fish; similarly, gill nets select those fish whose gills are caught in mesh of a particular size. Only fish of particular sizes are available to such fisheries.
- Gear saturation: this is a density effect—once the hooks on a long line have fish caught on them, other fish cannot be caught on those hooks; once a net fills up, fewer additional fish will be caught because the gear behaves less efficiently.
- Behaviour of fish: e.g. some seine net fisheries operate on moonless nights because the fish may see the nets in daylight or by moonlight. The schools of fish may have behaviour patterns enabling them to dive below the reach of nets if they see them.

For these reasons, to estimate fish abundance it is necessary to use a quantitative measure of the availability of fish to the fishery, the catchability coefficient:

$$F = q \times f \qquad (8.2)$$

or (fishing mortality rate) = (catchability coefficient) × (fishing intensity), where fishing intensity is the effort expended divided by the area fished, and fishing mortality is proportional to the number of fish caught.

Rearranging eqn (8.2),

$$q = F/f \qquad (8.3)$$

or (catchability coefficient) = (number of fish caught)/(fishing intensity). The catchability coefficient thus gives a measure of the efficiency of the fishery; if the coefficient is large, more fish are caught for the same effort spent fishing.

The above problems associated with using a relative index of fish abundance, such as the catch rate (CPUE), and with the reliability of measures of fishing effort, indicate that other measures of fish abundance need to be used, preferably ones that are independent of the fishery statistics of catch and effort.

8.2.2 Independent measures of fish abundance

8.2.2.1 Echo-sounder (hydro-acoustic) surveys

Traditionally, echo-sounders have been used in navigation to measure the depth of water, based on the speed of sound in sea-water and the time it takes for sound waves to penetrate the water from the vessel to the sea floor and be echoed back to an instrument in a ship's hull. Fishermen soon noticed smudges on their echo-sounder traces, which indicated sound-reflecting objects in the water, such as the swim bladders of fish, which often moved up at night and down during the day, in line with their ideas about vertical migration of fish (and plankton). Techniques have now been greatly improved and specialized fishing echo-sounders have been developed; skilled operators can identify many species of fish from the nature of echo received, as the shape of the marks on echo screens and charts is very characteristic for many species. These echoes can also be calibrated and counted electronically, using sophisticated echo-integrators, which are used in scientific surveys to quantify the number of fish detected. Of course, the method becomes unreliable with fish that live flat on the sea bed, such as soles, plaice or halibut, and with fish that live a few metres below the surface, above the transducer in the bottom of the ship. However, the main problem facing fishery scientists becomes one of proper statistical design of hydro-acoustic surveys by research ships to obtain representative samples of the fish stock in question.

8.2.2.2 Egg and larval surveys

Plankton surveys can be used to obtain an estimate of the number of eggs or larvae spawned. If representative samples are obtained during the spawning season, then the number of eggs can be used to back-calculate the number of spawning females if the average number of eggs produced per female is known. This, and the development rate of eggs and larvae at different temperatures, can be measured

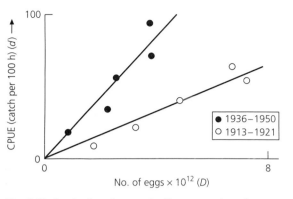

Fig. 8.12 Graph of catch per unit effort vs. number of eggs spawned in North Sea plaice fishery. Steeper lines represent more efficient fleets in 1936–1950 as a result of improved technology compared with the 1913–1921 period. (Slopes of lines = d/D catchability = q.)

experimentally. The method has been applied in the North Sea plaice fishery (Fig. 8.12). Here, the CPUE, which is one index of fish density, is plotted against another index, the estimated number of eggs spawned. The slopes of the lines give estimates of the fishing efficiency, or catchability coefficient (q) (see eqn (8.3)). The steeper slope of the line after the mid-1930s gives a measure of the improved fishing efficiency for plaice with the advent of steam trawlers.

8.2.2.3 Mark–recapture experiments

These can be useful for estimating the abundance of fish that can be caught, tagged and released without severe injury, provided the tagged fish mix randomly with the population again after tagging. There also needs to be a mechanism for detecting the tags and for reporting the finds to an appropriate fisheries authority, which may be difficult to achieve in international fisheries. The method is difficult to use in deep-living fish, such as cod and hake, which often suffer trauma caused by swim bladder expansion in trawl nets coming up from depth, and in fin-fish that readily shed their scales when handled, such as anchovy. The technique is good for crabs, lobsters, sardines and many flatfish. The equation used in estimating population size from mark–recapture experiments is well known in population dynamics and is dealt with in Section 8.5.2.

8.2.2.4 Visual surveys

A number of techniques using either scuba diving or remotely operated vehicles (ROVs) has been developed to count, photograph or video fish, crabs and lobsters in particular. This has been very useful for tropical and coral reef fish, which live in clear water with good visibility and in habitats not accessible to remote sampling techniques such as trawls.

8.3 Fish growth

8.3.1 Biological considerations

In nature, fish need to grow large as quickly as possible to avoid predation. Large fish have fewer fish larger than themselves preying on them. Thus a small increment in growth leads in a small decrement in mortality. Eggs and larvae, the stages which are most numerous in a fish's life history, have the fastest mortality rates (see Table 8.1). The balance between growth (leading to high fecundity) and mortality is what determines population stability. In many instances, especially for small pelagic fish such as sardine and anchovy, environmental conditions such as the temperature and abundance of food, play a large part in growth and mortality rates, particularly at the vulnerable egg and larval stages.

Most fish grow by a factor of 10^6 or 10^7 from egg to adult size because their eggs are so small. As they grow the growth rate declines with age, but so does the daily ration of food required, as a proportion of body weight, and assimilation efficiency declines even more. Our understanding of the physiological basis of fish growth owes much to the pioneering experiments of V.S. Ivlev (1961), a Russian fisheries biologist. He found that for each fish stage, there is an optimal prey size. Ivlev showed that giving fish different densities of food of the appropriate size, there is a threshold density of food above which growth can occur. Below this, all the food ingested (the daily ration) is used to fuel metabolism; above the threshold, the surplus food is available for growth. These experiments lead to the concept of growth efficiency, the amount of weight gained per unit of food eaten.

This was found to be higher in juvenile fish than in adults.

8.3.1.1 Density dependent growth

In fish there is little conclusive direct evidence for density dependent growth, which is a natural mechanism for population regulation. Indirect evidence is provided by examples such as the South African sardine, whose age at first spawning reduced from 4 to 2 years after the stock had declined to less than 10% of its former numbers during the mid-1960s. This suggests that sardines grew faster at the reduced population size, thus reaching the size for sexual reproduction sooner. The suggestion is, therefore, that the younger stages of fish may show density dependence in their growth rates, but there is no evidence for this in older fish, which continue growing after reaching sexual maturity. An important consideration in fisheries is that older fish spawn many more eggs that do smaller, younger fish. Thus a large mature cod may spawn as many eggs as 10 or 20 smaller adult cod. Furthermore, the eggs spawned may be larger and more viable than those of younger fish. Thus increasing size confers not only increased fecundity but also a reduced mortality rate of the eggs and larvae spawned. In general, fish have very variable growth rates and can respond to changes in the biotic or physical environment with increased growth rates.

8.3.2 Estimating growth rates

Because growth is so important in fish population dynamics, we need to obtain measures of the growth rates of fish to model the potential yield of fisheries. Several methods may be used, depending upon the characteristics of the fish stock.

8.3.2.1 Analysis of age marks

This technique has many variations, but they all apply when growth is interrupted annually, seasonally or by some other physiological rhythm such as spawning. The principle is exactly the same as in laying down rings in the trunks of trees. In the oceans, the changes in growth rate may be caused by changes in ocean temperature, changes in food supply, or by spawning, all of which affect the rates of physiological processes including the laying down of hard parts such as bone. Examples include the following.

• Otoliths (ear bones): These are commonly used to age fish such as cod, hake, sardine and many other fish. The otoliths are dissected from the head of fish, sectioned and examined on a microscope slide. Recently, computer image analysis has been used to automate the recognition and counting of rings. In some cases it is possible to detect daily growth rings. Figure 8.13 shows rings in the otolith of a South

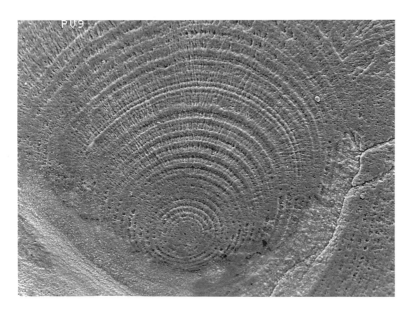

Fig. 8.13 Rings in the otolith of a South African sardine.

African sardine. For accurate analysis it is necessary to calibrate the rings to verify their periodicity and to ensure that they are indeed annual, or whatever the case may be—this may be done experimentally in aquaria or by marking animals, injecting them with tetracycline, which stains the hard part laid down, and then recapturing the animal after some growth has occurred, to see how many rings have been laid down in the interval between the marking and recapture.

• Scales: Annual rings in their scales have been used to age herring in the North Sea for many years.
• Spines and vertebrae have been used for ageing dogfish, sharks and rays.
• Teeth: The teeth of marine mammals, such as seals and toothed whales, are sectioned and rings counted.
• Ear plugs: Many baleen whales have 'ear plugs', which have rings that can be counted if the plugs are removed and sectioned.
• Mollusc shells: Many bivalves, especially, have concentric rings laid down in the shells. These often carry tidal, seasonal or annual growth signals, which can be used to age the bivalves.
• Beaks and statoliths of squids: Calcium carbonate is laid down in both the beaks and also the organs of balance (statoliths) in squids.

8.3.2.2 Mark–recapture experiments

Individuals are measured, marked (tagged) and released. Upon recapture, the animals are measured and the growth rate obtained from the difference in size and time interval. There can be expensive and difficult operations, and the return rate of tagged fish ranges from less than 1% for some sardines to as much as 30% for intensively fished flatfish, so thousands of fish may have to be tagged to obtain enough recaptures for statistically valid results.

8.3.2.3 Size–frequency analysis

This is often called Peterson analysis after the Danish marine biologist Johannes Peterson, who employed this method at the turn of the century. This method is applicable only for fish that have markedly seasonal spawning periods. By measuring a large sample of fish, and arranging the data in a size–frequency histogram, it is often possible to pick out peaks corresponding to year-classes. If the

process is repeated for the same stock over a number of years, one can follow the progress of a peak from year to year along the length axis, and measure the growth rate. If these peaks vary markedly from year to year, the task is made easier, as in the example of North Sea herring (Fig. 8.14). ELEFAN (Electronic LEngth Frequency ANalysis) and MULTIFAN are widely used variations on the theme of size–frequency analysis, in the form of computer packages. These are used to divide up a many-peaked size–frequency distribution into a series of normal curves by making some assumptions about the likely spread of size classes around each age-group.

8.3.2.4 In situ measurements

In the case of sedentary organisms, it may be possible to identify individuals such as barnacles which fix themselves onto rocks, and measure their growth by photography at regular intervals. Similarly, some territorial limpets can be identified individually from their position on a rock and their growth followed photographically.

8.4 Models of fish growth

To manage a fishery, one needs to know the growth, recruitment and mortality rates of the population. To predict the size of an average fish of any particular age in the stock, various models are used. The data required for models to describe growth and predict the size of fish are length-at-age data, and in the example shown in Table 8.2 ages were estimated by reading rings on otoliths (see Section 8.3.2). In general, the length increments are larger for young hake than for older ones. Thus length is not linearly related to age, but growth tends to slow down as the fish become older.

Three models are commonly used to describe growth, as follows.

8.4.1 Logistic curve

This is an S-shaped curve, with the two tails of the S being symmetric. The symmetric curve seldom fits data on growth in length or weight very well, but it is nevertheless a general model often used in population dynamics and fisheries biology. Most commonly it is used to describe the growth in

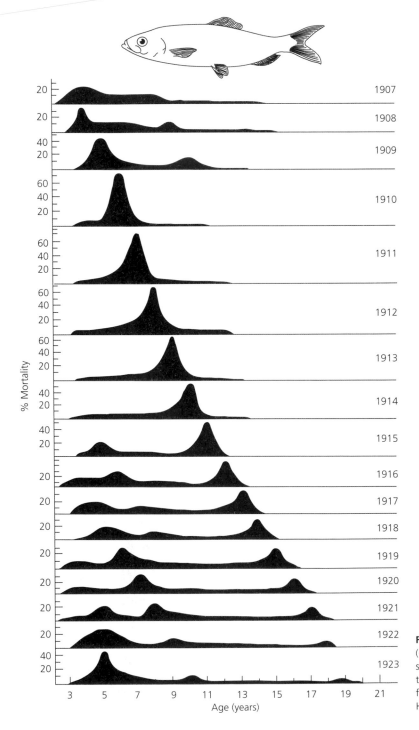

Fig. 8.14 Domination of a herring (*Clupea harengus*) population—as sampled by the Norwegian fishery—by the year-class which first entered the fishery in 1908 when 4 years old. (After Hardy, from Barnes & Hughes 1988.)

Table 8.2 Example of length-at-age data for Cape hake. Ages were estimated from otolith rings.

	Age (years)									
	2	3	4	5	6	7	8	9	10	11
Mean length (cm)	27.1	36.5	43.5	52.1	59.6	65.3	72.1	76.9	83.8	86.0
Length increase (cm)	—	9.4	7.0	8.6	7.5	5.7	6.8	4.8	6.9	2.2

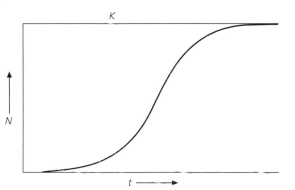

Fig. 8.15 Logistic curve showing population growth in numbers.

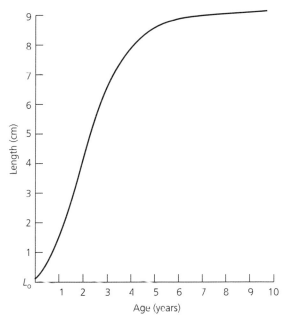

Fig. 8.16 Gompertz growth curve fitted to length-at-age data for clams *Mactra glabrata*. The clams were aged using growth rings in their shells. The curve has the equation $L_t = 9.0$ $(0.015)^{0.43^t}$ cm, and L_0 is 0.13 cm (see text for explanation).

population numbers rather than the growth in size of the average individual. The equation for logistic growth is given by

$$dN/dt = r(K - N)/K \qquad (8.4)$$

where N is the population size, r the intrinsic rate of natural increase, and K the carrying capacity. Figure 8.15 depicts a logistic curve showing population growth in numbers.

8.4.2 Gompertz curve

This is also an S-shaped curve, but it is asymmetric, with a much shorter tail for young individuals at the bottom left. The equation can be fitted only by using fairly sophisticated numerical methods on a computer, so has come into common use only in recent years. This is the most realistic curve for most fish growth, but often we do not have data for the very young stages, so cannot take advantage of the more complex form of model. Figure 8.16 shows a Gompertz growth curve fitted to length at age data for the clam *Mactra glabrata*. The bottom left-hand tail of the curve shows the slow growth of small

young individuals, starting from a very small size. In the middle of the S-shaped curve, growth is at its fastest; all energy is going into increasing body size. Towards the top right of the curve, growth slows down, usually as the animal becomes older and more energy is channelled into gonad output and less into somatic growth. In larger animals the metabolic rate also slows down. The Gompertz growth model has the following form:

$$L_t = L_\infty(a)^{b^t}$$

where L_t is length at age t, L_∞ is the maximum (asymptotic) length, a is the ratio of birth size to maximum size (L_0/L_∞), and b is an index of the steepness of the curve.

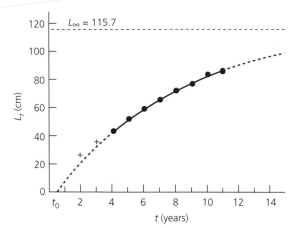

Fig. 8.17 Growth curve of Cape hake. The curve has the equation $L_t = 115.7\{1 - \exp[-0.13(t - 0.44)]\}$ cm.

8.4.3 Von Bertalanffy growth curve

This is the most commonly used growth model in fisheries biology because it fits into models used to predict fish catch, and because terms in the equation can be estimated graphically, so it has a long history of use. The curve has an asymptotic form, reaching the maximum size (L_∞), after which no growth is detectable. However, it does not describe early

Fig. 8.18 Mortality curves. (a) Typical mortality curve obtained by following a cohort of fish through time. The broken line depicts the increase as fish recruit into the fishery. The age at recruitment is here called t_r. (b) The same data plotted logarithmically. The slope of the logarithmic line, $-Z$, gives the instantaneous mortality rate.

growth well, having no left-hand 'tail'—but we can seldom obtain data for the young stages anyway.

The equation has the following form:

$$L_t = L_\infty\{1 - \exp[-K(t - t_0)]\}$$

where L_t is length at age t, L_∞ is the maximum (asymptotic) length, exp is the base of natural logarithms, K is the growth constant (or steepness of the curve) and t_0 is the theoretical age at which the length would have been zero, if the model were true at the youngest stages.

The terms in the Von Bertalanffy equation can be estimated numerically or graphically. Figure 8.17 shows a Von Bertalanffy curve fitted to the hake length-at-age data given in Fig. 8.2.

The parameters of the Von Bertalanffy growth equation for hake are

$$L_t = L_\infty\{1 - \exp[-K(t - t_0)]\}$$
$$L_t = 115.7\{1 - \exp[-0.13(t - 0.44)]\} \text{ cm.}$$

This allows us to predict the size of hake at any age, but caution should be used in extrapolating beyond the range of the data.

8.5 Mortality rates

Growth tells us about population increase, both in the biomass of the average fish and in the numbers of fish, whereas the mortality rate tells us about the rate of decrease in numbers of each cohort, or age-class of fish spawned in the same season. Figure 8.18 shows how the numbers in a cohort change logarithmically. (Note that the numbers (N) are given a

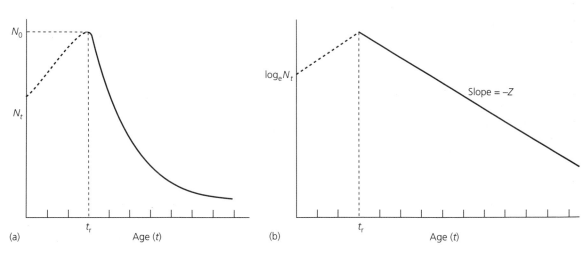

Table 8.3 Mortality rates estimated from a tagging experiment by Newman (1970). Natural mortality is calculated by subtraction.

Year	Total mortality (Z) =		fishing mortality (F) +		natural mortality (M)
1957–60	0.80	=	0.06	+	0.74
1963–66	0.71	=	0.12	+	0.59

subscript (t) to remind us that they change with age (t), usually expressed in years.) The broken line indicates the process of recruitment, when numbers in the cohort are increasing as they first grow to a size that is catchable by sampling or fishing gear. Once recruitment of a cohort is approximately complete, the numbers reach a maximum and thereafter mortality takes its toll. The instantaneous rate of decrease of a cohort is the mortality rate, Z, which is most easily measured as the slope of the graph of log (N_t) plotted against age (t).

Expressed as an equation, the slope of the line

$$Z = [\ln (N_0) - \ln (N_t)]/\Delta t$$

or

$$Z = 1/\Delta t \times \ln (N_0/N_t).$$

To find the number in a cohort at time, t, we integrate

$$N_t = N_0 e^{-Zt}. \tag{8.5}$$

It should be noted that Z is an instantaneous rate, it is the exponent by which numbers decline logarithmically.

8.5.1 Total mortality, fishing mortality and natural mortality

The mortality rate can be partitioned into two components, that due to fishing (F) and that due to all the natural causes such as disease, predation, etc., natural mortality (M). It is particularly important to be able to separate these two components, because it is precisely the fishing mortality that we are interested in as fisheries managers. Intuitively, fishing mortality (F) should be related to fishing intensity (f) and this is expressed as a simple proportionality constant, the catchability coefficient (q) (see eqn (8.2)). Fishing intensity is simply related to fishing yield (Y) and catch per unit effort (CPUE):

$$f = Y/\text{CPUE}.$$

Thus, substituting for f,

$$F = q \times Y/\text{CPUE}.$$

We can thus calculate the fishing mortality rate (F) from the catch statistics (Y and CPUE) if we know the catchability coefficient, q.

8.5.2 Separating fishing mortality from natural mortality

Two main techniques are available for separating fishing mortality from natural mortality. It should be remembered that these are instantaneous rates.

(a) We can follow the history of a fishery over varying levels of fishing intensity (effort). By calculating the total mortality each year,

$$Z = 1/\Delta t \times \ln (N_1/N_2) \tag{8.6}$$

where $\Delta t = 1$, because it is calculated each year, and $Z = (F + M)$. Therefore $(F + M) = \ln (N_1/N_2)$ and substituting for F,

$$(q \times f + M) = \ln (N_1/N_2).$$

Figure 8.19 shows total mortality (Z) plotted against effort each year. The result is a straight line

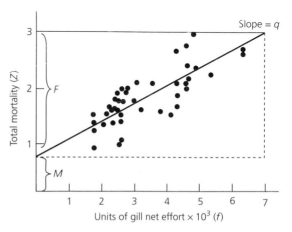

Fig. 8.19 Fraser River sockeye salmon: graph of mortality vs. fishing effort over the history of the fishery. (See text for details.)

whose slope is q, the catchability coefficient. The line gives the increasing mortality caused by fishing (F) as effort (f) is increased; if this is extrapolated back to zero fishing effort, it follows that the intercept is the average natural mortality over the period, with the scatter of data being due to variations in mortality from year to year. To obtain a good regression line, one needs to follow the history of a fishery over a wide range of fishing intensities, preferably right from the start of a new fishery.

(b) Tagging experiments can be used to estimate fishing mortality. The principle was first developed by Johannes Peterson in 1896. A good example of a tagging experiment to estimate the population size of bluegill sunfish in a fresh-water lake was given by Cushing (1968): 140 fish are tagged, by attaching numbered tags to their bodies (number marked, $N_m = 140$), and a total of 727 fish is caught (yield in numbers, $Y_n = 727$) and 28 of the marked fish are caught (number returned, $N_r = 28$). We can calculate that the proportion that die from fishing is: $N_r/N_m = (28/140) = 20\%$. The fishing mortality rate is the logarithm of this ratio:

$$F = \ln (N_r/N_m). \tag{8.7}$$

In reality, the picture is more complex and we also need to calculate the total mortality rate from changes in the stock size from year to year.

Thus, the estimated stock size in year 1 is

$$N_1 = N_m(Y_n + 1)/(N_r + 1) \tag{8.8}$$

where 1 is a statistical correction factor. Substituting for N_m, Y_n and N_r,

$$N_1 = 140 \times (727 + 1)/(28 + 1)$$
$$= 3515 \text{ fish in the population.}$$

Repeating the experiment the following year, we can obtain an estimate of the number in the population in year 2, N_2, and hence calculate the total mortality rate from the two estimates:

$$Z = \ln (N_1/N_2).$$

There are a number of complications such as emigration/immigration of fish from/to the fishing area, death caused by tagging, etc., which are beyond the scope of this book but have been explained more fully by Cushing (1968).

A real-life example of a tagging experiment is found in a paper by Newman (1970), who tagged a

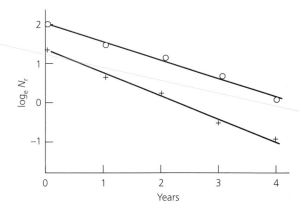

Fig. 8.20 Results of tagging sardine (pilchard) off Namibia in 1957–1960 (+) and 1963–1966 (o). The logarithm of numbers of returns is plotted against time since tagging.

large number of sardine off Namibia. The processing factories were fitted with machinery to detect the tagged fish magnetically and they were thrown off the conveyor belt for measurement and counting. Newman obtained the following graphs from the two tagging experiments, each of which was followed up for 4 years as the young fish grew and died. Figure 8.20 shows the decline in numbers of each cohort, the slope of the line giving the total mortality rate (Z). It should be noted that the fishing mortality rate obtained from eqn (8.7) was twice as large in the second period as in the first (Table 8.3). There is also considerable variation in the total and natural mortality rates, some of which may be due to fish emigrating out of the fishing area and appearing as natural mortality. This example re-emphasizes the importance of having good statistics of fishing effort.

8.5.3 Catch curves

These are based on the assumption that recruitment is constant from year to year. Catch curves are used to estimate the mortality rate of a population of fish by ageing the fish from even a single trawl (or catch), or by pooling the results of many trawls, and arranging the results into an age–frequency distribution. The log of numbers in each age-class ($\ln N_t$) is plotted against the age (t). Figure 8.21 shows that the graph is analogous to following a cohort from year to year, only here one follows the population

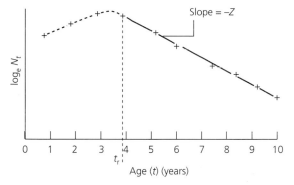

Fig. 8.21 Catch curve showing log numbers in each age-class of hake plotted against age. It should be noted that recruitment is complete only at age 4 (t_r).

from age-class to age-class, assuming that the starting point of each age-class (recruitment) was the same. In this example, 2–3-year-old fish are still recruiting into the fished population, so those data points are ignored and a straight line fitted to the descending limb of 4–10-year-old fish. The slope of this exponential decay line is Z, the instantaneous total mortality rate. If catch curves are plotted for virgin fisheries which have not been fished to any extent before, then the fishing mortality has been zero, and the total mortality is all due to natural mortality:

$$Z - F + M, \qquad (8.9)$$

and, as $F = 0$,

$$Z = M. \qquad (8.10)$$

Thus, this is a good way to estimate M, if the assumption of constant recruitment is true. This example illustrates the importance of obtaining good data right from the start of a new fishery. Even an approximate estimate of the natural mortality, based on far-fetched assumptions, is better than no estimate at all, because so many models need an estimate of M.

8.5.4 Virtual population analysis, VPA (also called cohort analysis)

Virtual population analysis is a technique used to estimate the size of a population and its fishing mortality rate, using catch data and samples of the catch for age estimation. The name 'virtual population' signifies all the fish that are destined to be

caught eventually. The method is developed from the catch-curve equation, i.e. exponential mortality. Figure 8.22 illustrates the principle, which is to follow a cohort from year to year, down the diagonal of a matrix of age against time, and to use simultaneous equations to obtain a best estimate of the average values of mortality rates.

Knowing that all fish die in their last year (by definition), we calculate the total mortality rate (Z) and the exploitation rate (E).

The exploitation rate (E) has two definitions, a very useful fact:

$$E = C_t/N_t \qquad (8.11)$$

and

$$E = F/Z \qquad (8.12)$$

where C_t is the number of fish of a cohort caught in year t and N_t is the number of fish actually in the cohort in that year. Thus, if the exploitation rate is 70%, then we can estimate the number in the cohort from the numbers caught:

$$N_t = C_t/0.7$$

and we can work out the numbers of each cohort from the catches.

Figure 8.22 depicts what one might find in a hypothetical VPA of Portuguese sardine. The first step is to divide the total catch for a year into age-classes, based on their proportions in samples taken from caught fish. The weights (which is how catches are normally reported) in each age-class (W_t) are then converted to estimated numbers caught (C_t), by dividing by the mean weight-at-age of each age-class. The second step is to convert the numbers caught into numbers in the cohort, using the exploitation rate ($N_t = C_t/0.7$). The third step is to check on the result for each year, by back-calculating the numbers in the cohort to those of the same cohort the previous year:

$$N_{(t+1)} = N_t e^{-Zt}. \qquad (8.13)$$

Thus,

$$N_t = N_{(t+1)} e^{+Zt}.$$

Often it is assumed that the exploitation rate is constant for any particular year, i.e. E is constant along one row of the matrix, and the natural mortality is assumed to be constant for any age-class (M

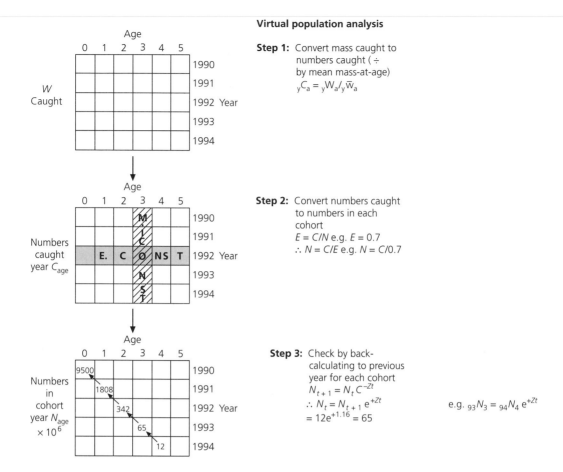

Virtual population analysis

Step 1: Convert mass caught to numbers caught (÷ by mean mass-at-age)
$$_yC_a = {}_yW_a/{}_y\overline{w}_a$$

Step 2: Convert numbers caught to numbers in each cohort
$E = C/N$ e.g. $E = 0.7$
$\therefore N = C/E$ e.g. $N = C/0.7$

Step 3: Check by back-calculating to previous year for each cohort
$$N_{t+1} = N_t C^{-Zt}$$
$$\therefore N_t = N_{t+1} e^{+Zt}$$
$$= 12e^{+1.16} = 65$$

e.g. $_{93}N_3 = {}_{94}N_4 e^{+Zt}$

Fig. 8.22 VPA—diagrammatic representation of the principle of virtual population analysis.

is assumed to be constant down one column). Using simultaneous equations one can then obtain the best solution for fishing mortality, total mortality and hence, natural mortality. For short-lived fish such as anchovy, the instantaneous natural mortality rate, M, is usually of the order of 0.8, and for longer-lived sardine, M is about 0.5.

Using the mortality equation, we can follow a cohort down a diagonal of the table as it ages and decreases, or we can back-calculate the number of recruits that entered the cohort by working backwards up the diagonal. This is one of the most powerful methods of estimating recruitment strength but it requires intensive length-at-age data. By adding up the numbers in all the cohorts of increasing ages

along a row, we can obtain an estimate of the total population size. This is therefore a useful method of stock assessment, provided there are several age-classes to give enough simultaneous equations for good estimate checks. Thus, the method is useful if the fish can be reliably aged, and if they live for at least several years. Anchovy are too short lived for VPA to be useful, but sardine, cod, flatfish, etc., all live long enough for good VPA.

8.6 Descriptive stock production models

A descriptive model investigates what happens to a stock without understanding or postulating a mechanism for the changes that occur. The first such model applied to fisheries is attributed to Graham (1935). He used the logistic curve to describe the biomass of plaice in the North Sea.

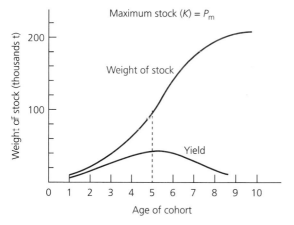

Fig. 8.23 Logistic stock model. Maximum yield is found at the point of inflection of the curve of increasing weight of stock.

8.6.1 Logistic stock model

Figure 8.23 describes the changes in biomass of plaice with age of the cohort in the S-shaped top curve, and the yield to fisheries in the lower curve. The familiar symbols used in ecology are modified to suit fisheries:

$$dN/dt = rN(K - N)/K \qquad (8.14)$$
$$dP/dt = rP(P_{max} - P)/P_{max} \qquad (8.15)$$

where P is the stock size, P_{max} is the carrying capacity, and r the intrinsic rate of natural increase. This is no longer used, but is included to show the origin of the Schaefer model below.

8.6.2 Schaefer production model

The logistic model was improved and adapted for fisheries by Schaefer (1954). It is particularly useful for predicting the biomass yield of stocks that cannot easily be aged, such as tuna. The principal disadvantage is that one cannot analyse reasons for failure of a fishery from the model, because it is purely descriptive:

$$dP/dt = aP(P_{max} - P) - FP. \qquad (8.16)$$

Here, a replaces r/P_{max} in eqn (8.15) and a new term for the fisheries yield is introduced, Yield = FP, where F is the fishing mortality, and $F = fq$ (eqn (8.2)). By rearranging the terms in the equations we can plot various very useful graphs for fisheries

management purposes. We can predict yield by plotting yield against stock size (P), or yield against fishing effort (f). Figure 8.24 shows predictions of yellowfin tuna catch using a Schaefer production model. The most useful graph for management purposes shows yield plotted against fishing effort (f) (Fig. 8.24a). Although the quantity of fish yield is not very reliable, it is useful to know the optimal fishing effort that should not be exceeded. Figure 8.24b shows the catch per unit effort plotted against effort, depicting the decline in stock size with heavier fishing. In plotting yield against stock size (Fig. 8.24c), the Schaefer stock production model has a parabolic curve, with the peak yield at exactly half the maximum stock size (P_{max}). This has been criticized as unrealistic for statistical reasons (the tails of fish age distributions are not symmetrical or bell shaped, as assumed by the model).

8.6.3 Dynamic stock production models

These have been developed by modifying the Schaefer model to take account of the criticism mentioned above, and give an asymmetric parabolic curve with the peak of yield at less than half the maximum stock size. There is a variety of such stock production, or surplus production models in this family of models developed by Pella and Tomlinson, Fox, and others. The details of these are beyond the scope of this book, but suffice it to state that they are widely used in fisheries management where ageing is difficult or unreliable (see Cushing (1975) or King (1995) for more details). Examples include fisheries for hake, tuna, anchovy, crabs and many others. Their main use is to predict the optimal level of fishing effort (F), which is very important in management.

8.7 Analytical stock models

These analyse the factors responsible for stock-size changes. They can therefore be used to analyse reasons for fluctuations in stock sizes, provided the assumptions of the model hold true. In most cases the models assume that the fish stock has settled down to a steady-state situation, often with constant recruitment. These assumptions are rarely met, so the models have limitations.

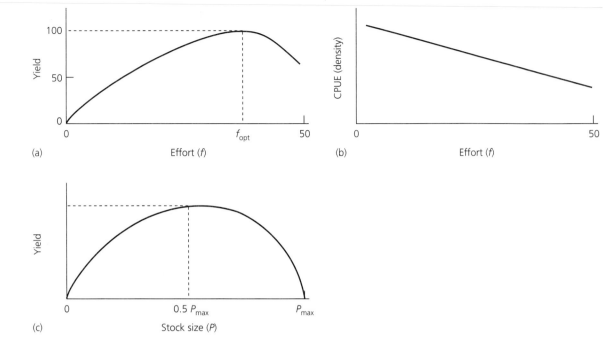

Fig. 8.24 Schaefer stock production model can be plotted in three different ways. (a) Yield vs. fishing effort, showing the optimal fishing effort (in thousand fishing days), f_{opt}; (b) catch per unit effort (CPUE) vs. effort (f), showing the decline in the index of stock density; (c) yield vs. stock size (P), showing the maximum yield at half the maximum stock size. (After Schaefer 1957.)

8.7.1 Russell's equilibrium yield model

The first analytical fisheries model was devised in 1931 by Fredrick Russell, who compared the biomass of a fish stock in two successive years:

$$P_2 = P_1 + (R' + G') - (F + M) \qquad (8.17)$$

where P_2 and P_1 are stock biomasses at years 2 and 1, respectively. Factors that increase the stock biomass are recruitment (R') of fish into the fishery, and growth (G') of those already recruited. Factors that decrease the biomass are mortality caused by fishing (F) and natural mortality (M). Russell was able to predict sustainable yields at different fishing intensities, assuming the fishery remained in an equilibrium state (i.e. $P_2 = P_1$).

In an unexploited stock the biomass may stay fairly constant, $P_2 = P_1$, thus, $R' + G' = M$. In an exploited stock that has stabilized to equilibrium, $P_2 = P_1$, then $R' + G' = F + M$.

Russell could explain and predict different age distributions and yields, assuming equilibrium conditions. Figure 8.25a shows a scenario of heavy fishing pressure, with an exploitation rate of 80% (or $E = 0.8$); 800 of the recruits would be caught in their first year, 20% survive fishing and put on weight each year so that the total yield in weight per 1000 recruits is about two-thirds that of the other scenario of 50% exploitation rate. The numbers of fish caught are given above each bar and the biomass caught is given by the area of the bar itself. Thus after 6 years all 1000 recruits would have been caught under the heavy fishing scenario and 969 under the moderate fishing scenario. It should be noted that more older fish are left in the stock to be caught under moderate fishing than under heavy fishing, which is what is commonly observed in real fisheries: the sizes of fish in heavily exploited stocks decrease compared with unfished or moderately fished stocks.

8.7.2 Underfishing vs. overfishing

Russell's model can also be used to predict maximum sustainable yield (MSY), when most fish are

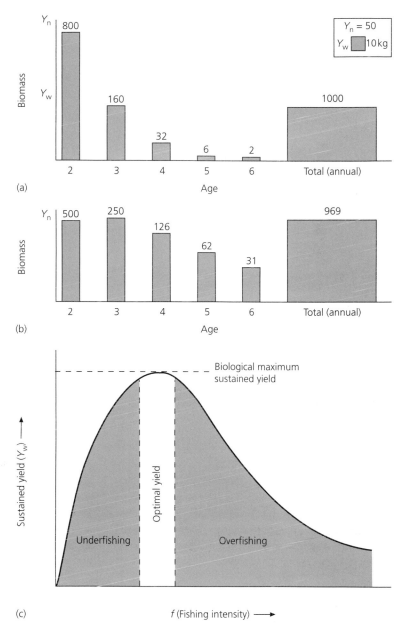

Fig. 8.25 Russell's equilibrium yield predictions. (a) Intensive fishing; (b) moderate fishing; (c) yield vs. fishing effort. (From Russell-Hunter 1970.)

left in the sea long enough to grow at a fast rate, but not so long that their average growth rate slows down. We can, in the sense of getting the most out of the sea, think of 'underfishing' as the situation in which the average fish is left in the water until it reaches an age when it is approaching max-´mum size and its growth rate slows down. In these conditions, most fish are large and slow growing

and the average productivity is slow—one is fishing at a low level of effort and not catching as much as might be obtained—a very unusual situation these days. At the opposite end of the spectrum is the overfishing scenario, when most fish are caught while they are young and before they have a chance to put on weight. This is the right-hand side of the curve and depicts overfishing. Russell's model also predicts that most of the fish caught will be small.

8.7.2.1 Growth overfishing

This occurs when the stock is reduced by catching most fish before they have put on weight. It applies to fish that grow considerably after they recruit into the fishery, such as plaice, cod, hake, haddock and many long-lived fish. Russell's model applies only to predicting the maximum sustainable yield of stocks of this sort, because it is based on the assumption that recruitment is constant. The model is not able to predict a spawning or recruitment failure.

8.7.2.2 Recruitment overfishing

This occurs when stock size is limited by the number of recruits entering the fishery. Here, too many fish are caught before they have a chance to spawn, reducing their reproductive output and hence recruitment. It seems to apply to stocks that grow rapidly before they recruit into the fishery, such as anchovy, sardine and herring. Russell's model does not apply to recruitment overfishing, as it assumes constant recruitment.

8.7.3 Yield-per-recruit models

Yield-per-recruit models were developed by Beverton & Holt (1957), based on the same principles as the Russell model. The biomass caught is the product of the numbers caught (N_t) and their average weight (W_t), where the subscript t indicates that they are age dependent. Algebraically,

$$Y_w = \int F\, N_t\, W_t\, dt. \qquad (8.18)$$

Figure 8.26 depicts a hypothetical curve of yield in biomass (Y_w) vs. age, from the age of recruitment (t_r) to the age at which all are caught (t_{ext}).

From the equation we see that, to predict yield per recruit, we need to know the fishing mortality (F); the numbers in each cohort (N_t) obtained from the total mortality rate, hence one also needs to ascertain the natural mortality rate (M); the weight of the average fish (W_t) is obtained from the Von Bertalanffy growth curve, so one needs the growth parameters L_∞ and K. To unscramble the problem from that of recruitment variation, it is usually

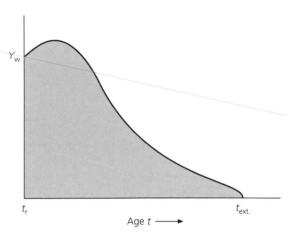

Fig. 8.26 Yield per recruit curve. This is dependent upon age at recruitment (mesh size).

expressed as yield per recruit (Y_w/R') and this is plotted against fishing effort (f), which in turn is proportional to fishing mortality (F).

8.7.3.1 Maximum sustainable yield (MSY)

Figure 8.27a shows a yield-per-recruit curve calculated for plaice at different levels of fishing effort. It should be noted that the peak maximum sustainable yield (MSY) is predicted to occur at fairly low levels of fishing mortality (F_{opt}). This kind of curve is common for fairly long-lived fish such as cod, haddock, halibut and hake. These all have a large maximum weight (W_∞) and a slow growth rate (K). These fish put on a lot of weight after recruitment to the fishery. The catch is the product of numbers and mean weight: if many fish are caught before they put on weight, there is a marked peak and then a decline with increasing fishing mortality. The more fish can grow after entering the fishery, the sharper the peak.

On the other hand, Fig. 8.27b shows a yield-per-recruit curve for herring, which is very different in shape. This curve is typical for fish such as herring, sardine and anchovy, with a fast growth rate (large K) and small W_∞. Herring do most of their growing before recruitment, so the curve is asymptotic, with no peak. This should not be interpreted as indicating that herring can be exploited indefinitely without collapsing, but simply that this model predicts yield per recruit and not yield—it is assumed that ther

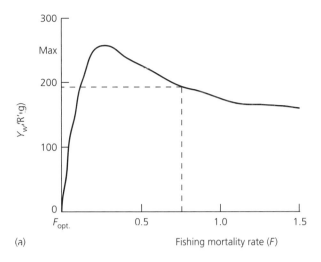

(a)

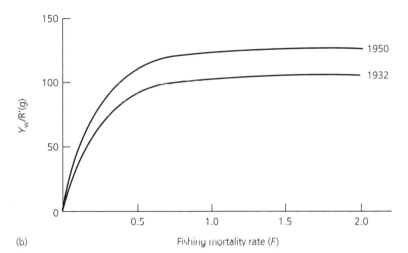

Fig. 8.27 (a) Yield per recruit curve for plaice. (From Beverton & Holt 1957.) ($W_\infty = 2867$ g, $K = 0.095$) (b) Yield per recruit curve for herring. (From Cushing & Bridger, 1966.) ($W_\infty = 210$ g, $K = 0.4$).

(b)

is constant recruitment each year. In other words, MSY is predicted only for stocks liable to growth overfishing and not recruitment overfishing, just as with Russell's model.

8.7.3.2 MSY and different levels of natural mortality

Figure 8.28a shows that the higher the natural mortality rate, the less peaked the yield-per-recruit curve. For very high values of M (e.g. $M = 0.5$) the curve tends towards an asymptote, as for herring. In other words, the loss in numbers by natural death (regardless of F) is so large that it nullifies any gain

in weight by individuals which might give a peak to the curve.

8.7.3.3 MSY and age at recruitment (effect of mesh size)

The yield-per-recruit curve is affected by age at recruitment (t_r), which is linked to the start of the integral of eqn (8.18) (Fig. 8.26). If the integration starts at a later age, then the numbers are likely to be smaller (N_t) through mortality, but the average weight per individual in the cohort (W_t) will be larger through growth. By plotting yield against t_r, one can find the peak yield and read off the optimal age of recruitment

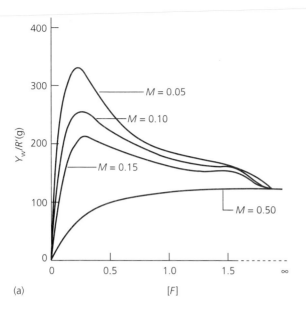

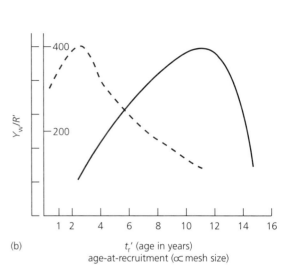

(a)   [F]

(b) t_r' (age in years)
age-at-recruitment ($\propto$ mesh size)

Fig. 8.28 (a) Yield curve and natural mortality for plaice. (b) Yield curve and age at recruitment for plaice (continuous line) and herring (dashed line).

(Fig. 8.28b). In a net fishery this can be related to net mesh size, and net mesh regulations fixed accordingly.

8.7.3.4 MSY and regulation of fishing effort

So far it has become clear that by controlling the fishing effort one can optimize the yield per recruit (e.g. plaice: $F = 0.2$ for $M = 0.05$). We have also seen that the curve is affected by age at recruitment (t_r)—in practice, this can be regulated by controlling mesh size, allowing small fish to pass through the nets. These two effects can be represented simultaneously in a three-dimensional graph. The section at $F = 0.73$ in Fig. 8.29 gives the plaice curve in Fig. 8.27a. Well-spaced contours indicate the gentle increase in yield to $t_r = 9$ years. Close contours beyond $t_r = 12$ years show the sharp decline in yield. If fish are recruited too old, the average fish dies naturally instead of being caught.

8.7.4 Recruitment and parent stock size

Figure 8.30 shows the history of the Californian sardine (= pilchard) stock from 1932 to 1950. There is

a clear relationship between the size of the parent stock and fishing effort: as fishing effort increased, stock size decreased. After the fishery collapsed and became uneconomic in 1950, fishing effort reduced dramatically. Simultaneously, there was a complete recruitment failure and instead of recovering back up the stock effort line, the stock density remained extremely low, despite the lack of fishing. There is still controversy as to whether the recruitment failure was due to natural causes or to overfishing, although the low stock of the late 1940s suggests a link with heavy exploitation. The modern view is that, in all probability, the collapse was brought about by a combination of overfishing and environmental factors (see Section 8.7.5). A similar fate befell the Peruvian anchoveta in the 1970s, which recovered to catches of over 12×10^6 tonnes in 1994.

It appears that in some fish (e.g. cod, salmon, haddock) there are compensatory mechanisms that adjust the recruitment in a density-dependent fashion—Ricker has suggested that this may be by cannibalism or other predation on the young stages when they are very dense (Fig. 8.31a). In these stocks the recruitment declines at high stock densities, providing a natural density-dependent check on population growth. A statistical trick used to linearize the relationship is to plot log R/P against P (Fig. 8.31b).

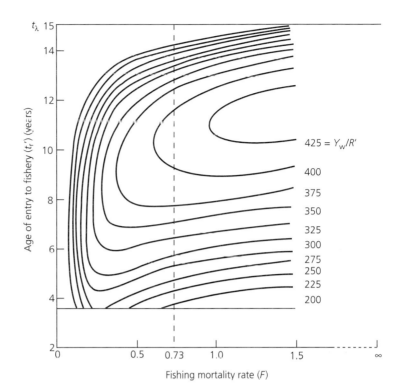

Fig. 8.29 Yield curve: effect of recruitment age and fishing effort for plaice. (From Beverton & Holl 1957.)

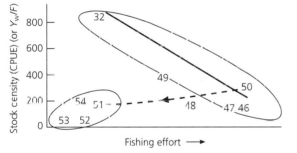

Fig. 8.30 Decline of the Californian sardine stock from 1932 to 1950 showing the declining stock density with increasing fishing effort.

Plaice, on the other hand, show little density-dependent mortality, with a flat-asymptotic recruitment vs. parent stock curve (Fig. 8.31c).

These curves (with wide scatter though) enabled Cushing (1975) to calculate actual yield curves (Y_w) as opposed to yield-per-recruit curves (Y_w/R'). He estimated the number of likely recruits from the stock–recruit curve and multiplied these by the yield per recruit to give the expected yield (Fig. 8.32). Where there is density-dependent mortality (e.g. cod) the actual yield curve shows much lower optimal fishing effort than the yield-per-recruit curve. For plaice, however, the actual yield curve gives a similar optimal fishing effort to the Y_w/R' curve, because the stock–recruit curve is asymptotic and not domed. It should be noted that for very small stock sizes (P), the compensatory relationship does not apply and there is still the danger of recruitment failure to the left of the 'optimal' stock size.

Table 8.4 summarizes the main types of model used in fisheries biology, their assumptions, data requirements and what they predict. There are many variations on these themes and new, more complex models are being developed every year.

8.7.5 The link between recruitment and environment

Cushing (1995) has analysed the recruitment of nine well-known stocks from which there are data for over 30 years from both Pacific and Atlantic oceans: these range from Skeena sockeye salmon and halibut, to various cod and herring stocks. Two

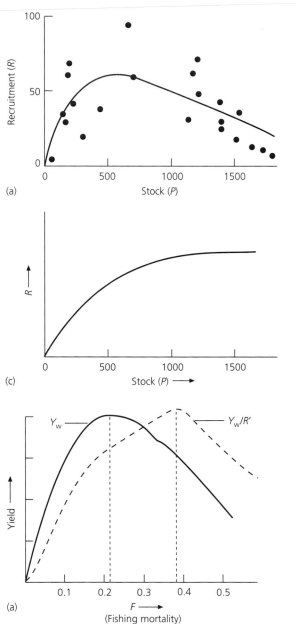

(a)

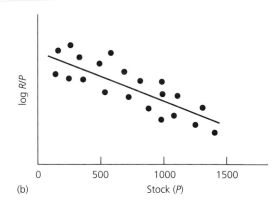

(b)

(c)

Fig. 8.31 Parent stock and recruitment. (a) Ricker model showing strong density-dependence (e.g. cod). (b) Linearized Ricker model (log. R/P vs. P). (c) Beverton and Holt model showing weak density dependence (e.g. plaice).

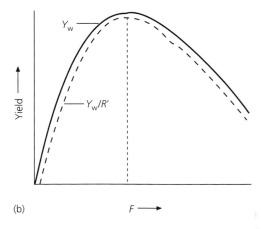

(a)
(Fishing mortality)

(b)

Fig. 8.32 Yield curves incorporating both yield-per-recruit and stock-recruit models. (a) Cod and (b) Plaice. (From Cushing 1975.)

examples are shown here in Figs 8.33 and 8.34. A remarkable feature is that there are large peaks in recruitment, and that these occur, on average, about once in 10 years. This suggests that large-

scale physical forcing may be creating favourable conditions for recruitment periodically. In the case of the North Atlantic stocks the decadal periodicity appears to be connected to the low-frequency variations of the North Atlantic Oscillation, which affects both atmosphere and ocean. It also affects the biota from phytoplankton to fish. The periodically good recruitment of various fish stocks caused

Table 8.4 Summary of the main classes of model used in fisheries management.

Model	Assumption	Data requirement	Output
Surplus production	Constant recruitment	No ageing, biomass index (e.g. CPUE)	MSY
Yield per recruit	Constant recruitment Equilibrium	Growth curve, natural mortality	Optimal effort, size at entry
Yield	Equilibrium	As above plus stock-recruitment	MSY
VPA	Natural mortality Exploitation rate	Catch at age Several age-classes	Stock size Recruitment

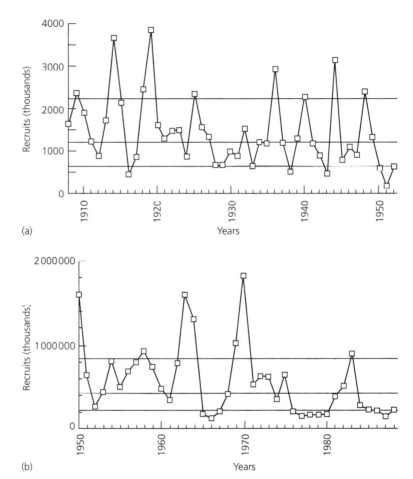

Fig. 8.33 (a) Recruitment of Skeena sockeye salmon. (From Cushing 1995.) (b) Recruitment of Arcto-Norwegian cod. (From Cushing 1995.)

Cushing to propose the match–mismatch hypothesis (discussed by Cushing (1975)). The essence of this is that if the timing of the zooplankton maximum coincides with the development of the fish larvae that feed on them, this will give rise to a strong year-class. Conversely, if the timing is not matched, the year-class will be average to poor. Serial spawning in sardine and anchovy populations (they may spawn up to once a week for months over the breeding season) may be an adaptation to ensure that

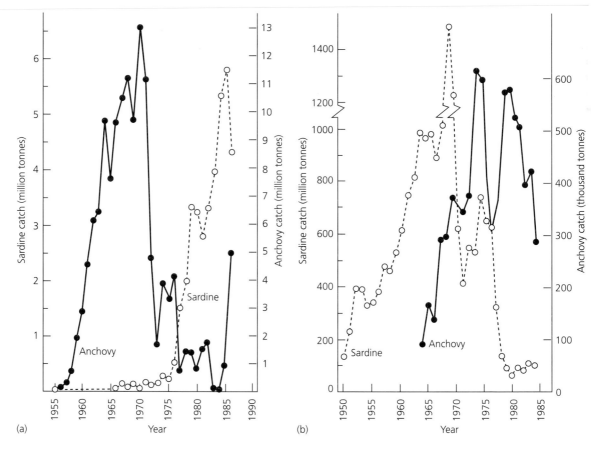

Fig. 8.34 (a) Humboldt sardine and anchovy catches, Pacific Ocean. (b) Benguela sardine and anchovy catches, South Atlantic Ocean (From Lluch-Belda *et al.* 1989).

at least some larvae are matched with food production in fluctuating upwelling systems. In contrast, herring spawning is less spread out in time and may be geared to specific events (e.g. the spring bloom of phytoplankton).

On a larger scale, stocks of sardine and anchovy seem to alternate in synchrony right across the Pacific Ocean from Japan to Chile. Figure 8.34 shows alternating peaks in sardine and anchovy catches; a peak anchoveta collapse and a peak sardine catch in the mid-1980s. Simultaneously, there appear to be peaks in anchovy catches in the southeast Atlantic off South Africa and Namibia when sardines peak in the Pacific, i.e. a similar 30-year period but of opposite phase. This suggests that there may be global-scale environmental 'tele-connections' causing strong

and weak recruitment in different ocean basins over decades. Evidence that such changes are environmentally induced, and not solely caused by overfishing, comes from studies of fish scales in anoxic sediments off California and Namibia, where periodic oscillations, averaging about 30 years, of sardine and anchovy fish scale abundance have been shown to occur many centuries before commercial fishing began. This has been discussed further by Cushing (1995) and Mann & Lazier (1996).

The moral of the story is that managers of fisheries cannot take good or even average recruitment for granted. There is now convincing evidence that the environment plays an important part both in strong year-classes and in recruitment failures. Most of the traditional models on the other hand assume constant recruitment, or are based on equilibrium conditions. When stocks are in decline through a series of poor recruitment years, managers need to be able to cut down on exploitation, so as to slow the decline of the stock and give it a chance to rebuild. On the other hand, when there is a particularly good year-class, there may be an opportunity to increase fishing effort. This is particularly true for short-lived fish, which usually have high natural mortality rates. These stocks can be exploited much better if we know that a good year-class is about to arrive. At present our understanding of the links between environment and recruitment are not good enough to be able to predict strong or weak recruitment from medium-term climate variations, such as El Niño. However, such possibilities are likely within the next decade and present exciting prospects for new research.

9 Ecology of life histories

9.1 Introduction

Amongst the bewildering diversity of marine organisms run some basic patterns of life history that are correlated with the physical and biological nature of the habitat. The present chapter briefly reviews such patterns, interpreting wherever possible the significance of particular phenomena in terms of natural selection. In making these interpretations, it is necessary to keep in mind the counterbalancing benefits and costs associated with most adaptations. Energy or material committed to one function may be unavailable for another, and the development of a certain attribute may restrict the development of others. Each organism represents a compromise to the conflicting demands of simultaneous evolutionary problems. Excellence in one field usually costs poorer performance in another, with the result that different organisms outclass one another in different circumstances. It should also be borne in mind that observed characteristics are not necessarily interpretable just in terms of current selection pressures, but may represent states that were fixed historically in ancestral lineages, so implicating phylogeny (see Harvey & Pagel (1991) and Section 9.3.2.7).

Early attempts to rationalize data on life histories often used MacArthur and Wilson's concept or r- and K-selection (e.g. Pianka 1974), where r denotes the intrinsic (unlimited or exponential) per capita rate of increase and K the carrying capacity (asymptotic population size) of the logistic growth equation. In the logistic model, growth rates of populations at densities well below the carrying capacity are influenced largely by the variable r, whereas growth rates of populations at densities close to the carrying capacity are greatly influenced by K. Populations kept at, or repeatedly reduced to, low densities by physical or biotic environmental factors will be influenced by a different set of selective pressures (r-selection) from the set influencing more stable populations at high densities (K-selection). r-selected species correspond to opportunistic, early successional organisms in which rapid growth, early sexual maturity, high fecundity and great dispersal abilities are often advantageous. K-selected species correspond to late-successional organisms occupying more permanent habitats, in which competitive ability, resistance to predation, greater longevity, greater investment per offspring and the capacity to reproduce repeatedly in successive seasons are at a premium. r-selected features tend to be correlated with small body size and K-selected features with larger body size. As body size is inversely proportional to population growth rate and hence to the ratio of production to biomass, there is a trend of decreased population turnover rate and productivity from more r-selected to more K-selected species (Fig. 9.1). The 'r–K continuum' has provided a framework upon which to build theories

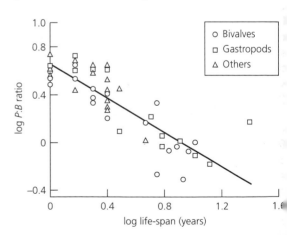

Fig. 9.1 The ratio of production (P) to biomass (B) with increasing life-span and hence with increasing body size. (After Robertson 1979.)

of life history, but the need for modification has become increasingly apparent. Some of the organisms discussed below (e.g. heteromorphic algae, benthic invertebrates with planktonic larvae) will be seen to possess a mixture of r- and K-features, whereas other properties, such as resistance to physiological stress, are not covered adequately by the 'r–K' concept. The latter is evidently too restrictive and is gradually being replaced by more comprehensive schemes of classification as knowledge and understanding of life history ecology develops. For example, Grime (1979) included physiological stress along with population density, and Sibly & Calow (1986) offered a general scheme, encompassing r- and K-selection, based on gradients of juvenile growth rate and the ratio of juvenile to adult survivorship.

9.2 Algae and higher plants

9.2.1 Benthic forms

Benthic vegetation extends to depths where the sea bed coincides with the lower limit of the photic zone. Representatives from the wide range of algal types, encompassing unicellular, filamentous, encrusting and foliose forms, are potentially able to colonize any kind of physical substratum. Because of their small size, rapid population growth and physiological robustness, unicellular and filamentous algae are able to survive under harsh or frequently disturbed conditions, such as the splash-zone of rocky shores, heavily grazed sublittoral rock surfaces or on the surfaces of mobile sediments. In more equable circumstances, foliose and encrusting macroalgae establish themselves on hard substrata, forming the dominant vegetation, e.g. intertidal fucoids or sublittoral kelps (Chapter 5).

Macroalgae are not so successful at colonizing sedimentary substrata. Fucoids and kelps may be found attached to shells or small stones and may continue growing even when detached from the substratum in very sheltered localities. Indeed, some members of the Fucales, such as certain morphs of *Ascophyllum nodosum* and the well-known oceanic rafts of *Sargassum* in tropical seas, can lead an entirely pelagic life. However, in general, angiosperms replace macroalgae as the dominant vegetation of shallow-water sediments, e.g. salt-marshes, sea-grass

meadows, mangrove-swamps (Chapter 4). Features giving higher plants the ability to dominate terrestrial habitats also enable them to reinvade these aquatic habitats with equal success. Sophisticated skeletal and stomatal systems keep emergent vegetation erect and protected from desiccation while allowing efficient gaseous exchange. Roots and rhizomes provide anchorage, access to sedimentary nutrients and may serve as perennating organs protected from some of the physical hazards occurring above-ground. Angiosperm root-systems, however, are unable to cope effectively with hard surfaces, which therefore provide a competitive refuge for the macroalgae.

Macroalgae occur in an impressive variety of forms, but these can be grouped into several broad categories such as filamentous, membranous, finely branching, coarsely branching, calcareous and encrusting forms (Fig. 9.2a). Life cycles may involve very different growth forms, e.g. minute gametophytic and large sporophytic phases of kelps (Fig. 5.15a), or the frondose gametophytic and encrusting sporophytic phases of certain red algae (for a review of red algal life cycles, see Searles (1980)). Large physiological differences also occur among algae: some are quick growing but flimsy and short lived, others slow growing, robust and long lived; some contain bacteriocidal and herbivore-repellent chemicals; some are resistant to considerable physiological stress and others less so.

These features represent counterbalancing capacities for rapid growth, reproduction, environmental tolerance, resistance to predation and competition for resources (nutrients, space, light). Among the possible combinations of such properties, certain groupings are particularly common, and many of these can be understood in terms of the type of habitat in which the algae are found. The role of physical and biological disturbances in determining community structure were discussed in Sections 5.1, 5.2 and 6.4, from which it is evident that there exists a continuum from frequently disturbed habitats, such as heavily grazed, rocky surfaces, to relatively undisturbed habitats, such as the fucoid zone on sheltered shores. Combinations of features expected to be advantageous in species exploiting young, temporally fluctuating communities of disturbed habitats (corresponding to an r-selected regime, Section 9.1)

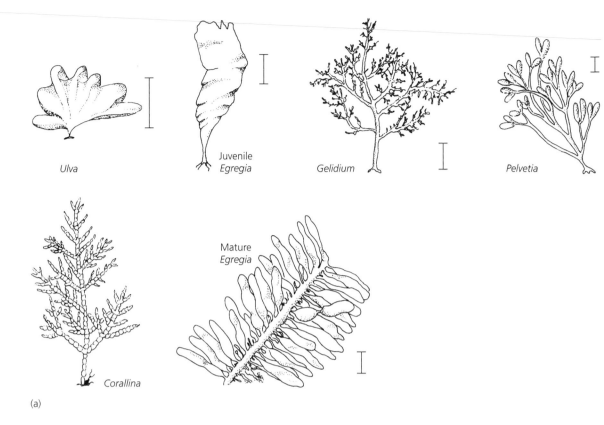

(a)

Fig. 9.2 (a) A range of algal morphologies. Scales represent 2 cm. The segment of *Egregia* is from a frond about 4 cm long. (*Facing page*) (b) Frond-toughness. Percentage of structural (non-photosynthetic) material in the frond, and resistance to wave shock among a series of Californian intertidal algae; (c) mean net primary productivities, energy contents, and palatabilities of the algal species as in (b). (After Littler & Littler 1980.)

are compared in Table 9.1 with those expected to be advantageous in species from mature, temporally constant communities from undisturbed habitats (similar to a *K*-selected regime). The counterbalancing benefits and costs ('trade-offs') associated with these features are listed in Table 9.2. To a large extent, theoretical expectations are borne out in nature. Thus, membranous forms such as *Ulva* and *Porphyra* are commonly the earliest colonizers of disturbed, rocky intertidal substrata, and if succession is allowed to proceed, the early colonists are replaced by more robust species (Fig. 9.2). Opportunistic species have higher net productivities than later colonists, which invest more material and energy

into non-photosynthesizing structural or defensive materials. This trend is reflected by the ratio of productivity to total biomass (P/B), which decreases in the order of membranous, finely branching, coarsely branching and encrusting forms. The high net productivities of membranous algae such as *Ulva*, *Porphyra* and *Enteromorpha* are partly attributable to the extremely thin construction of the thallus and the large size of the cells, which result in relatively little self-shading by non-photosynthesizing materials. Late successional species grow more slowly and reproduce less profusely than opportunistic species, but their investment in non-photosynthesizing materials gains them greater life expectancy in at least three ways. First, non-photosynthesizing supporting tissue enables the algae to attain large sizes and to adopt overtopping growth forms that impart a competitive advantage over more delicate forms. The dominance of many sheltered, rocky shores by fucoids may be largely attributable to this factor. Secondly, the supporting tissue increases algal resistance to physical damage by waves, currents and scour. Thirdly, a tough

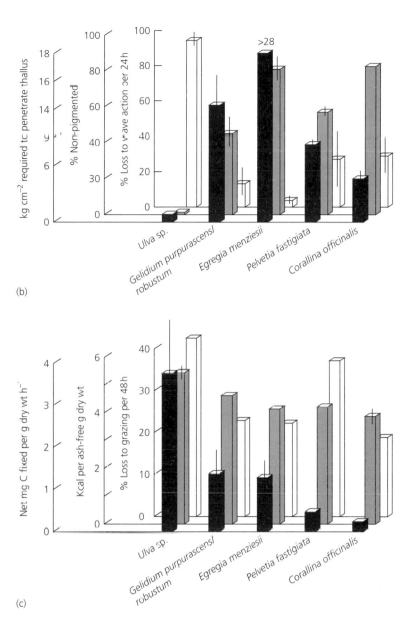

(b)

(c)

Fig. 9.2 *Contd*

epidermis, calcareous matrix, low energy content or presence of toxic chemicals reduce the acceptability to grazers.

Tolerance to physiological stress, such as desiccation and extremes of temperature or salinity, may or may not be associated with reduced productivity. High-shore fucoids such as *Pelvetia canaliculata* are more resistant to desiccation, but are slower growing than lower-shore fucoids (see Fig. 5.6). Some long-lived, slow-growing, encrusting algae

such as *Hildenbrandia* are highly resistant to physiological stress (Section 5.1.4.2). On the other hand, opportunistic macroalgae such as *Ulva* and *Enteromorpha*, together with unicellular blue–green and diatomaceous algae, are not only fast growing, but also are among the most resistant algae to physiological stress.

A number of red algae form epilithic, encrusting sporophytes that are long lived, apparently having evolved under selective pressures from intense

Table 9.1 Hypothetical *a priori* survival strategies available to opportunistic macroalgae representative of stressed* communities vs. macroalgae characteristic of nonstressed† communities. (After Littler & Littler 1980.)

Opportunistic forms	Late successional forms
1 Rapid colonizers on newly cleared surfaces	**1** Not rapid colonizers (present mostly in late seral stages) and invade pioneer communities on a predictable seasonal basis
2 Ephemerals, annuals, or perennials with vegetative short-cuts to life history	**2** More complex and longer life histories; reproduction optimally timed seasonally
3 Thallus form relatively simple (undifferentiated) and small with little biomass per thallus; high thallus area to volume ratio	**3** Thallus form differentiated structurally and functionally with much structural tissue (large thalli high in biomass); low thallus area to volume ratio
4 Rapid growth potential and high net primary productivity per entire thallus; nearly all tissue is photosynthetic	**4** Slow growth and low net productivity per entire thallus unit because of respiration of non-photosynthetic tissue and reduced protoplasm per algal unit
5 High total reproductive capacity with nearly all cells potentially reproductive and many reproductive bodies with little energy invested in each propagule; released throughout the year	**5** Low total reproductive capacity and specialized reproductive tissue with relatively high energy contained in individual propagules
6 Calorific value high and uniform throughout the thallus	**6** Calorific value low in some structural components and distributed differentially in thallus parts. May store high-energy compounds for predictable harsh seasons
7 Different parts of life history have similar opportunistic strategies; isomorphic alternation; young thalli just smaller versions of old	**7** Different parts of life history may have evolved markedly different strategies; heteromorphic alternation; young thalli may possess strategies paralleling opportunistic forms
8 Escape predation by nature of their temporal and spatial unpredictability or by rapid growth (satiating herbivores)	**8** Reduce palatability to predators by complex structural and chemical defences

*Young or temporally fluctuating.
†Mature, temporally constant.

grazing and stressful physical conditions, e.g. wave action, sand scour. They also form short-lived, upright gametophytes that are faster growing but less robust than the sporophytes. When first describing these 'heteromorphic' forms, taxonomists mistook them for separate species, and although their identity is now recognized, the nomenclature is retained. For example, a common heteromorphic red alga on western North American rocky shores is the encrusting sporophyte known as *Petrocelis middendorffii*, which alternates with the foliose gametophyte known as *Gigartina papillata*. The sporophyte grows slowly, increasing in area by about 4% a year, or even shrinking in some years, but is generally long lived, an average-sized individual being anywhere from 25 to 90 years old. The gametophyte is probably annual, is more productive (0.64 vs. 0.09 mg C g^{-1} per day) and is over twice

as palatable to grazers as the sporophyte. Other algae, e.g. *Chondrus crispus* or *Corallina officinalis*, may have extensive crustose basal or holdfast systems from which grow erect thalli, and perhaps these algae invest differentially in the two structures according to environmental conditions. Among brown algae, kelps have heteromorphic life cycles (Fig. 5.15a), the tiny, ephemeral, opportunistic gametophytes alternating with the large, longer-lived sporophytes. However, even the sporophytes of some kelps may change morphologically as they grow. The young sporophytes of *Egregia* spp. (Fig. 5.15b) are thin, membranous forms with high growth rates enabling them to recruit effectively during early successional stages, but with increasing age the sporophytes invest more energy and material into supporting tissues, thereby reducing productivity but at the same time increasing competitive

Table 9.2 Hypothetical costs and benefits of the survival strategies proposed in Table 9.1 for opportunistic (inconspicuous) and late successional (conspicuous) species of macroalgae. (After Littler & Littler 1980.)

Opportunistic forms	Late successional forms
Costs	
1 Reproductive bodies have a high mortality	**1** Slow growth, low net productivity per entire thallus unit results in long establishment times
2 Small and simple thalli are easily outcompeted for light by tall canopy formers	**2** Low and infrequent output of reproductive bodies
3 Delicate thalli are more easily crowded out and damaged by less delicate forms	**3** Low surface to volume ratios relatively ineffective for the uptake of low nutrient concentrations
4 Thallus is relatively accessible and susceptible to grazing	**4** Overall mortality effects are more disastrous because of slow replacement times and overall lower densities
5 Delicate thalli are easily torn away by the shearing forces of waves and abraded by the sedimentary particles	**5** Must commit a relatively large amount of energy and materials to protecting long lived structures (energy that is thereby unavailable for growth and reproduction)
6 High surface to volume ratio results in greater desiccation when exposed to air	**6** Specialized physiologically and thus tend to be stenotopic
7 Limited survival options because of less heterogeneity of life history phases	**7** Respiration costs high because of the maintenance of structural tissues (especially unfavourable growth conditions)
Benefits	
1 High productivity and rapid growth permits rapid invasion of primary substrates	**1** High quality of reproductive bodies (more energy per propagule) reduces mortality
2 High and continuous output of reproductive bodies	**2** Differentiated structure (e.g. stipe) and large size increases competitive ability for light
3 High surface to volume ratio favours rapid uptake of nutrients	**3** Structural specialization increases toughness and competitive ability for space
4 Rapid replacement of tissues can minimize predation and overcome mortality effects	**4** Photosynthetic and reproductive structures are relatively inaccessible and resistant to grazing by epilithic herbivores
5 Escape from predation by nature of their temporal and spatial unpredictability	**5** Resistant to physical stresses such as shearing and abrasion
6 Not physiologically specialized and tend to be more eurytopic	**6** Low surface to volume ratio decreases water loss during exposure to air
	7 More available survival options due to complex (heteromorphic) life-history strategies
	8 Mechanisms for storing nutritive compounds, dropping costly parts, or shifting physiological patterns permit survival during unfavourable but predictable season

ability and physical robustness (Table 9.3). This shift in thallus form allows *Egregia* to compete effectively with opportunistic species for newly available space as well as to persist among late successional competitors for light and space.

The distribution of green, brown and red algae is very loosely related to depth. Green algae tend to be abundant intertidally, brown algae flourish both intertidally and sublittorally, and red algae are most numerous sublittorally. This distributional sequence used to be interpreted as chromatic adaptation to prevailing light conditions. Blue and green light is least

Table 9.3 Comparative values for juvenile and mature individuals of *Egregia menziesii* used to test the shifting-strategy hypothesis. (After Littler & Littler 1980.)

Thallus used	Net productivity (mg C fixed g^{-1} (dry wt h^{-1})	Toughness (kg cm^{-2} to penetrate thallus)	Time of appearance on successional plots (months)
Juvenile	2.51 ± 0.13	3.60 ± 0.33	3.0
Mature	1.26 ± 0.52	> 28 (off scale)	10.0

absorbed by water, whereas longer wavelengths of red, orange and yellow light are strongly absorbed (see Sections 1.2 and 2.1). Phycobilins and carotenoids of complementary colour to underwater light absorb the latter more efficiently than chlorophyll, on to which they are able to pass the excitation energy for photosynthesis. These accessory pigments give brown and red algae their characteristic colour, which has therefore been regarded as adapted to changing light quality with increasing depth. Accumulating documentation of distributional exceptions and conflicting physiological data, however, throw increasing doubt on the general applicability of the chromatic adaptation hypothesis. For example, green algae survive at the expense of brown and red algae when heavily grazed by sea-urchins in deep sublittoral areas (Section 5.2.4.1). Chromatic adaptation could still be important in allowing the red algae to outcompete the green algae when undisturbed, or to survive under the canopy of tall brown algae that also flourish at these depths, but its role is far from clear. Physiological experiments, moreover, have shown algal morphology to be at least as important as colour in determining potential vertical distribution (Ramus *et al*. 1976), and the actual limits to range are probably set by such factors as desiccation, grazing, competition and availability of sites (Sections 5.1 and 5.2).

9.2.2 Phytoplankton

Phytoplankton is the most abundant pelagic algal life and is virtually ubiquitous in the surface waters of the seas. Floating rafts of macroalgae, such as *Sargassum*, are very restricted in geographical distribution and are confined to the air–water interface.

By existing as small suspended particles, phytoplankton gains access to subsurface nutrient supplies through transportation by vertical mixing. Small size is advantageous in a planktonic existence for reasons considered in Chapter 2.

Shape is also important: flattened discs, long cylinders or filaments sink more slowly than spheres of similar volume. Spheres and discs absorb nutrients more rapidly per unit mass than long filaments of similar diameter when sinking at the same rate. Elaborations in the form of spines or hairs (Fig. 9.3) may reduce sinking rates in some instances, but because of the relatively high density of cell wall material, pronounced ornamentation could increase sinking rates in other cases. Large, spiny diatoms and dinoflagellates frequently contain oil droplets and large vacuoles that increase their buoyancy, and it would appear that the function of spines and other processes is frequently to deter zooplanktonic grazers such as copepods. Some diatoms are enclosed in gelatinous capsules of only very slight excess density. The gelatinous capsules may protect the diatoms from predation either by making them too big for zooplanktonic grazers to handle or by allowing them to pass unharmed through the guts of their predators. Chain-formation may decrease or increase sinking rate according to whether the aggregative surface area to volume ratio is increased or decreased (Fig. 9.3) and may deter some predators. Size and shape may therefore represent an evolved compromise to the effects of sinking rate, efficiency of nutrient uptake, predation and possibly other factors (Hutchinson 1967).

Although planktonic algae exist as single cells or small cellular aggregates it is important to realize that the cells divide mitotically under favourable circumstances to produce clones of cells with identical genotypes. Each clone is genetically equivalent to the multicellular body of a nonclonal alga such as a fucoid or a kelp, and in this sense the clone is the genetic 'individual' and the component cells are spatially scattered modules of its 'soma' (for a discussion of clonal population ecology, see Harper (1977)). Advantages gained by dividing the body into a large number of detached modules include: (1) a high productivity facilitated by high ratios of absorptive surface area to cytoplasmic volume and of photosynthetic to structural materials; (2) the potential ability to disperse modules among

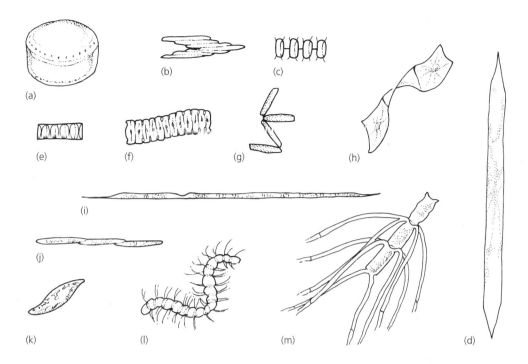

Fig. 9.3 Adaptive shapes and ornamentations of planktonic algae: (a) *Coscinodiscus concinnus*; (b) *Bacillaria paradoxa*; (c) *Thalassiosira gravida*; (d) *Rhizosolenia styloformis*; (e) *Paralia sulcata*; (f) *Bellarochea malleus*; (g) *Thalassiothrix nitschioides*; (h) *Streptotheca thamensis*; (i) *Rhizosolenia hebetata*; (j) *Nitschia seriata*; (k) *Gyrosigma* sp.; (l) *Chaetoceros curvisetus*; (m) *Chaetoceros convolutus*. (After Hardy 1962.)

different water masses, thereby increasing the chance of encountering favourable growing conditions; (3) the ability to quickly exploit localized resources by rapid (exponential) clonal growth; (4) lessening the probability of extinction of the genotype by spreading the risks of mortality among many independent modules.

9.3 Animals

Animals lead more varied lives than algae and higher plants, largely because of the more diverse sources of energy and nutrients that they exploit. The richness of life histories of marine animals could exceed the capacity of even a large textbook, but some important general features will be discussed here under the arbitrary headings of feeding and reproduction.

9.3.1 Feeding

9.3.1.1 Filter feeders and deposit feeders

Just as benthic algae and higher plants can acquire energy by remaining stationary and intercepting sunlight, so numerous animals can meet their energy requirements by remaining attached to the substratum and intercepting water-borne food particles: either planktonic organisms or particulate organic matter. Filter feeders sift food particles still in suspension, whereas deposit feeders gather food particles that have fallen out of suspension into the sediment. Filter feeders are represented among diverse phyla (Fig. 9.4 and Plates 23 and 24, facing p. 208), being most numerous where currents bring food particles from large catchment areas. Many filter feeders employ ciliary tracts or sticky mucus, or both, to trap food particles and transport them to the mouth. Arthropods lack cilia and use meshes constructed from interlocking hairs and bristles (Fig. 9.4g). The filtration device is either external, as with the radially symmetrical crown of tentacles or bristles of coelenterates, bryozoans, phoronids, polychaetes, arthropods, echinoderms and hemichordates, or internal, as with the choanocyte chambers of sponges, lamellae of

Fig. 9.4 Filter-feeding benthic animals: (a) the sponge *Amphilectus*; (b) the bryozoan *Bugula*; (c) the hydroid *Algaophenia* with its flat surface at right angles to the prevailing current; (d) the bivalve *Cerastoderma edule*; (e) the polychaete *Bispira volutacornis*; (f) the tunicate *Clavellina* *lepadiformis*; (g) the barnacle *Semibalanus balanoides*; (h) the amphipod *Haploops tubicola*; (i) the brittlestar *Ophiothrix fragilis*; (j) the anemone *Metridium senile*. (After Hughes 1980b.)

bivalves, lophophores of brachiopods and pharyngeal chambers of tunicates. Radially symmetrical discs, cups or fans formed by tentacles or intermeshing bristles optimize combinations of surface area, mechanical strength, metabolic operational costs and trapping efficiency. The last may also be increased by adjusting the orientation of the filter so that it always faces the water currents. This may involve movement of the animal, as in barnacles and polychaetes, or a modification of colonial growth as

in feather- or leaf-shaped hydroids, bryozoans and gorgonians that grow with their broad surfaces at right angles to the current (Fig. 9.4c).

Cnidarians, through their possession of stinging nematocysts, are potentially able to catch large prey in addition to small suspended particles. Some anemones, such as the western North American *Anthopleura elegantissima*, which subsists largely on dislodged mussels, can cope with prey almost as large as themselves. Other filter feeders, e.g. stony

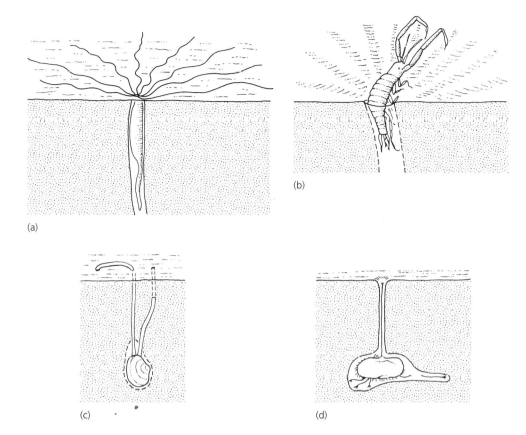

(a)

(b)

(c)

(d)

Fig. 9.5 Deposit feeders: (a) the polychaete *Amphitrite*; (b) the amphipod *Corophium volutator*; (c) the bivalve *Scrobicularia plana*; (d) the burrowing urchin *Echinocardium cordatum*. (After Hughes 1980b.)

corals, soft corals, giant clams (*Tridacna*) and certain ascidians living in shallow, sunlit waters, rely to a greater or lesser extent on the photosynthates of symbiotic zooxanthellae (dinoflagellates, or in the case of tunicates, blue–green algae) as a source of energy (Section 6.10.1).

Deposit feeders also form a diverse taxonomic assemblage (Fig. 9.5 and Plates 25 and 26, facing p. 208). Whereas filter feeders tend to live on hard substrata or in sediments where there is little silt to clog the delicate filtering mechanisms, deposit feeders are able to cope with silty sediments, which they ingest to extract the micro-organisms associated with detrital or mineral particles. Because deposit feeders need to process copious quantities of sediment to extract sufficient food and because they also tend to live in

unstable substrata, they are generally more mobile than filter feeders, showing greater morphological adaptation for locomotion and lacking the radial symmetry and colonial existence of many filter feeders.

9.3.1.2 Predators

In an ecological context, herbivory and carnivory can be regarded as types of predation, the only essential difference being the trophic level of the prey. Predators may consume entire prey organisms, as with fish eating zooplankton or dogwhelks eating barnacles, or they may consume only part of the prey organism without killing it, as with herbivorous grazers of benthic macroalgae or carnivorous grazers of sedentary colonial animals. Both grazers and non-grazers show a wide range of feeding behaviour from extreme dietary specialization to extreme generalism. Dietary specialization is more appropriate where specific prey are predictably abundant. Such is often the case with prey that have defence mechanisms effective against predators, resulting in low overall predation pressure. Examples are plentiful among sedentary animals. Sponges are pervaded by spicules that not only make sponge tissue unpalatable to many predators, but also lower its energetic value as food. In tropical to warm temperate regions where grazing fish are numerous, sponges living on open surfaces contain toxic chemicals that repel fish; for example, in the Red Sea, the sponge *Latrunculia magnifica* contains a cholinesterase inhibitor. In the same regions, sponges occupying protective microhabitats within crevices or beneath ledges do not possess toxins, which presumably are manufactured at some cost. Cnidarians are charged with stinging nematocysts that are defensive as well as offensive in function. The sea anemone *Anemonia sulcata* also contains toxic polypeptides that are lethal when experimentally injected into fish and crustaceans, and which presumably repel predators in nature. Organisms protected from the majority of predators present a potentially rich food source for any animals that develop means of breaking through the defence mechanisms. Not surprisingly therefore, virtually all plants and animals armed with antipredation devices are exploited by a few coevolved predators, specialized to deal with these defences. Certain nudibranches and cowries specialize on sponges; for example, the nudibranch *Archidoris pseudoargus* feeds entirely upon the sponge *Halichondria panacea*, and the nudibranch *Aeolidia pappilosa* feeds on sea anemones.

Generalist predators can be expected to contract or expand their diets according to the relative abundances of more of less preferable prey. The western North American starfish *Pisaster ochraceus* prefers mussels but will also feed on barnacles and then gastropods as the preferred prey become scarce. Dietary preferences sometimes reflect the profitabilities of different prey. Profitability will be a complex function of the likelihood of capture, yield of food material and the time and effort needed to handle the prey. In other instances, dietary preferences may reflect the limited time available for foraging or the need to minimize the time spent foraging, during which the predator may itself be at risk to predation or to other mortality factors.

Predators can be classified as searchers, pursuers and ambushers (Fig. 9.6). Searchers (Plate 27, facing p. 208) spend most of their foraging time locating prey which, once encountered, are caught and eaten relatively quickly; these predators will tend to be generalists, feeding opportunistically on any easily caught prey. Pursuers (Plate 28, facing p. 208) feed on prey that take a relatively long time to catch and subdue, so that a large proportion of the foraging time is spent pursuing individual prey; these predators will tend to specialize on prey with high profitabilities, perhaps becoming morphologically adapted to capturing particular prey efficiently. Ambushers wait for prey to come close enough to be caught by surprise or by blundering into a trapping mechanism; these predators do not actively search for prey or pursue them, but some considerable time may be spent subduing and digesting very large prey. Ambushers cannot determine which prey shall be encountered and so will tend to accept all capturable prey, sometimes being able to tackle prey as large as or even larger than themselves.

As with most biological classifications, this categorization of predators is arbitrary and many animals will have intermediate predatory behaviour or may even switch from one category to another according to circumstances. The shore crab, *Carcinus maenas*, is a most opportunistic feeder as it migrates with the tide to forage on the shore (Fig. 9.6a). When feeding among an abundance of mussels, however, *Carcinus* displays a surprising ability to modify its predation technique to deal specifically with mussels and tends to choose the most profitable mussel sizes (Fig. 9.7). The dogwhelk, *Nucella lapillus*, feeds almost entirely on barnacles and small- to medium-sized mussels. Although these sedentary prey are not pursued, dogwhelks spend a very long time handling them (over 10 h to drill and over 15 h to consume a medium-sized mussel) and because of this, dogwhelks fit into the 'pursuer' category. However, like most predators, dogwhelks will feed opportunistically on any moribund prey, such as a freshly killed fish. Ambushers are often morphologically constrained to adopt the one predation method. Anemones such as *Anthopleura elegantissima* or *Actinia equina* that feed on macroscopic food are unable to pursue prey, but their stinging nematocysts and numerous tentacles enable them to subdue large prey which slip through the wide pharynx, lubricated with mucus, into the distensible sacklike stomach. Deep-sea fish such as gulper eels (Fig. 9.6f) and angler-fish are ambushers *par excellence*. At these great depths prey are scarce indeed (see

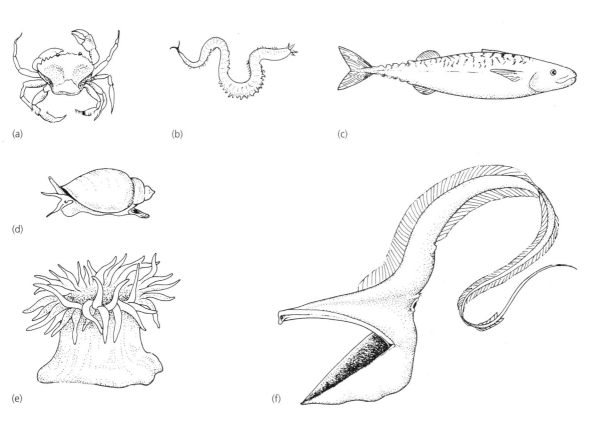

Fig. 9.6 Kinds of predators. Searchers: (a) crab, *Carcinus maenas*; (b) polychaete, *Nereis* sp. Pursuers: (c) mackerel, *Scomber scombrus*; (d) dogwhelk, *Nucella lapillus*. Ambushers: (e) anemone, *Actinia equina*; (f) deep-sea gulper, *Eupharynx pelecanoides*. (a and c after Hughes 1980b; d and e after Hughes 1980a; f after Briggs 1974.)

Section 1.3), so that pursuit and selective feeding would be uneconomical. Gulper eels, angler-fish and others have become little more than suspended traps, most of the body muscles and skeleton having atrophied except for the jaw apparatus, so that sustained searching or pursuing would be impossible. The mouth has an enormous gape and in some cases can even be unhinged, and the stomach is hugely distensible so that prey considerably larger than the predator can be swallowed.

Copepods, which account for a large proportion of the zooplankton, are important filter feeders on phytoplankton (Chapter 2) and as with all arthropods, the filter is constructed from intermeshing bristles borne on modified limbs. Size selection of algal cells

occurs as a mechanical consequence of the mesh diameter of the filter, but whether mesh size can be modified according to the availability of different algal types remains in debate. Algal quality also is important in addition to size. *Acartia clausi* avoids the dinoflagellate *Ceratium tripos*, which is armoured with cellulose plates, but accepts unarmoured larger and smaller algae. Copepods are sometimes able to grasp individual items with the mandibles. *Acartia tonsa* can use the grasping mechanism to prey upon nauplii of other copepods while simultaneously filtering algal cells.

9.3.2 Reproduction and dispersal

9.3.2.1 Asexual vs. sexual reproduction

Asexual reproduction occurs among most phyla and may take place by two fundamentally different processes. First, embryos may develop parthenogenetically from unfertilized eggs and secondly, the body may divide as it grows. Parthenogenesis has evolved

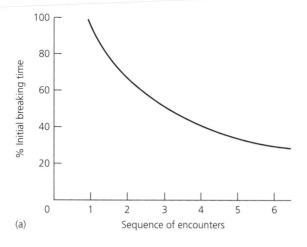

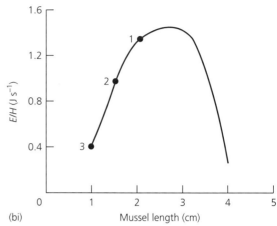

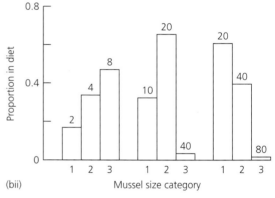

Fig. 9.7 (a) The time taken by the crab *Carcinus maenas* to open a mussel, *Mytilus edulis*, decreases as the crab becomes more experienced at handling the prey. (After Cunningham & Hughes 1984.) (bi) The profitability (energy yield per handling time) of mussels to foraging crabs peaks at intermediate mussel size; and the blue crab, *Callinectes* *sapidus*, opening a marsh mussel, *Geukensia demissa*. (bii) The diet of shore crabs fed on three sizes of mussels where the profitability is 1 > 2 > 3. Numbers offered appear over the histograms. As the most profitable mussels became more abundant the crabs fed disproportionately upon them. (After Townsend & Hughes 1981.)

recurrently, and continues to do so in many independent lines, especially in those occupying terrestrial and freshwater habitats. Some parthenogenetic lines have lost the ability to reproduce sexually, and although benefiting in the short term from a higher potential rate of increase than sexual relatives (by omitting males that cannot give birth), these parthenogenetic animals lack the evolutionary potential associated with sex and so are much more prone to extinction than sexual lines. Other animals incorporate both parthenogenetic and sexual phases into their life cycles. Cyclical parthenogens usually reproduce sexually when growing conditions begin to deteriorate. The parents die and the progeny endure the ensuing harsh periods as dormant zygotes, or early embryos, protected within resistant capsules. When favourable growing conditions return, each embryo develops into a parthenogentic female. This female reproduces parthenogenetically throughout the favourable growing period, forming a clone of genetically identical modules. Such a life cycle is similar to that of many unicellular algae (Section 9.2.2) and is associated with similar advantages, i.e. the maintenance of a favourable surface area to volume ratio by splitting the 'body' into modules, the ability to disperse modules over a wide area in search

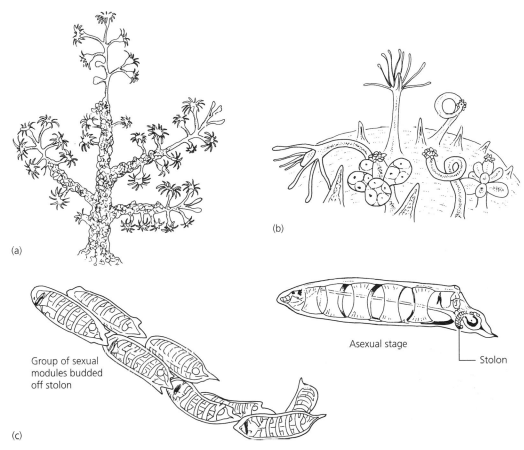

(a)

(b)

Group of sexual
modules budded
off stolon

Asexual stage

Stolon

(c)

Fig. 9.8 Clonal animals. (a) When modules (zooids) of a
growing clone remain united they form a colony, which is
usually a benthic sedentary form as in this example of the
hydroid *Bougainvillia*. (b) The clonal modules are genetically
identical but may become specialized for different functions
as in *Hydractinia echinata* with feeding (long tentacles),
defensive (cluster of knob-like tentacles) and reproductive
(large vesicles, left and right of the colony) zooids.
(c) Modules of a growing clone may become detached as
with most pelagic tunicates. Depicted is the salp *Iasis* (*Salpa*)
zonaria, which alternates between an asexual, clone-forming
stage and a sexual, outbreeding stage. The asexual stage
buds off chains of molecules from a stolon. The modules
remain attached in groups for a while, during which time
they are functional females. After giving birth to a single
asexual young, each female module changes sex and
becomes detached from the others, shedding sperm into the
sea-water and perhaps fertilizing another clone. (After Hardy
1962.)

of unpredictably located patches of resource, the
rapid exploitation of local patches of resource, and
the lessening of the probability of extinction of the
genotype by spreading the risk of mortality among
many independent modules. Cyclical parthenogens
include aphids, rotifers and cladocerans, but whereas
rotifers and cladocerans are common in fresh water,
they are relatively scarce in the sea, being oversha-
dowed by copepods. Why neither marine nor freshwater copepods have developed any parthenogenetic
lines remains a tantalizing, unanswered question.

Asexual reproduction by budding or fission of
the body does not involve gametogenesis and is
therefore different from parthenogenesis. Both processes, however, result in the formation of a clone
of genetically identical modules. Budding or fission
is common among many invertebrate phyla and the
resulting clonal modules may remain attached to
each other to form colonies, e.g. colonial hydroids,
zooanthids, corals, bryozoans and colonial tunicates
(Fig. 9.8a), or they may become detached to form
a clone of aggregated or dispersed modules, e.g.

solitary hydroids, scyphozoans, anemones, certain polychaetes and most pelagic tunicates (Fig. 9.8c). Division of the growing body into a clone of modules has several advantages. If the modules are dispersed the clone gains similar advantages to those of parthenogenetic animals. *Thalia democratica*, for example, is a warm-water, pelagic tunicate that completes the alternation of asexual budding and sexual reproduction within 2 days. It has about the shortest generation time of any metazoan and is thus well able to exploit local phytoplankton blooms (Heron 1972). If the modules form an aggregation or an organically united colony, the dispersal ability of the clone is severely reduced, but several other advantages remain that are of great significance among sedentary animals. First, filter-feeding devices such as the tentacular crowns of coelenterates, bryozoans and polychaetes work efficiently only below a certain size. Modular iteration preserves the optimal size of the feeding apparatus while allowing a continued increase in biomass. Secondly, modules can become specialized for different functions. Such 'division of labour' is seen most often in organically united colonies, e.g. the hydroid *Hydractinia echinata*, in which different polyps are specialized for feeding, reproduction and defence (Fig. 9.8b), but may occur in modular aggregations, as with the anemone *Anthopleura elegantissima* in which peripheral polyps forgo sexual reproduction and use the energy to defend the reproductive polyps within. Division of labour and co-operation among modules of a clone are not altruistic, as only a single genotype is involved, but are analogous to the division of labour among the organs of a non-clonal animal. Thirdly, a very flexible growth form is achieved, which can be modified to suit local conditions, e.g. topography of the substratum, direction of prevailing currents (Fig. 9.4c), or presence of other organisms. Fourthly, sedentary animals are at risk of predation by carnivorous grazers. Modules escaping predation can sustain damaged clone-mates until these are regenerated or replaced. Fifthly, the ability to compete for and retain space on the substratum is increased by the collective effort of the modules. Reviews of clonal biology have been given by Jackson *et al.* (1985), Harper *et al.* (1986) and Hughes (1989).

Among the great variety of forms exhibited by sedentary modular colonies, Jackson (1979) recognized six basic shapes: runners (linear or branching encrustations), sheets (two-dimensional encrustations), mounds (massive, three-dimensional encrustations), plates (foliose projections from the substratum), vines (linear or branching, semi-erect forms with restricted zones of attachment) and trees (erect, usually branching projections) (Fig. 9.9). Runners and vines are the most opportunistic forms, advancing quickly in a linear fashion to make temporary use of unoccupied space before being outcompeted by other growth forms that show increasing commitments to survival within their own areas of settlement (sheets < mounds < plates < trees). For example, the bryozoan *Electra pilosa* adopts the runner growth form and opportunistically colonizes unoccupied patches on the surface of *Fucus spiralis*, but is eventually overgrown by *Alcyonidium hirsutum* and *Flustrellidra hispida*, which grow more slowly in any single direction but also advance two-dimensionally to form sheets over the substratum (Section 5.1.7.6). When two similar growth forms meet, the competitive outcome may depend on subtle factors such as whether two growing edges meet or whether one growing edge impinges on a non-growing edge.

Sometimes, colonies react allelochemically to competitors. When different, and therefore genetically distinct, clones of the sponge *Hymeniacidon* grow into contact, they produce a substance that interferes with cellular adhesion. By means of this chemical warfare, one colony usually dominates the other, in a fashion reminiscent of the competitive hierarchies by mesenterial digestion found among corals (Section 6.4.1). Allelochemical interactions are common among sponges, bryozoans and tunicates inhabiting crevices, caves and shaded overhangs on coral reefs. Coexistence among these animals apparently is facilitated by the presence of competitive loops rather than hierarchies. For example, species A might overgrow species B which might overgrow species C, but species C is able to suppress species A by an allelochemical interaction. The occurrence of competitive loops in a community of potential competitors forms what Buss & Jackson (1979) called a competitive network, the complexity of which increases with the frequency of loop formation. Among the inhabitants of cryptic coral reef habitats, competitive loops are common and often permanent, whereas among the epiphytes of *Fucus serratus* they are infrequent and temporary (Section 5.1.7.6), and are probably absent among corals.

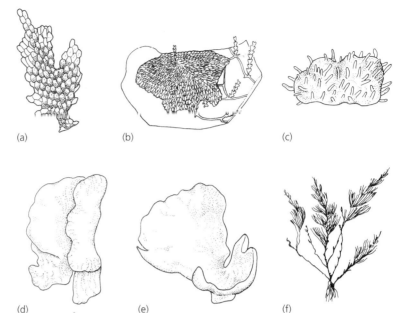

Fig. 9.9 Growth forms of sedentary modular colonies. (a) Runner: bryozoan *Electra pilosa*. (b) Sheet: bryozoan *Alcyonidium polyoum*, overgrowing vine: hydroid *Dynamena pumila* (after Stebbing 1973). (c) Mound: sponge *Polymastia* sp. (after Bowerbank 1874). (d) Vertical plate: sponge *Phakellia robusta*. (e) Horizontal plate: sponge *P. ventilabrum* (after Bowerbank 1874). (f) Tree: bryozoan *Bugula plumosa* (after Ryland 1962).

9.3.2.2 *Larval life*

Sexual reproduction in most marine animals results in the formation of larvae that persist for varying lengths of time before metamorphosing into the pre-adult stage. The range of larval sizes is narrow, invertebrate larvae being about 0.5–1.5 mm in overall diameter and fish larvae perhaps an order of magnitude larger. Various modes of larval life are possible. Larvae may be pelagic (planktonic) or non-pelagic (benthic), feeding (planktotrophic) or non-feeding (lecithotrophic), brooded or non-brooded, and intermediate or mixed modes also occur. It is difficult to generalize about the relative advantages of the different modes of larval life because many interacting selective forces may be involved, and because the experimental testing of ideas is hindered by the difficulty of measuring larval survivorship and dispersal in the field. At present it is possible only to list plausible hypotheses. Before doing so, it will be helpful to consider some general biological relationships.

1 The larger the egg the longer is the embryonic development time from fertilization to hatching (Fig. 9.10). There may be two contributing factors to this correlation. First, larger eggs contain more yolk, which retards cleavage; some animals reduce this effect by providing extra-embryonic yolk supplies to the embryos, e.g. the non-developing 'nurse eggs' contained within the egg capsules of dog-whelks (*Nucella lapillus*). Secondly, the longer the pre-hatching development time, the more advanced is the developmental stage at hatching. Animals requiring developmentally more advanced hatchlings must therefore provide the eggs with more yolk, thereby making them larger. The largest eggs have sufficient yolk to nourish the embryo through to metamorphosis before hatching (direct development), whereas the smallest eggs contain yolk sufficient only for limited embryonic development, so that the hatching larvae must spend some time feeding in the plankton (planktotrophic) to complete their development. Eggs of intermediate size hatch into 'lecithotrophic' larvae, endowed with sufficient yolk to sustain them during their brief motile existence and to enable them to complete metamorphosis. Postlarval juveniles hatching from larger eggs with direct development are slightly more advanced or larger than newly metamorphosed postlarvae from smaller eggs with lecithotrophic development. Among direct developing eggs, larger eggs produce larger hatchlings.

2 Because the energy available to a parent for egg production is limited, the larger the egg the smaller is the clutch size (Fig. 9.11a).

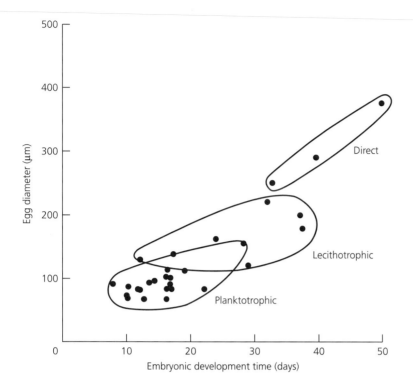

Fig. 9.10 The time taken by nudibranch eggs to develop and hatch as veligers from the egg capsules increases in larger eggs. (After Todd & Doyle 1981.)

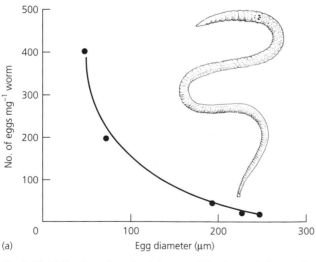

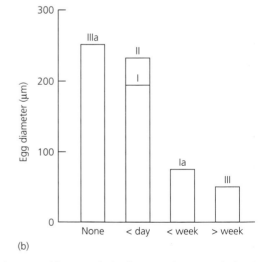

Fig. 9.11 (a) The fecundity of polychaetes within the *Capitella capitata* species-complex decreases as larger eggs are produced. (b) Egg size, and hence the amount of yolk reserve, is negatively correlated with the time that larvae spend in the plankton. *Capitella* species IIIa has direct development without a pelagic phase; species I, Ia and II have lecithotrophic larvae with short pelagic phases; and species III has planktotrophic larvae with a longer pelagic phase. (After Grassle & Grassle 1977.)

Table 9.4a Maximum time of survival in days at an assumed oxygen consumption of 5 ml g^{-1} (dry wt) h^{-1}. (After Crisp 1976.)

Source of energy	Percentage of tissue weight devoted to energy store				
	5	10	25	50	75
Lipid	0.8	1.7	4.2	8.3	12.5
Protein	0.5	1.0	2.5	5.0	7.5
Carbohydrate	0.3	0.7	1.7	3.3	5.0
Metabolic rate of non-storage tissue ml O$_2$ g^{-1} h^{-1}	5.3	5.6	6.7	10.0	15.0

Table based on $t + \dfrac{q \times x}{24 \times r}$ days,

where q is oxygen requirement in ml g^{-1} of metabolite, x is fraction of body tissue devoted to reserves and r is respiration rate in ml g^{-1} h^{-1}.

Table 9.4b Oceanic diffusion. (After Crisp 1978.)

Pelagic life	Log$_{10}$ (probable distance transported in cm)	Order of magnitude	Whether likely to be exceeded by tidal currents
3–6 h	4	100 m	Yes
1–2 days	5	1 km	Yes
7–14 days	6	10 km	?
14 days–3 months	7	100 km	No
1 year	8	1000 km	No

It follows from **1** and **2** that planktotrophic larvae are produced in greater quantities per brood than lecithotrophic larvae which are, in turn, produced in greater quantities than direct developing larvae.

3 Released larvae are at constant risk to predation, and if pelagic, also to transportation by currents into unfavourable areas. The cumulative risk of mortality therefore increases exponentially with increased duration of the mobile phase.

4 Dispersal ability increases with increased duration of the mobile phase. This relationship is not simple, however, because in nearshore waters tidal currents are usually stronger than residual currents, with the result that the average distance transported would increase to a first maximum after 6 h in the plankton, beyond which it would increase only very slowly. Hence a pelagic phase exceeding 6 h duration would achieve little additional dispersal while greatly increasing the cumulative risk of mortality. Significantly increased dispersal would only be achieved by greatly prolonging the pelagic phase so

that larvae could be transported long distances by non-tidal currents (see Crisp 1976).

With these generalizations in mind, some guesses can be made about the ecological significance of various modes of larval development.

Dispersal hypothesis. Animals with limited powers of postlarval dispersal, e.g. most benthic invertebrates, must rely on pelagic eggs and/or mobile larvae for dispersal. As with seeds transported by the wind, the small size of larvae enables them to use passive transportation (by water currents) as an energetically cheap means of dispersal. Also like seeds, larvae have some capacity to delay metamorphosis until a suitable settlement site is encountered. Among non-feeding larvae, energy reserves dictate that metamorphosis can be delayed only for a short while (Table 9.4) (this restriction is much less severe in cold, polar regions where metabolic rate is reduced; see Environmental constraints), after which settlement becomes more indiscriminate and usually proves fatal. Feeding larvae can delay settlement for longer, but never as long as many seeds, which can remain dormant and viable for several years. Larvae on the other hand have the additional feature of small-scale locomotive powers, so that the encountering of suitable places for settlement and metamorphosis is less haphazard than with seeds.

Increased dispersal ability is achieved at the cost of increased larval mortality (see 3 above). Long-distance dispersal therefore has a reasonable chance of success only if very large numbers of larvae are released, and because of energetic constraints (see 2 above), these must be planktotrophic. As long as food is available to them, planktotrophic larvae theoretically could spend any amount of time drifting and, indeed, some Pacific echinoderm larvae spend as much as 36 weeks in the plankton, giving them sufficient time to traverse the ocean in favourable currents. If long-distance dispersal is unnecessary, an animal could gain by producing fewer larger eggs, each of which has a greater chance of surviving to metamorphosis, a trend which reaches its peak with direct development and elimination of the mobile larval stage. The optimal dispersal ability will depend on the spatial distribution and degree of permanence of suitable sites for colonization. If these are unpredictably located, then long-distance dispersal may be advantageous, whereas if they are predictably located nearby, then more limited dispersal will suffice and greater larval survivorship will be more advantageous. Also, predictably located settlement sites are more likely to be encountered by larvae than unpredictably located sites, so that competition among larvae and postlarvae may be more severe in the former than in the latter. Competitive ability will increase with advanced embryonic development or with larger size, so that increased egg size and associated reduced clutch size may be expected among animals successfully exploiting predictably located settlement sites. It is becoming increasingly apparent that larval recruitment tends to be local, often close to the parent, and that many benthic colonial species release larvae when tidal currents are minimal, or have larval behaviour which favours local settlement.

Some support for the dispersal hypothesis is to be found in the reproductive modes of sibling species of polychaete comprising the *Capitella capitata* species-complex. These sibling species are morphologically so similar that their identities were only

ascertained when their isozymes were examined electrophoretically by Grassle & Grassle (1977). Their morphological similarity and recent common ancestry makes these polychaetes ideal subjects for comparing variations in reproductive mode, because the influences of morphological and phylogenetic constraints can be discounted. Among the sibling species of *Capitella*, decreasing egg size is correlated with increasing clutch size, with increasing duration of the pelagic phase and hence with increasing powers of dispersal (Fig. 9.11b). Species IIIa has non-pelagic (benthic) larvae with very limited dispersal potential, species I and II have non-feeding (lecithotrophic), pelagic larvae, and the remainder have feeding (planktotrophic) larvae that remain in the plankton for several days (species Ia) to 2 weeks (species III) and are capable of long-distance dispersal. All the *Capitella* species can be regarded as opportunistic (Section 9.1), typically colonizing fine sediments that have been denuded of competitively superior species by environmental disturbance such as pollution. Some of the *Capitella* species, however, behave more opportunistically than others. Species I, II and IIIa are the most opportunistic, settling soon after hatching as relatively advanced larvae. Larval survivorship is high and these species can rapidly colonize local areas after disturbance has eliminated competitors. The continued existence of species I, II and IIIa depends on the frequent occurrence of disturbed patches of habitat, which they are ready to exploit at any time by virtue of their continued reproduction throughout the year. Species Ia and III occur in less variable subtidal habitats and because of the longer pelagic phase, they are slower to colonize new habitats than the more opportunistic species. The relatively wide larval dispersal of species Ia and III, however, enables them to select potentially more favourable habitats, which will be rarer and less predictably located than the frequent local disturbances exploited by species I, II and IIIa. The persistence, however, of invertebrate species with direct development along with others possessing a pelagic larval phase in the Antarctic, where ice scour repeatedly decimates shallow benthic populations, shows that larval dispersal is not a prerequisite for coping with disturbed habitats.

A serious problem with the dispersal hypothesis is that the successful establishment of an individual following dispersal depends to a great extent on colonizing ability and this may be uncorrelated, or even negatively correlated with dispersability. The production of planktotrophic or lecithotrophic larvae, or the brooding of young may represent the action of various selection pressures, some of which are discussed below, or may simply be inherited from distantly ancestral lineages without any relation to present conditions. If so, dispersability is merely a secondary consequence of developmental mode.

Moreover, the highest levels of dispersability, conferred by planktotrophic development, may hinder colonization of isolated habitats that do not lie in major pathways of larval transportation. This is because planktotrophic larvae of a new colonist are likely to be carried away by currents, leaving the parent in reproductive isolation. On the other hand, brooded young, or even lecithotrophic larvae, will tend to remain in the parental habitat, potentially establishing a reproductive population. This principle may explain the absence of the planktotrophic *Littorina littorea* from Rockall and the successful colonization of this small, offshore island by the ovoviviparous *Littorina saxatilis* (Johannesson 1988). The relationship between developmental mode and geographical distribution is discussed further in Section 10.3.1.

Size-threshold hypothesis. Some researchers have postulated that to produce sufficient numbers of lecithotrophic or direct developing larvae, an animal must attain a threshold size, at which the body will contain sufficient energy reserves to manufacture the large eggs. Below the threshold, the body contains insufficient stored energy to make lecithotrophy or direct development worth while, because the resulting clutch sizes would be too small. Todd and Doyle used this hypothesis to explain lecithotrophy in the larger nudibranch *Adalaria proxima* and planktotrophy in the smaller nudibranch *Onchidoris muricata*.

Conversely, others have postulated that planktotrophy will not be feasible below a threshold body size because, even though the eggs are small, restricted bodily energy reserves could not produce enough of them to compensate for the larval mortality incurred by a long pelagic phase. Tiny bivalves belonging to such genera as *Gemma* and *Mysella*, which live in shallow-water sediments, or *Lasaea* and *Turtonia*, which inhabit small intertidal crevices, all brood their embryos and release them at the post-larval stage. *Lasaea rubra*, for example, incubates 12–22 embryos in its suprabranchial chambers and releases them as 0.5–0.6 mm juveniles, even though the parent is itself only 2–3 mm in length. Perhaps, with such small energy reserves, the potential clutch size of these tiny animals is so restricted that larval mortality must be avoided altogether. This also may be the case among meiofaunal animals, ranging from 50 µm to 3 mm in length and adapted to life in the interstices of sedimentary particles. Ninety-eight per cent of interstitial species lack pelagic larvae and many brood their young to an advanced developmental stage, whereas others protect their eggs in sticky cocoons that readily adhere to sand grains. Clutch sizes are correspondingly small, two to three eggs being normal and seldom exceeding 10. An exception 'proving the rule' is to be found among tubicolous polychaetes in the genus *Dodecaceria*. Most species form clones by fission and although each worm is diminutive, producing only a few small eggs, its larvae are planktotrophic. But as all clonemates share the same genome, they are collectively equivalent to one individual of an aclonal species, and together they produce multitudes of larvae.

Energy-subsidy hypothesis. The plankton is a food resource used by animals ranging from small invertebrates to the largest vertebrate (Chapter 2). During phytoplankton blooms this food resource is particularly rich and it is conceivable that it may pay some species to exploit the plankton not only as food for themselves but also as food for their progeny, thereby subsidizing the energetic cost of reproduction. Fecundity would be increased because the parent need make only a small energetic investment per larva (Fig. 9.11b). Larval growth, and hence chance of successful recruitment, would be enhanced when planktonic food is particularly abundant, allowing the parent to increase its reproductive value in years with a good bloom. The converse, however, would be true in years with a poor bloom, and it has not yet been demonstrated that there is a net gain in energy as a result of feeding in the plankton. It is certain that the spawning of some animals is timed so that their larvae can exploit the phytoplankton bloom. For example, increased phytoplankton concentration in the spring triggers spawning in the chiton *Tonicella lineata* on western North American shores.

Food-niche hypothesis. It has been suggested that by exploiting a different food resource, planktotrophic larvae avoid competition with the parental generation or with meiofaunal species similar in size to the macrofaunal larvae (Warwick 1989), but such competition could equally be avoided by the production of non-feeding larvae. Feeding larvae may need to forage in the plankton to encounter suitably small food organisms, but again this constraint could be avoided by provisioning the young with yolk. The food-niche hypothesis therefore cannot explain the relative selective advantages of feeding and non-feeding larvae among different animals.

Although larvae are chiefly pelagic in low latitudes, where planktonic productivity is less seasonal, direct development is prevalent in polar regions, where planktonic production is highly seasonal and irregular, making planktotrophy too risky (see Section 3.2.4).

Environmental constraints. In his classic work on the latitudinal distribution of larval forms, Thorson (1946) concluded that pelagic larvae rarely occur in polar regions because the seasonal pulse of primary production is too short to sustain planktotrophy. Thorson's observation was based mainly on tropical–Arctic comparisons for prosobranch gastropods, and later was corroborated for Antarctic prosobranchs. In general, prosobranch species follow a cline from the almost complete absence of pelagic forms in polar seas to some 95% with pelagic larvae in the tropics. Other taxa, however, do not necessarily show this trend. Echinoderms, for example, are represented predominantly by species with pelagic larvae at all latitudes. In this case a different cline exists, with planktotrophic species predominating at lower latitudes and lecithotrophic species in polar regions (Pearse 1994). Recent year-round surveys have revealed an unexpectedly high diversity of pelagic larvae in Antarctic waters, comparable with values for temperate latitudes (Stanwell-Smith *et al.* 1997). Some taxa showed seasonal peaks in summer (molluscs and annelids presumably exploiting the microphytoplankton bloom) or winter (echinoderms and nemerteans), whereas others were present throughout the year. We must conclude, therefore, that although environmental constraints may significantly influence latitudinal variation in the relative abundance of larval types, observed trends are not entirely explicable in these terms. The lower diversity of pelagic larvae in the Arctic Ocean probably originates from the geological youth of this region rather than from environmental constraints operating on an ecological time scale.

9.3.2.3 Parental investment

Planktotrophic, lecithotrophic and direct developing larvae represent increasing levels of energetic investment per offspring by the parent. Some of the possible selective advantages of these different levels of parental investment are discussed in the previous section, but the picture is left far from complete. Among species with direct development, parental investment varies considerably according to the amount of yolk provided per egg, the amount of encapsulating material per egg and the amount of parental care in the form of brooding or guarding the eggs and young (Fig. 9.12). Parental investment beyond the minimum required to produce a viable egg is worth while only if juvenile survivorship is significantly increased and outweighs the concomitant losses in potential fecundity or dispersal potential. For example, the winkle *Littorina saxatilis* retains its young within a brood chamber, which is a modified jelly gland, until the shell is well developed and the young are able to lead an independent existence on the shore. *Littorina compressa* has a functional jelly gland, which it uses to provide a gelatinous protective capsule round the egg mass placed beneath stones or within crevices on the midshore. Because *L. compressa* does not brood its eggs, these can be produced more quickly, so that the fecundity of *L. compressa* is about three times that of *L. saxatilis* on the same shore. *L. compressa* therefore thrives well at midshore levels where the survivorship of egg masses is high, but at higher shore levels egg masses would become desiccated and it is here that the brooding behaviour of *L. saxatilis* places the latter species at an advantage. Greater parental investment therefore enables *L. saxatilis* to reproduce in harsher environments (high shore levels, salt-marshes) than related species which do not brood their young.

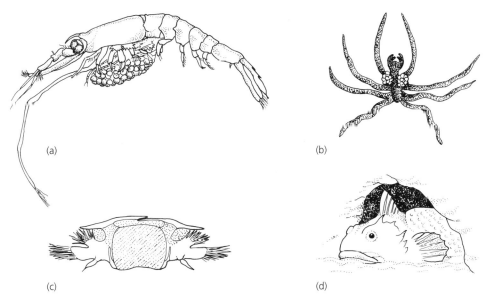

Fig. 9.12 Parental care. (a) As is typical among decapods, the euphausiid, *Nematoscelis difficilis*, carries its eggs on modified thoracic limbs. (After Briggs 1974.) (b) In pycnogonids (Arachnida), females transfer their eggs to males, which carry the eggs on modified limbs. (After Nakamara & Sekiguchi 1980.) (c) The scale worm, *Harmothoe imbricata*, broods its eggs on its back, where they are covered by plate-like processes from the parapodia (After Daly 1972.) (d) The sea scorpion (sculpin), *Cottus bubalis*, guards its egg mass. (After Hughes 1980b.)

9.3.2.4 Semelparity and iteroparity

Organisms may reproduce sexually once and then die (semelparity) or reproduce more than once, or continually, over a protracted period (iteroparity). Care should be taken not to confuse semelparity and iteroparity, which are defined relative to the lifetime of the organism, with the terms ephemeral, annual and perennial, which are defined relative to the year. Nudibranchs, such as *Onchidoris* spp. are annuals that die after spawning and are therefore semelparous. Eels are perennials that also die after spawning and are semelparous. Winkles, *Littorina* spp. (Section 9.3.2.3), are perennials that are iteroparous, whereas some bryozoans, such as *Celleporella hyalina*, can be ephemeral and iteroparous.

Whether semelparity or iteroparity is the more advantageous depends very much on the ratio of juvenile to adult survivorship, as determined by the morphologies and habitats of the two stages. Semelparity relies on the survival of the young to maturity whereas iteroparity relies more on survival of the adult. The benthic adults of marine, sedentary invertebrates are less at risk to mortality than the dispersing larvae, and iteroparity predominates among these organisms. Apart from some very small species, most prosobranch gastropods are well protected by their shells and have a much higher life expectancy as adults than as juveniles, and are correspondingly iteroparous. Nudibranchs lack a shell and even though they use other protective devices such as acidic secretions, nematocyst-bearing dorsal cerata or crypsis, they are likely to have lower life expectancies during the benthic stages than have the armour-plated prosobranchs. All nudibranchs are semelparous. These rationalizations should be regarded with caution, however, because natural magnitudes of juvenile or adult survivorship necessary for a sound comparison have not been measured.

Semelparity in migratory fish such as salmon, eels and lampreys probably has a different evolutionary history. The life history of salmon and lampreys is partitioned between marine habitats that are highly productive, promoting growth and hence fecundity of the adults, and freshwater habitats that are much less productive, but safer nursery grounds for the young. The migration from feeding to nursery grounds, however, demands so much energy and entails such high risks that it pays only to do it once, committing all the remaining energy into a suicidal bout of reproduction. The reverse migration of eels, from freshwater feeding grounds to marine spawning grounds, is more difficult to understand. Eels often exploit productive, relatively safe habitats such as eutrophic muddy ponds, but the conditions on the nursery grounds in the Sargasso Sea remain unknown. Perhaps it is significant that eels have a recent marine ancestry and salmon a freshwater one.

9.3.2.5 Hermaphroditism; why combine sexes?

In most actively mobile animals male and female functions are performed by separate individuals, termed gonochorists (equivalent to dioecious plants), and in a minority by single individuals, termed hermaphrodites (equivalent to monoecious plants). In simultaneous hermaphrodites male and female functions are performed together, whereas in sequential hermaphrodites the male function may precede the female function (protandry) or vice versa (protogyny). How can these sexual permutations and combinations be accounted for in terms of natural selection?

Gonochorism owes its origins to the evolution of anisogamy. Primitive sexual organisms probably had one type of gamete (isogamy) as does the unicellular alga *Chlamydomonas*. Fertilization and zygotic development require mobility, whereby unrelated gametes can encounter one another, but also energy reserves to fuel subsequent development. Mobility is facilitated by small size but hindered by bulky yolk, hence the evolutionary division of labour (anisogamy) into smaller motile sperm whose function is to seek, and larger non-motile eggs whose function is to nourish. It will often be advantageous for division of labour to be carried through to the parents, males being specialized for seeking females and perhaps defending them from other males (ensuring paternity), females being specialized for nurture of the young. Different adult morphological features are often involved, especially with regard to the reproductive organs. The combination of sexual roles within a single individual could therefore be disadvantageous because of incompat-

ible morphology and because of the extra energetic investment in two sets of reproductive apparatus. Hermaphrodites are, however, widespread in nature, so that in certain circumstances the advantages of gonochorism must be outweighed by other factors.

A clue to one selective advantage of hermaphroditism is given by the greater preponderance of sedentary than of mobile hermaphrodites. Among gonochoristic, sedentary animals, there is a risk that neighbours may be of similar sex and unable to fertilize each other, a risk that is avoided by hermaphroditism. Similarly, mobile animals that normally occur at low densities are sometimes hermaphroditic, e.g. nudibranchs, thereby ensuring that when two mature individuals meet, they are capable of crossfertilization. An interesting alternative solution to this problem is found among several gonochoristic, oceanic angler-fish, e.g. *Ceratias holbolli*, which live at exceedingly low population densities: the dwarf male becomes organically attached to the female, deriving nourishment from her and remaining with her throughout reproductive life. The rate of encounter between sexes is so low that once contact has been made, it is advantageous to maintain it. Once having evolved in copulating animals, simultaneous hermaphroditism may be unlikely to revert to the ancestral state of separate sexes because of the stabilizing influence of gamete trading. Life-history traits usually make one or other sexual mode more advantageous in terms of potential genetic contribution to future generations. During mating, therefore, individuals should try to 'cheat', by performing only the preferred sexual role. Gamete trading probably has evolved to prevent such cheating. It involves bouts of limited performance by each individual in the non-preferred sexual role, so forcing the partner to reciprocate. Egg trading occurs, for example, in the hamlet fish *Hypoplectrus nigricans*, and sperm trading in the sea slug, *Navanax inermis* (Leonard & Lukowiak 1991).

Whereas hermaphroditic animals that copulate such as barnacles and nudibranchs, function simultaneously as males and females (Fig. 9.13e), sedentary animals that liberate sperm into the sea-water are usually sequential hermaphrodites, e.g. the protogynous tunicate *Botryllus schlosseri* (Fig. 9.13a) thus preventing self-fertilization and loss of fitness

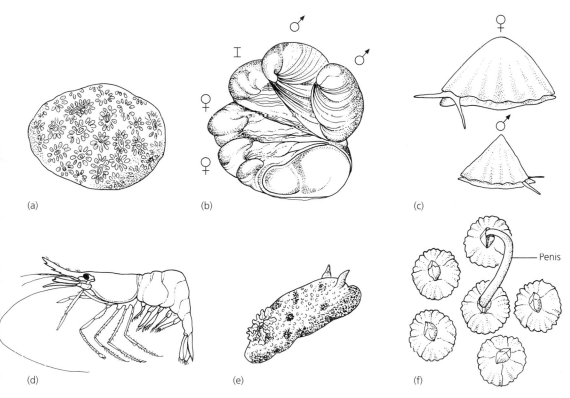

Fig. 9.13 Hermaphrodites. (a) A proptogynous sequential hermaphrodite, the colonial tunicate, *Botryllus schlosseri*. (After Millar 1970.) (b) A protandrous sequential hermaphrodite, the slipper limpet, *Crepidula*. (c) A protandrous sequential hermaphrodite, the limpet, *Patella* *vulgata*. (d) A protandrous sequential hermaphrodite, the shrimp, *Pandalus montagui*. (After Smaldon 1979.) (e) A simultaneous hermaphrodite, the nudibranch, *Archidoris pseudoargus*. (f) A simultaneous hermaphrodite, the barnacle, *Semibalanus balanoides* (showing copulation).

cause by excessive homozygosity. The cyclical, sequential hermaphroditism of sedentary animals, however, is rather different from the permanent, sequential hermaphroditism of certain mobile animals. In populations of the limpet *Patella vulgata* most individuals become male at a relatively small size and switch to being female as they grow larger (Fig. 9.13c). A size threshold has been postulated, rather like that for planktotrophy (see Size-threshold hypothesis, Section 9.3.2.2.), below which energy reserves are inadequate to produce a sufficient number of eggs (that hatch into planktotrophic larvae). Sperm, however, are energetically cheap to produce, so that small individuals can function adequately as males. Difficulties arise with this interpretation because of intermale competition. Bigger males would be more fecund than smaller males, and as limpets are external fertilizers, the larger males should outcompete smaller ones to fertilize eggs. Indeed, a small proportion of *P. vulgata* do remain male throughout life.

Protogyny is common among coral-reef fish. In some wrasses and parrot-fish, only territory-holding males are accepted as mates by the females. Male reproductive success therefore depends on being able to defend a territory from competing males and this can only be achieved by large, strong fish. Young, small fish could not compete for territories, so they function as females until the critical size is reached.

Sometimes, sex is determined environmentally. The coral-reef fish *Anthias squammipinnis* lives in schools within territories on the reef. Each school is composed of females attended by a male. If the male is removed, one of the females changes sex, evidently in response to visual behavioural stimuli.

The value of this particular sex ratio, however, remains to be elucidated. Some species of shrimp, belonging to the genus *Pandalus*, are protandrous (Fig. 9.13d), but can adjust the age of sex change according to the age composition of the population, which fluctuates from year to year because of irregularities in recruitment. This flexibility is selectively advantageous because the value to an individual of being male rather than female increases when males are rarer. As first pointed out by Fisher (1930), on average the rarer sex contributes genetically to more zygotes than the commoner sex.

9.3.2.6 Reproductive effort

Different modes of larval development and parental care represent various ways in which an animal can spend the resources (energy, nutrients, time) allotted to sexual reproduction, but what factors determine the total amount of reproductive expenditure? For simplicity and comparability energy can be regarded as the resource of overriding importance. Energy assimilated from the food is accumulated as body tissues and gametes, dissipated as heat resulting from metabolism, or lost in nitrogenous excretions, leakage of dissolved organic matter, or in the production of mucus. Usually over 50% of assimilated energy is lost as heat, but the amount varies according to the level of metabolic activity, and may exceed 80%. Muscular activity and physiological homeostatic mechanisms such as osmoregulation are metabolically costly. Because, in the long term, an animal assimilates energy at a fixed rate, increased metabolic costs must reduce somatic growth or gametogenesis, and vice versa. The proportion of assimilated energy devoted to reproduction is termed reproductive effort, and this may be expected to vary according to the life history and the degree of environmental stress. In the strict sense, reproductive effort includes not only the energy accumulated in gametes and associated structures such as egg capsules, but also the metabolic energy spent on all reproductive activities such as searching for mates, copulation, defence of young and other kinds of parental care. In practice, reproductive metabolic costs are extremely difficult to measure, so that estimates of reproductive effort are usually approximations in the form of the proportion of assimilated

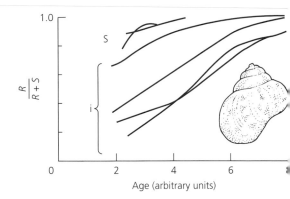

Fig. 9.14 Reproductive effort (the production of assimilated energy used in spawn production) is greater in semelparous (S) than in iteroparous (i) marine snails. (From Hughes & Roberts 1980.)

energy channelled into eggs or even just the ratio of egg production to somatic biomass. The pattern emerging from such measurements is complex, but there is a marked tendency for total lifetime reproductive effort to be greater in semelparous than in iteroparous animals, and for 'instantaneous' reproductive effort to increase with increased age and decreased life expectancy in iteroparous species (Fig. 9.14). Explanations for these trends are similar to those for semelparity and iteroparity (Section 9.3.2.4). If the body is unlikely to survive beyond a single reproductive season, then once sexual maturity is reached, all available energy should be committed to reproduction, depleting the body reserves to a lethal level. If the body is likely to survive beyond a single reproductive season, it becomes worth investing a greater proportion of assimilated energy in bodily maintenance. This investment provides the chance to reproduce more than once, a valuable strategy if reproductive seasons vary greatly in quality. It also takes advantage of the general tendency for fecundity to increase as the body gets bigger (Fig. 9.15). The likelihood of investment in the body paying off, however, declines as life expectancy decreases, so that less energy should be gambled on the body and more should be firmly committed to reproduction as the animal ages. More opportunistic (*r*-selected) species may be expected to have higher reproductive efforts than more stable habitat (*K*-selected) species (see Section 9.1 and Pianka (1974)).

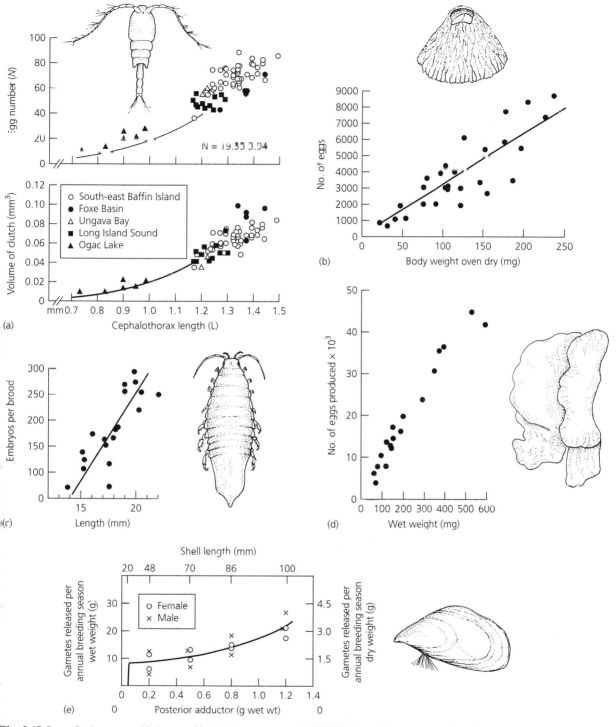

Fig. 9.15 Fecundity increases with increased body size.
(a) The copepod *Pseudocalanus*. (After McLaren 1965.)
(b) The barnacle *Tetraclita rufotincta*. (After Achituv & Barnes
1978.) (c) The isopod *Idotea baltica*. (After Strong & Daborn
1979.) (d) The polychaete *Harmothoe imbricata*. (After Daly
1972.) (e) The mussel *Choromytilus meridionalis*. (After
Griffiths 1977.)

9.3.2.7 *History and phylogeny*

Because nature is so complex and dependent on contingency, most ecological theories will have only limited applicability. For example, the diversity of life histories shown by rough periwinkles in the North Atlantic probably can be understood more through history and phylogeny than in terms of r- and K-selection. Rough periwinkles form a group of closely related species and ecotypes derived from an oviparous ancestral form that invaded the North Atlantic from the Bering Strait some 30 million years ago (Section 10.2.1). Two species, *Littorina arcana* and *L. compressa*, retain the ancestral trait of oviparity, laying jelly-coated egg masses beneath stones and within crevices. The jelly gland of a third species, *L. saxatilis*, is modified into a series of brood chambers, where the eggs develop until hatched. This difference in reproductive mode is uncorrelated with any aspect of demography that has been studied and so is uninterpretable in terms of life-history theory. The relatively recent (within 3–4 million years) evolutionary transformation of the jelly gland into a brood chamber, however, has enabled *L. saxatilis* to colonize habitats such as salt-marshes, estuaries, mobile pebble beaches and silty shores that would be lethal to egg masses. The presence of oviparous and ovoviviparous forms on European shores therefore represents a phase in adaptive radiation, probably having little, if anything, to do with demographical regimes envisioned by life-history theory.

Rough periwinkles also differ in size at birth and at sexual maturation. Despite the claims of certain researchers, there is no conclusive evidence linking any such differences with factors that could be interpreted in terms of r- and K-selection. The sizes of hatchlings and adults are more likely to reflect the diverse nature of mortality factors, such as desiccation, dislodgement, mechanical impact and predation operating in different microhabitats and on different shores (Hughes 1995).

10 Speciation and biogeography

10.1 Introduction

Within an ecosystem, such as a coral reef, rocky shore, or tract of sediment, the numbers and kinds of organisms present are determined by the availability of resources, competition, predation, environmental disturbances, local colonizations and extinctions, all acting on an ecological time scale and therefore having an immediate influence on the biota (Chapters 1, 3 and 5–7). The possible combinations of species present, however, will depend on the pool of potentially available species in the geographical area. Guilds of herbivorous grazers on rocky shores are dominated by periwinkles, top shells and limpets in temperate regions, but by a wider range of gastropods together with grapsid crabs on tropical shores (Section 5.1.2.2). Shallow, sublittoral, rocky substrata are colonized by kelps in temperate latitudes and by corals in the tropics (Figs 5.14 and 6.3). Geographical faunistic and floristic variations are determined not only by immediate ecological and environmental factors, but also by processes of speciation, colonization and extinction acting on an evolutionary time scale. Of course, ecological and evolutionary time scales merge, and at a detailed level are meaningful only with reference to the generation times of particular organisms. Fruitful generalizations can be made, however, and the following examination of speciation and biogeographical patterns forms a natural extension of the ecological discussions in previous chapters.

10.2 Speciation

10.2.1 Allopatric speciation

The number of species living on Earth represents a balance between rates of speciation and extinction. Causes of extinction are little understood, but somehow must be related to changes in the physical environment, to changes in the biological environment as competitors, predators and pathogens evolve, or, in small isolated populations, to accidents of chance. From the fossil record, it appears that major episodes of extinction tend to occur in waves rather than gradually, suggesting the influence of climatic changes.

The influence of ecological factors is illustrated by the recent extinction of the limpet, *Lottia alveus*. This limpet lived and grazed entirely upon the blades of sea-grass, *Zostera marina*, in the western North Atlantic. When populations of the host plant were eradicated by a pathogen in the 1930s, the limpet became extinct.

The influence of climatic factors is illustrated by increasing rates of extinction and speciation accompanying the acceleration of glacial cycles in late Pliocene and Pleistocene time. Beginning some 2.7 million years ago, these cycles brought about changes in sea-level of up to 100 m, changes of up to 2–6°C in sea surface temperature, together with fluctuations in upwelling, current strength and wind patterns. In the Caribbean, there was a wave of extinction of planktonic foraminiferans, molluscs and corals, and this was followed by rapid speciation, producing the present fauna of very young species (Jackson 1994).

Speciation occurs when an ancestral gene pool diverges either as a result of genetic drift or in response to natural selection. Divergence is facilitated by the spatial isolation of gene pools. If by subsequent migration the populations meet, interbreeding may combine incompatible genes, resulting in low fitness. In this case, reproductive isolating mechanisms will evolve, creating separate species.

Incipient reproductive isolation is occurring among populations of the small harpacticoid copepod *Tisbe clodiensis* from different parts of the Mediterranean and Atlantic coasts of Europe. Individuals from most populations interbreed in the laboratory, but this results in progeny of reduced viability. Evidently the geographically isolated gene pools have become modified by local conditions so that when alien gametes combine, the finely tuned genetic architectures break down, resulting in hybrids of low fitness. Reproductive isolating mechanisms have not evolved between these disjunct populations, and indeed would be expected to occur only if the populations had become sympatric.

Such an event has evidently happened among German populations of small littoral isopods belonging to the *Jaera albifrons* species complex. By allowing individuals access only to mates from other species of the *albifrons* complex, hybridization can be induced in the laboratory and although the F_1 progeny are viable, the F_2 progeny show reduced viability. In nature, the frequency of hybrids is less than 1%, because of the operation of behavioural, reproductive isolating mechanisms between sympatric populations. Mating is preceded by courtship, and females refuse males presenting alien stimuli.

The formation of species from a geographically discontinuous, ancestral gene pool is called allopatric speciation. Geographical discontinuity can result from environmental forces that fragment the ancestral population, a process known as vicariance. For example, fossils show that when the Bering Strait opened during Late Pliocene time, some 3.5–4.0 million years ago, *Littorina squalida* extended its range from the northern Pacific, across the Arctic Ocean and into the Atlantic, probably following an eastward route (Reid 1996). Climatic cooling towards the end of Pliocene time then eliminated the periwinkle from the Arctic Ocean. Isolated from the ancestral population in the Pacific, the Atlantic population diverged to become *L. littorea*.

Geographical discontinuity also can result from the colonization of isolated habitats peripheral to the central population (peripheral isolates mechanism). Typically, colonizing populations are small, perhaps even just a single fertilized female, with the result that rapid genetic differentiation occurs through founder effects, inbreeding, drift or selection. A possible example of this concerns *L. horikawai*, confined to a relatively small area off Kyushu. The sister species, *L. sitkana*, is widespread in the northern Pacific, but the nearest population to Kyushu is some 1200 km distant, suggesting a rare colonizing event (Reid 1996).

When peripheral populations become larger relative to the central population, the situation increasingly resembles vicariance, the difference between vicariance and peripheral isolates mechanisms of allopatric speciation being only one of degree. As an arbitrary criterion for the peripheral isolates mechanism, it has been proposed that the ratio of distributional areas between sister species should exceed 20 : 1.

If, as is thought to be the case, allopatric divergence of gene pools is the principal mechanism of speciation, then the rate of speciation ought to be highest in the most heterogeneous environments, where opportunities for spatial isolation and divergent selection are greatest. These opportunities are less likely to arise in the water column than on the sea bed, and accordingly only about 2% of marine species are entirely pelagic. Environmental conditions within the interstices of sediments are more monotonous throughout the world than are those on exposed surfaces. Consequently, geographical variations are far more pronounced among epifaunas than among infaunas. Gastropods, for example, tend to have restricted geographical distributions and entirely different species live in the tropics than in temperate regions. Infaunal polychaetes on the other hand, tend to have wide geographical limits and many are cosmopolitan.

Not only the physical nature of the environment, but also the dispersability of organisms will influence the opportunity for speciation. Gene flow is more restricted among animals with direct development than among those with pelagic larvae, and more restricted among those with short-lived than among those with long-lived, pelagic larvae. For example, periwinkles in the *Littorina saxatilis* species complex brood their young or lay benthic egg masses from which crawling young emerge. Dispersal is therefore almost entirely by crawling and perhaps occasionally by rafting on floating objects, so that gene flow between distant populations must be very slight indeed. On European shores, at least three species and several other morphs of undetermined taxonomic status exist within the *saxatilis* complex. Two geographically outlying populations

Plate 23 Permanently anchored to hard substrates, solitary sea squirts such as *Polycarpa aurata* and sea squirt *Rhopalaea crassa*, draw water currents in and out through their siphons. Food particles are then strained through their unique pharyngeal filtering mechanism (Madang, Papua New Guinea). (L. Newman & A. Flowers.)

Plate 24 Like a bunch of grapes, these colonial ascidians, *Clavelina*, are joined with a common stolon at the base of the colony (Heron Island, southern Great Barrier Reef). (L. Newman & A. Flowers.)

Plate 25 Sea cucumber *Holothuria forskali* (Scarba, West Scotland). (C. Howson.)

Plate 26 Spaghetti worms, *Reteterebella queenslandica*, normally dwell in soft mud-sand tubes within coral rubble, however, they often slip out of their tubes when disturbed. Long stickly feeding tentacles are used to sweep food particles off lagoonal sandy bottoms at Heron Island (southern Great Barrier Reef). (L. Newman & A. Flowers.)

Plate 27 Red mangrove crab, *Sesarma meinerti*, an omnivorous grapsid foraging at low tide on a South African shore. (R.N. Hughes.)

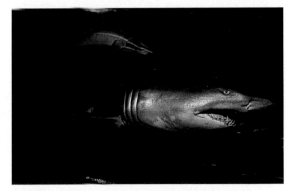

Plate 28 Looks can be deceiving. This inoffensive grey nurse shark, *Eugomphodustaurus*, looks menacing but is quite harmless unless provoked. Cruising different depths is easy as these sharks gulp air at the surface and hold it in their stomach to attain neutral buoyancy. Adjusting their position is accomplished by releasing bubbles from their cloaca. (L. Newman & A. Flowers.)

Plate 29 *Littorina saxatilis* from Langebaan lagoon, South Africa. (R.N. Hughes.)

Plate 30 *Littorina saxatilis* from Knysna lagoon, South Africa. (R.N. Hughes.)

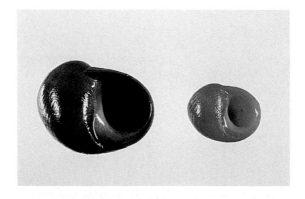

Plate 31 Sibling species *Littorina obtusata* (left) and *Littorina fabalis* (right). (photo courtesy of G. Williams.)

Plate 32 Gravel fauna in an area of seabed (Georges Bank, Northwest Atlantic) undisturbed by fishing gear. Note the prevalence of emergent fauna and associated fish and crustacean fauna (depth 84 m). (D. Blackwood and P. Valentine, US Geological Survey.)

Plate 33 A gravel seabed at the same depth as in Plate 32, but in an area of intensive scallop dredging with a distinct absence of emergent sessile fauna. (D. Blackwood and P. Valentine, US Geological Survey.)

occur in South Africa, probably having been transported as juveniles accidentally lodged on the bodies of palearctic waders during the winter migration. Because of evolutionary processes such as founder effect, drift and possibly selection influencing small, isolated breeding groups, these populations have become phenotypically (Plates 29 and 30, facing p. 208) and genetically distinct from northern populations (Reid 1996). As expected, electrophoresis of allozymes shows that there is considerable genetic variation even among local populations of the *saxatilis* complex, but that populations of *L. littorea*, which has planktotrophic larvae, are genetically more homogeneous even from both sides of the Atlantic Ocean.

Larval dispersal, however, does not necessarily result in genetic homogeneity. The oyster, *Crassostrea virginica*, has a larval phase lasting several weeks and, as expected, populations in estuaries along the coast of Florida are homogeneous for allozyme markers. On the other hand, genetic markers on mitochondrial and nuclear DNA reveal differences north and south of a point in mid-Florida. Reeb & Avise (1990) suggested that the allozymes are subject to balancing selection, making them similar among populations, whereas the DNA markers are selectively neutral, showing a genetic pattern that has not yet reached equilibrium, but reflects the effects of historical events. Northern and southern populations may have been isolated when large coastal estuaries drained during the Pleistocene lowering of sea-level. Genetic divergence resulting from this vicariant event still persists, even though subsequent rise in sea-level has restored panmixia.

The production of pelagic larvae, however, does not necessarily lead to thorough genetic mixing. Inside Long Island Sound is a population of *Mytilus edulis* genetically distinct from populations outside, as revealed by the electrophoresis of allozymes (Lassen & Turano 1978). The Sound has a typical estuarine hydrography, whereby less saline water flows out on top of a more saline, compensating bottom current. Larvae of resident mussels are carried out of the Sound in the surface current, but oceanic larvae fail to penetrate the Sound because when freshwater discharge is low during the summer breeding season, the weakened bottom current fails to flow over a sill near the mouth of the Sound. Genetic isolation of the Sound population is therefore imposed by hydrographical conditions.

A consequence of reduced dispersability and hence reduced interpopulation gene flow is that gene pools can become closely adjusted through natural selection to local conditions, but this is also likely to make them more vulnerable to environmental changes and prone to extinction. Enhanced dispersability and extensive interpopulation gene flow prevent localized differentiation of gene pools, so that species with long-lived, widely dispersing larvae must be broadly adapted to a range of environmental conditions and therefore less liable to extinction. It is sometimes possible to deduce the larval mode of life from the larval shell of fossilized gastropods. Scheltema (1978) thus deduced that the mean evolutionary longevity of species with teleplanic larvae (i.e. living several months to a year in the plankton) is about 19 million years, whereas that of species with direct development is only about 3 million years.

10.2.2 Sympatric speciation

At least in theory, geographical isolation is not always a prerequisite for speciation. When different selective forces act on different parts of a continuous population, they may, if strong enough to override the effect of gene flow, cause divergence within the gene pool, eventually leading to reproductive isolation. Perhaps the easiest kind of situation to envisage is where an ancestral gene pool diverges in response to the extreme conditions at opposite ends of a strong environmental gradient.

A difficulty with the concept of sympatric speciation is that very high selection pressures would be needed to counteract the homogenizing effect of gene flow. Attempts to cause sympatric, genetic divergence in laboratory cultures of *Drosophila* have met with varied success, and only rarely has convincing evidence of sympatric, genetic divergence and reproductive isolation been found in the field (e.g. copper-tolerant plants on copper-mine tailings). Genetic responses to strong selection pressures associated with environmental gradients are frequently found in nature, but gene flow usually persists, preventing speciation and resulting in gradual changes in gene frequency (clines) along the gradients. For example, Schopf & Gooch (1971)

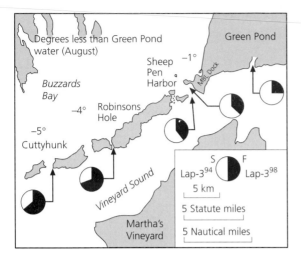

Fig. 10.1 Cline in frequencies of 'slow' (S) and 'fast' (F) alleles at a leucine-amino-peptidase locus of *Mytilus edulis* associated with a temperature gradient. 'Slow' and 'fast' refer to the relative speeds at which the proteins marking the different alleles move during electrophoresis. (After Schopf & Gooch 1971.)

found that the frequencies of two alleles at a leucine-amino-peptidase (Lap) locus in the bryozoan *Schizoporella unicornis* changed along the coastline of Cape Cod, paralleling a gradient in summer water temperature (Fig. 10.1).

Williams (1977) interpreted the data differently, proposing that there was probably sufficient mixing of larvae by currents to prevent any local genetic responses within the coastal population of *S. unicornis*. Larvae with a similar range of genetic qualities would be produced all along the coast each year. The clinal changes in gene frequencies could then only be caused by recurrent environmental elimination of locally inappropriate genotypes.

Without knowing the details of evolutionary history it is difficult to judge whether closely related, coexisting species have evolved sympatrically or have evolved allopatrically and subsequently become sympatric by range extension. However, the rapidly increasing number of sympatric 'sibling' or 'cryptic' species that are being discovered with the aid of electrophoresis of allozymes and other detailed taxonomic methods suggest that allopatric speciation cannot account for all cases. Sibling or cryptic species are groups of species that are morphologically difficult or impossible to distinguish

within each group, yet do not interbreed. Examples are bryozoans of the *Alcyonidium gelatinosum* species complex, discovered from differences in allozyme markers among samples taken from the Bristol Channel (Thorpe *et al.* 1978), two forms of the subtidal nudibranch *Doto coronata*, which, in the Irish Sea, feed on two different hydroids and also differ in allozyme profiles (Morrow *et al.* 1992), and the flat periwinkles *Littorina obtusata* and *L. fabalis* (formerly *mariae*) (Plate 31, facing p. 208), the former living mid-tidally on *Ascophyllum nodosum* or *Fucus vesiculosus* and the latter living on *F. serratus* or kelp at lower levels on British shores. Sibling speciation in marine invertebrates has been reviewed by Knowlton (1993). (For a general discussion of speciation in marine habitats, see Palumbi (1994).)

10.2.3 Environmental harshness and instability

Whether allopatric or sympatric processes are involved in speciation, genetic diversity will be enhanced by environmental heterogeneity. But other factors can also be important. Harsh environments (e.g. extremes of temperature or salinity) and fluctuating environments will cause strong selection for physiological robustness or phenotypic plasticity rather than genetic variability. Genetic variability within several species of crustacean on European shores decreases from marine, brackish-water to rock-pool habitats as the environment becomes harsher and more variable (Table 10.1).

Environmental variability hinders the fine tuning of genotypes to local conditions and is of particular significance with regard to fluctuations in food supply. Stable food resources allow competition to generate selection for trophic specialization. Fluctuating food resources render trophic specialization less feasible because animals will need to become more opportunistic in times of food shortage. Among species of krill, levels of genetic variation are correlated with the trophic stability of the environment. Heterozygosity increases in the order *Euphausia superba* from highly seasonal Antarctic waters, *E. mucronata* from Chilean waters with moderately irregular upwellings, and *E. distinguenda* from the trophically stable waters of the equatorial eastern Pacific. *E. superba* is apparently represented by relatively homogeneous generalist genotypes and *E. distinguenda* by more heterogeneous specialist genotypes.

Table 10.1 Mean genetic variation in species in three different environments. (After Battaglia *et al.* 1978.)

Variable	Environment and species		
	Marine *Tisbe clodiensis* *Tisbe holothuriae** *Tisbe biminiensis*	Brackish water *Tisbe holothuriae*† *Gammarus insensibilis*	Rock pools *Tigiriopus brevicornis* *Tigiriopus fulvus*
Mean number of alleles per locus	1.982	1.788	1.297
Mean percentage of loci polymorphic (0.99)	56.84	42.03	18.27
Mean percentage of loci polymorphic (0.95)	46.78	38.89	16.13

*Banyuls-sur-Mer.
†Sigean.

10.3 Biogeography

10.3.1 Geographical range

In addition to historical events (examples in Section 10.2.1), factors determining geographical range include dispersal ability, physiological tolerances and habitat requirements.

In some animal groups such as cone shells, geographical range is correlated with developmental mode, those species with planktotrophic larvae having the greatest ranges (Kohn & Perron 1993). In other groups such as species of *Littorina*, there is no such correlation (Reid 1996). Factors contributing to such lack of correlation include colonizing ability, discussed in Section 9.3.2.2, and the possibility of long-distance dispersal by post-larval or even adult stages. Small gastropods and bivalves with direct development can produce mucous threads that are caught by currents, so effecting dispersal (Martel & Chia 1991). A wide range of animals are dispersed by rafting on flotsam such as dislodged algal fronds, pieces of wood, pumice and, in recent times, on plastic and glass containers. Pumice probably has been an important rafting medium throughout evolutionary time, as it is continually produced in all oceans and is abundantly recorded in marine rock formations back to Precambrian time (Jokiel 1990). Pumice stranded on tropical beaches commonly bears coral colonies that have established themselves on the 'raft' as planula larvae (Section 6.3.1). Animals also may 'board' a raft by clinging to it, e.g. amphipods, or swimming within its shelter, e.g. reef fish (Myers 1994).

The dispersal potential of rafting is very great. Tree trunks and pumice washed ashore at Cocos Atoll in the early 1900s had been drifting in the Pacific for some 20 years after the eruption of Krakatoa, 1000 km to the northeast. Over geological time, drifting coral skeletons probably have been an important means of rafting. A skeleton of *Symphyllia agaricia*, afloat on the Great Barrier Reef, was found to be covered in coralline and filamentous algae and also to support goose barnacles, decapod crustaceans, oysters, gastropods, foraminiferans and bryozoans. The sizes of these organisms indicated that the skeleton had been drifting for several months, deriving buoyancy from gas trapped in the septal chambers (De Vantlier 1992).

Physiological limitation of range by temperature is particularly pertinent to latitudinal distribution (also discussed in Section 10.3.6) and its effect sometimes is detectable in the fossil record. When the climate was cooler during glacial periods in Pleistocene time, *Littorina littorea* occurred on the coast of Morocco, 850 km south of its present limit. During warmer interglacial periods the range of *L. littorea* extended to Siberia, over 2000 km north of its present limit (Reid 1996). The general limitation of kelp forests to cooler waters and of coral reefs to warmer seas is discussed in Sections 5.2.1 and 6.1.

Habitat requirements may limit geographical range; for example, on the western Atlantic coast, shores south of Connecticut are almost entirely sedimentary and the southward distribution of rocky shore organisms is limited to isolated rocky outcrops, harbour walls and jetties. Similarly, reef corals are largely absent from the sandy coastline of West Africa (Section 6.2).

10.3.2 Latitudinal gradients of species diversity

The latitudinal decrease in species diversity from equatorial to polar regions was considered briefly in Chapter 3, but can now be considered in further detail in the context of the determinants of speciation.

10.3.2.1 Area effect and environmental heterogeneity

The diversity of habitats and opportunities for the genetic isolation of populations on continental shelves increases as the total area of shelf increases.

Therefore, the amount of speciation among shelf faunas ought to be higher in parts of the world with greater areas of shelf, assuming, as seems likely, that there is no compensating increase in extinction rate. The species diversity of bivalves and bryozoans is indeed correlated with continental shelf area, which decreases away from the equator, but there is much residual variation (Fig. 10.2). Schopf *et al.* (1978) suggested that the correlation between species diversity and shelf area would improve if total shelf area could be replaced by a measure of habitable shelf area. The proportion of shelf suitable for bivalves and bryozoans may vary independently of latitude. For example, the southeastern coast of North America is sandy and more suitable for infaunal bivalves than the southwestern coast, which has coarser sediments and more rocky areas favouring epibenthic animals such as bryozoans.

10.3.2.2 Trophic stability

Prosobranch gastropods are one of the most diverse groups of marine macroinvertebrates and they show a strong latitudinal decrease in diversity from equatorial to polar regions. This trend may partly

Fig. 10.2 (a) Latitudinal changes in species diversity of bryozoans and bivalves. (b) Latitudinal changes in a continental shelf area. (After Schopf *et al.* 1978.)

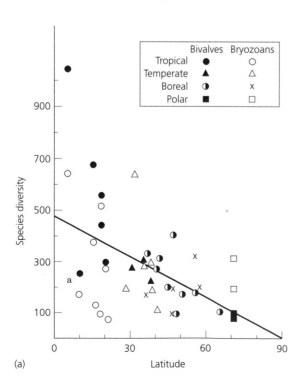

(a)

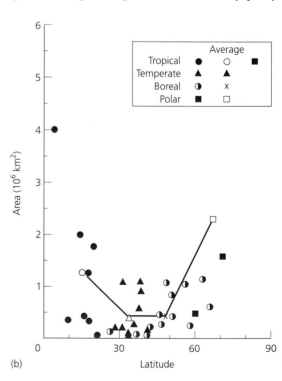

(b)

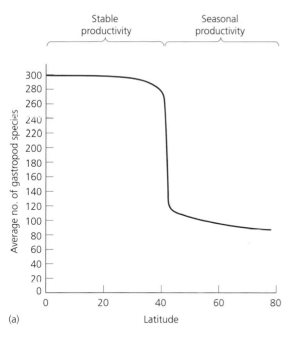
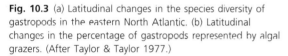
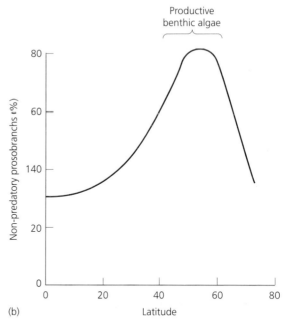

Fig. 10.3 (a) Latitudinal changes in the species diversity of gastropods in the eastern North Atlantic. (b) Latitudinal changes in the percentage of gastropods represented by algal grazers. (After Taylor & Taylor 1977.)

be due to latitudinal changes in the area of habitat, as with bivalves and bryozoans, but is also due to the trophic stability of the environment (Taylor & Taylor 1977). A sharp drop in the diversity of eastern Atlantic gastropods occurs round about latitude 40°N, coinciding with a change from a tropical oceanic regime in which primary production is continuous throughout the year, to a temperate oceanic regime in which primary production is seasonally pulsed. The trophic stability of tropical habitats allows specialized feeding behaviours to evolve, whereas at higher latitudes trophic instability has the opposite effect. Tropical families of predatory gastropods have narrower diets than families from higher latitudes. For example, the Cassidae feed on echinoids, the Tonnidae on holothurians, the Mitridae on sipunculan worms, most Conidae, Vasidae and Bursidae on polychaetes, and the Harpidae on decapod crustaceans. By contrast, the Buccinidae, which are abundant at high latitudes, are opportunistic feeders, whose diets include polychaetes, sipunculans, crustaceans, molluscs, echin-

oids and carrion. A given supply of food resource can either be partitioned among many species with narrower diets and smaller populations or among fewer species with wider diets and large populations, so that more species exist in low latitudes where diets are relatively narrower than at higher latitudes where diets are wider.

The Turridae are an exception which proves the rule of trophic stability. These predators are abundant at high latitudes yet appear to be specialist feeders on polychaetes. Their prey are, however, deposit feeding worms whose population dynamics are little affected by the seasonally pulsed primary production.

Quality of the food resource changes throughout the world and this influences faunal species composition rather than species diversity. The proportion of non-predatory gastropods rises between latitudes 40 and 60°N (Fig. 10.3). This is due to a surge in the number of algal-grazing species in response to the high benthic algal productivity of temperate regions (Chapter 5).

10.3.2.3 Diversity associated with coral reefs

Because of the physiology of the cnidarian–dinoflagellate symbiosis, reef-building corals are confined to shallow, clear waters within the 20°C winter isotherms (Chapter 6). The intricate three-dimensional architecture of corals and associated reef structures creates multitudinous microhabitats that support very rich faunas, an effect that is reinforced by the extremely high, stable productivities of coral-reef ecosystems. Coral reefs, therefore, account for a significant proportion of the high overall species diversity of tropical shelf faunas.

10.3.3 Oceanic differences in species diversity

The highest diversity of shallow benthic species in the world is in the tropical Indo-Pacific, followed by the tropical Pacific coast of America. The lowest tropical benthic species diversities occur on both sides of the Atlantic. Within the Indo-Pacific the highest diversity is centred on the Indo-Malayan region (Fig. 10.4), where most of the diversity-promoting factors previously described occur together. First, the equatorial location is associated with stable, non-seasonal, primary production (Section 10.3.2.2). This stability is enhanced by the numerous small islands and continental land masses surrounded by a large ocean, a geographical configuration that produces a very even maritime climate and stable water column. Secondly, the continental shelf is dissected by numerous deep basins between the islands and continents, providing good opportunities for the genetic isolation of populations (Section 10.3.2.1).

Fig. 10.4 Generic diversity of hermatypic corals in the Indo-Pacific. The centre of diversity is in the Indo-Malayan region. (After Stehli & Wells 1971.)

Thirdly, the coral reefs provide a spatially heterogeneous environment and stable high primary productivity (Section 10.3.2.3).

The reduced species diversity of shallow, benthic assemblages outside the Indo-Pacific region is partly correlated with the increased continentality of the climate. As a result of land having a lower specific heat than water, temperature fluctuations increase as the ratio of continuous land surface to sea surface increases. This ratio is least among the archipelagos and peninsulas of the Indo-Pacific, greater along the west American shelf where large continents border a large ocean, and greatest in the Atlantic where large continents face a small ocean.

Other theories attempting to explain the negative gradient in species richness radiating from the Indo-Malayan region are based on the geographical origin and dispersal of newly evolved species. The centre-of-origin theory proposes that speciation occurred principally within the Indo-Malayan region, perhaps on account of factors such as those described above, and that newly evolved species diffused outwards from this centre over geological time. The centre-of-accumulation theory proposes that speciation occurred principally in peripherally isolated archipelagos, from where species were transported to the central region by prevailing oceanic currents. Over geological time, this process would cause species to accumulate in the central, Indo-Malayan region. The provincial overlap theory proposes that the Indo-west Pacific is composed of several biogeographical provinces (including Pacific Ocean, Indian Ocean and Indonesian), each with different geological and evolutionary histories. These

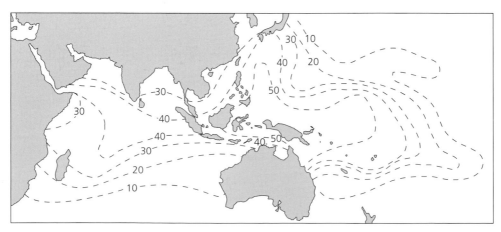

provinces overlap in the Indo-Malayan region, creating an area of exceptionally high species diversity.

The first two of these three theories can be tested, with the use of neutral genetic markers such as those on mtDNA, by examining patterns of allelic distribution in the central and peripheral regions. The centre-of-origin model predicts that movement to the periphery of the Indo-West Pacific is uncommon, each event involving relatively few colonists. These initially will carry with them a small, probably random, sample of alleles from the large, central population. If the peripheral region is reached relatively quickly, the randomly sampled alleles will be preserved. If, on the other hand, the periphery is reached slowly via a series of stepping-stone islands, alleles in the peripheral populations are likely to be highly modified compared with those in the ancestral gene pool. The centre-of-accumulation hypothesis predicts that, because new species originate at the periphery, the most ancient alleles should exist in peripheral populations, compared with those in the centre. Preliminary data for the sea-urchin, *Echinometra* sp. nov. A, show that an

outlying population at Tahiti has a cluster of highly derived mtDNA alleles. At Fiji, one step closer to the Indo-Malayan region, some of the alleles are evolutionarily basal to the Tahitian alleles, supporting the stepping-stone version of the centre-of-origin hypothesis (Palumbi 1996).

10.3.1 Geographical barriers

Physical features of the oceans and continents, such as unfavourable currents, large stretches of cold, deep water or shallow, warm water, and land barriers, may impede dispersal and colonization, with important consequences to the species diversity of shelf faunas. However, such barriers to range extension seldom are absolute over a geological time scale because very occasionally they may be crossed via some fortuitous event, such as rafting. The crossing of geographical barriers by species is termed jump dispersal.

The East Pacific Barrier (Fig. 10.5) is a wide stretch of deep water between Polynesia and America that has greatly restricted the spread of tropical, Indo-West Pacific, shallow-water species to the coasts of tropical America. Only about 6% of tropical, Indo-West Pacific, shore fish have managed to cross the East Pacific Barrier, as has a similarly small proportion of shallow-water invertebrates. Consequently,

Fig. 10.5 Zoogeographical barriers separating the tropical shelf regions. The arrows indicate the direction and approximate relative amount of colonization that has occurred. (After Briggs 1974.)

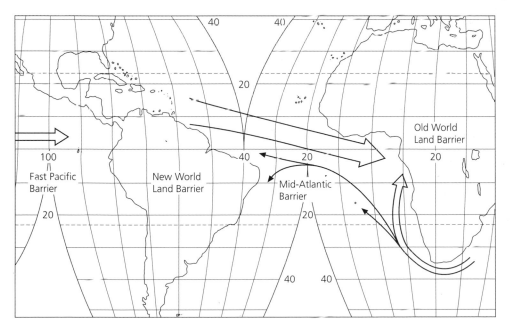

the tropical East Pacific shelf fauna is depauperated by the existence of the East Pacific Barrier. Because of the prevailing currents, no species originating in the eastern Pacific have managed to extend westwards across the East Pacific Barrier.

Other barriers with similar zoogeographical effects (Fig. 10.5) include the New World Land Barrier, which prevents movement of tropical marine species between the eastern Pacific and western Atlantic, the Mid-Atlantic Barrier, a broad stretch of deep water that separates tropical western and eastern shelf faunas, and the Old World Land Barrier, which separates the tropical eastern Atlantic and Mediterranean from the Red Sea and tropical Indian Ocean.

Warm, shallow, tropical shelf waters isolate the shallow, benthic faunas of northern and southern temperate regions. A number of species, however, such as the barnacle *Semibalanus balanoides* and the colonial tunicate *Botryllus schlosseri* (see Fig. 9.13a), are able to extend across the equator by living in deeper, cooler water, a phenomenon known as equatorial submergence.

Because the relative positions of land masses and associated shelf areas have changed drastically over geological time as a result of plate tectonics (Section 10.3.7), distances separating modern biogeographical areas may not resemble situations in the past. The occurrence of closely related taxa on either side of an apparent geographical barrier may have originated not by jump dispersal but by tectonic events, including the separation of land masses, or deep submergence of intermediate habitat such as islands, reefs and guyots, by subsidence of the ocean floor. Direct evidence of population splitting by tectonic processes is limited by the destruction of material at the subduction zones. Close to the Pacific subduction zones, for example, the ocean floor is only some 150 million years old, so the geological record does not extend beyond Jurassic times. Tectonic plates, on the other hand, have a much longer geological history, and so biogeographical inferences can be made by comparing taxonomic distributions among plates.

Many species, for example of fish, molluscs and crustaceans, show plate endemism, being confined to a particular plate even though geographical barriers to dispersal may appear weak. The circumtropical amphipod genus *Globosolembos* has nine species distributed among several tectonic plates. By comparing this geographical distribution with phylogenetic relationships among the species, it is possible to make inferences about how plate tectonics has influenced their biogeography. On a phylogenetic cladogram (Fig. 10.6) the first dichotomy separates four species on the African–Indian Ocean and India–Australian plates from five species on the Pacific, North American and African–Atlantic plates. This dichotomy corresponds to the 'Andesite' line, along which the rocks of the Pacific Basin are compositionally distinct from those of land masses to the west. The geological significance of this apparent barrier remains conjectural, but it is possible that species absent from the Pacific plate never have been present on it. The second dichotomy separates an Atlantic species from a group of Pacific plate and western Atlantic species, corresponding to the opening of the Atlantic in early Jurassic time some 165–180 million years ago. The third dichotomy separates two South Pacific species from a Hawaiian and a western Atlantic species, reflecting changing geography resulting from plate movements and volcanism (Myers 1994).

On the one hand, major differences in regional biotas may be interpretable in terms of plate tectonics. For example, the suturing of Africa with Europe in late Cretaceous time formed a physical barrier to dispersal of stenothermic, tropical shallow-water taxa across the Tethys Sea (see Fig. 10.8), resulting in great taxonomic differences between the floras and faunas of the Caribbean Sea and Indian Ocean. On the other hand, similar associations between corals, seagrasses and mangroves, occurring in widespread but disjunct areas of tropical ocean (Section 10.3.7), may be derived from an association that once extended throughout the Tethys Sea. Continental movements during Eocene–Miocene time led to extensive extinctions, notably in the Caribbean and eastern Pacific, so fragmenting the original distribution.

In summary, many large-scale biogeographical patterns can be understood in terms of Tectonic, Eustatic, Climatic and Oceanographic processes (TECO, Rosen 1984). TECO processes fall into several major categories: (1) splitting marine biotas by tectonic rifting, orogenic changes in direction of large rivers discharging into shelf areas, sea-level rise, and changes in ocean currents; (2) combining species pools through tectonic convergence and sea-level lowstands; (3) shifting, contracting and

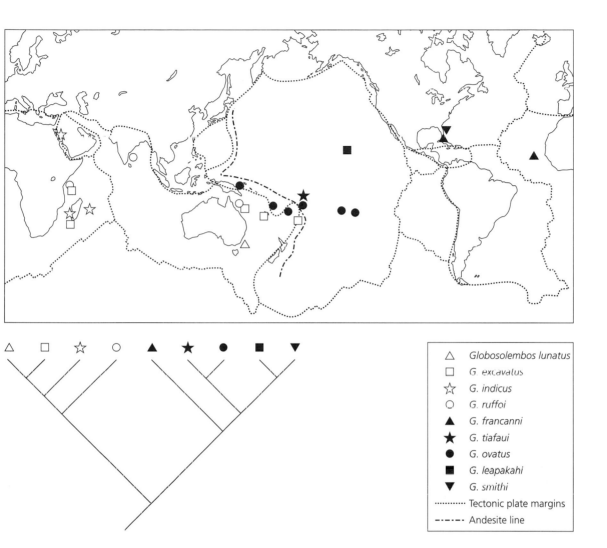

Fig. 10.6 Geographic distribution of amphipod crustaceans of the genus *Globosolembos*, with cladogram of phyogenetic relationships. (After Myers 1994.)

expanding species ranges through climate change and variation in continental shelf area; (4) controlling immigration rates through changes in distances between islands and in current regimes; (5) indirectly controlling extinction through climate change.

10.3.5 Deep-sea benthos

Deep-sea, benthic, faunal diversity is extraordinarily high (Chapter 7). Although the physical conditions are extreme compared with those of more familiar shallow habitats, they are stable. Relatively large particles, such as dead vertebrates, are quickly consumed by scavengers, as demonstrated by the rapid exploitation of bait lowered to the sea bed. Scavengers, mainly fish, disperse the organic matter as faeces, thereby evening out the spatial distribution of food for deposit feeders.

The predictability of the deep-sea environment would therefore seem conducive to specialization. However, the problem remains of explaining how competitive exclusions are prevented in such an apparently homogeneous habitat and how gene pools diverge in the apparent absence of barriers to gene flow. Dayton & Hessler (1972) proposed that generalist 'croppers' keep prey population densities sufficiently low to prevent competitive exclusions (Chapter 7). But would not the evolutionary arms race between predators and prey tend to offset this in such a stable environment? Hessler & Sanders (1967) suggested that the long geological history of stability in the deep sea has allowed the evolution of numerous species with narrow niches, but they did not indicate how the resources are partitioned or how genetic divergence and reproductive isolation take place. It has been suggested that many deep-sea species have not evolved from common ancestors *in situ* but are derived from ancestral immigrants from shallow waters. Comparative morphological studies of deep-sea isopods, however, indicate a long history of *in situ* evolution. The explanation of speciation and faunal diversity in the deep sea therefore remains an important problem, for the deep-sea environment accounts for a very large proportion of the biosphere.

10.3.6 Latitudinal zonation

Just as species are vertically zoned on shores in response to the terrestrial–marine environmental gradient (Section 5.1.1), so they also tend to be horizontally zoned on a geographical scale in response to a complex gradient associated with latitudinal changes in climate. The latitudinal climatic gradient is particularly pronounced on the east coast of North America. The south-flowing, cold Labrador Current and the north-flowing, warm Florida Current cause a very rapid change from arctic to tropical temperature regimes within 20° of latitude (Fig. 10.7). Not surprisingly, the latitudinal zonation of many eastern North American intertidal species is correlated with critical boundaries where temperatures become too extreme for reproduction or adult survival.

Along the western coast of North America the climatic gradient is much more gradual and factors other than temperature become important in determining latitudinal, zonal boundaries. Behrens-Yamada (1977) identified factors limiting the latitudinal distribution of two species of periwinkle by transplanting the snails beyond their normal ranges. The southern limit of *Littorina sitkana* and the northern limit of *L. planaxis* occur at about latitude 43°N. *L. planaxis* transplanted further north survived and reproduced for 4 years, but the potential northerly range of this species is curtailed by south-flowing currents that prevent the northward spread of the pelagic larvae. *L. sitkana* transplanted south of its natural range succumbed both to desiccation, because of a lack of damp microhabitats, and to predation by crabs, which are more abundant in the high intertidal zone to the south.

10.3.7 Plate tectonics and provincialization

Waves of extinction and taxonomic diversification are prominent features in the fossil record. Sometimes extinction coincided with a reduction in taxonomic diversity (diversity-dependent extinction), suggesting that the carrying capacity of the biosphere had somehow declined, but at other times waves of extinction did not affect taxonomic diversity (diversity-independent extinction), evidently because lost lineages were replaced quickly by new ones. The most likely cause of diversity-independent extinction is a change in climate that causes unadaptable species to die out but does not affect the carrying capacity of the biosphere, so that taxa soon evolve to utilize vacated niches.

A density-dependent wave of extinction occurred in Mid- to Late Permian time (just over 200 million years ago), reducing the diversity of most taxa to their lowest levels since Cambrian time, when fossils first became numerous. The cause of the Permian crash in taxonomic diversity remain in debate, but the surge of rediversification during the Mesozoic–Cenozoic period (180–120 million years ago) is much better understood.

Before the Mesozoic period the continents were grouped together, surrounded by the huge Pacific Ocean and partly divided by the Tethys Sea (Fig. 10.8). During the Mesozoic period the continents began to separate owing to sea-floor spreading and this must have altered the climate considerably (for an account of plate tectonics and its biogeographical implications, see Valentine (1973) and Gray & Boucot (1979)). The huge continuous area of the pre-Mesozoic supercontinent would have intensified temperature fluctuations, producing a more continental world climate (Section 10.3.3). Separation of the continents had several important effects. First, the increased coastline per unit area of land ameliorated the world climate, making continental shelf areas more stable and therefore more conducive to the evolution of species with specialized niches (Section 10.3.2.2). Secondly, separation of the continents promoted faunistic and floristic isolation,

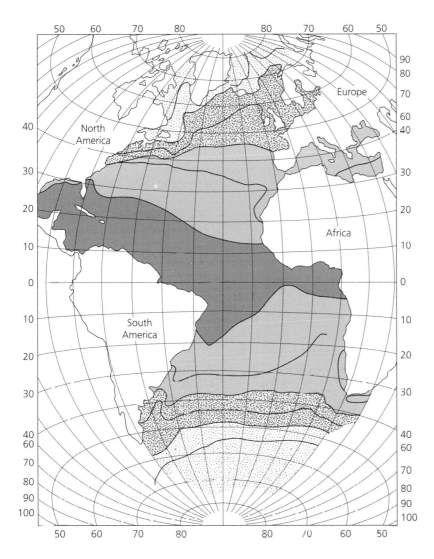

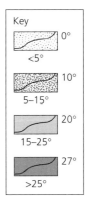

Fig. 10.7 Mean annual temperature isotherms of the surface water of the Atlantic, showing the compressed temperature gradient on the east coast of North America. (After Ekman 1967.)

increasing the opportunities for allopatric evolution. Consequently, biotas of geographically separate shelves developed different taxonomic compositions, or in biogeographical terminology, developed provincial differences. During early Jurassic time there seems to have been no provinciality at all, as ammonite and bivalve taxa were cosmopolitan. By Middle and Late Jurassic time there were two, well-defined faunal realms, and today over 30 marine provinces are recognized. Thirdly, the changing positions

of the continents altered the pattern of oceanic circulation so as to intensify the equatorial–polar temperature gradient. Owing to the isolation of the Arctic Ocean by continental topography and of the Antarctic seas by the Circumpolar Current, polar seas at present are as cold as they can ever be. The pronounced, latitudinal, temperature gradient led to latitudinal zonation (Section 10.3.6) and increased provinciality. The glacial cycle, which may have started in association with changing continental configuration and oceanic circulation, has not caused massive extinctions. This is because the predominantly north–south continental orientation allows species to move gradually with the latitudinal advance and retreat of cooler conditions.

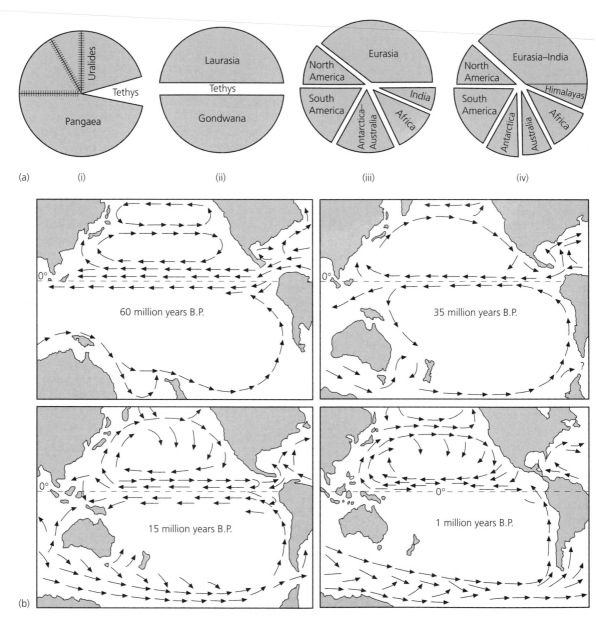

Fig. 10.8 (a) Inferred continental and oceanic configurations during the last 200–300 million years. (i) In Permian time, Pangaea is partly intersected by the Tethys Sea. (ii) In Early Mesozoic time, Laurasia and Gondwana are separated by the Tethys Sea. (iii) At the end of the Cretaceous period, Gondwana is highly fragmented. (iv) At the present, India is joined to Eurasia. (After Valentine 1973.) (b) Changing configuration of continents in the Pacific region during the last 60 million years and the associated changes in surface currents. (After Van Andel 1979.)

10.3.8 The effects of man

Current theories of plate tectonics and evolutionary ecology could not be expected to predict the small but significant effects that man himself has had on contemporary biogeography. Man's technology has resulted in the passive transportation of organisms throughout the world and even the opening up of new dispersal routes between previously isolated seas.

Organisms are dispersed attached to the hulls of ships or to commodities such as oysters. Most hitch-hikers will be killed in transit or will fail to establish themselves in foreign localities because conditions are unsuitable for their growth or reproduction, or because there are too few simultaneous arrivals to form a viable breeding group (propagule). Opportunistic (*r*-selected) species (Section 9.1), with their high potential population growth rates and ability to flourish from small initial densities, are the most likely organisms to spread by hitch-hiking. Examples include the Australasian barnacle *Elminius modestus*, which was first noticed in Chichester harbour, England, in 1945 and now thrives in moderately sheltered localities throughout Britain. Presumably *Elminius* was given the opportunity to breed and recruit young while ships were laid up during the Second World War. In 1890 a shipment of American oysters brought the slipper limpet *Crepidula fornicata* to Essex, since when the stowaway has become firmly established in the oyster beds of southwest Britain.

Successful immigrants can make an impressive ecological impact. In the early nineteenth century the cord grass *Spartina alterniflora* was introduced from America to Southampton Water, where it hybridized with the European species *S. maritima*, forming a sterile hybrid *S. townsendii*. This hybrid subsequently became polyploid, forming the sexual species *Spartina anglica*, which is so vigorous that it has spread throughout much of the British Isles, forming dense stands where once there were open mud flats. Animals such as *Scrobicularia plana* and *Macoma balthica* disappear as *Spartina anglica* monopolizes the substratum, but other species such as *Littorina rudis* and *Anurida maritima* move in.

Ecological impact of the shore (or green) crab, *Carcinus maenas*, is being studied in California. Although *C. maenas* has been transported about the world by human activities for over two centuries, it has only recently reached the North Pacific, first having been recorded in San Francisco Bay in 1989. By 1995, *C. maenas* had extended its range northward to Bodega Harbor, apparently the result of larval recruitment by a single cohort carried by the prevailing northward surface currents.

Enclosure experiments have shown that *C. maenas* significantly reduces densities of bivalves and infaunal crustaceans in sheltered bays, and it is predicted that the alien will have great impact on the benthic ecology of the region (Grosholz & Ruiz 1995).

As well as carrying organisms about with him, man has assisted their dispersal by connecting the Red Sea to the Mediterranean. Organisms began to disperse along the Suez Canal immediately after its opening in 1869. Currents flow from the Red Sea to the Mediterranean for 10 months of the year so that most invertebrate migration has been in this direction. At least 140 animal species, including fish, decapod crustaceans, molluscs, polychaetes, ascidians and sponges have reached the Mediterranean via the Suez Canal.

The Panama Canal, connecting the Pacific to the Atlantic, has a salinity of less than 1‰ and therefore can only be negotiated by a few euryhaline species such as the blue crab *Callinectes sapidus*. The proposal for a sea-level canal, if implemented, would probably have important biogeographical consequences. The eastern and western Central American shelf faunas contain approximately 8000 and 6000 species, respectively. Bringing these species into contact could cause the extinction of many of them.

Finally, man has severely depleted many of the populations he exploits. Stellar's sea-cow (Section 5.2.4.2) and the great auk were exterminated, and the large whales seem dangerously close to a similar fate. Ecological consequences of such overindulgences are not yet predictable, but could be far-reaching. Nevertheless, accelerating alteration of habitat and transportation of species around the world are beginning to have far-reaching effects. Regionally distinct biotas are threatened by the extinction of endemics and their replacement by opportunistic species that are good colonizers.

Human impact on the marine biota is, at present, miniscule compared with biogeographical events that have occurred in the geological past.

Useful texts on speciation, biogeography, and plate tectonics in the marine context include those by Ekman (1967), Valentine (1973), Briggs (1974), Gray & Boucot (1979), Myers (1994) and Palumbi (1994).

By now the broad outlines of interrelationships between different marine organisms and different marine habitats should be fairly clear. It is the purpose of this chapter to quantify some of the fluxes between the different compartments in the overall system and to draw together the various threads introduced in the earlier chapters. First, however, it may be helpful to summarize the broad outlines referred to above.

11.1 The general nature of interactions within the marine ecosystem

Excluding the centres of primary chemosynthesis on the sea bed, the initial production of organic compounds in the ocean is confined to the shallow, photic zone. In this surface water layer and in those littoral areas bathed by it, carbon dioxide is fixed by photosynthesis and nutrients are incorporated into living tissues. The dominant organisms achieving this primary production are the minute protistan phytoplankton together with, in the littoral zone, the related benthic species, some of which, in the form of seaweeds (or macrophytes), attain macroscopic size. Also in the littoral zone are semi-aquatic stands of plants of terrestrial affinity, which are highly productive but which are relatively little grazed by consuming species. Neither are the larger seaweeds consumed directly to any appreciable extent, except when at the sporeling stage, although their finer, filamentous epiphytes can be consumed as efficiently as the free-living protists. The uni- or non-cellular and smaller multicellular photosynthetic protists therefore provide a direct source of food for herbivorous species whereas the production of the coarser, macrophytic seaweeds and of the semi-terrestrial tracheophytes enters marine food webs largely as detritus. Together their biomass has been estimated to total some 4×10^9 t (dry weight) of organic

matter, and this produces annually some 55×10^9 t (dry weight), or about 14 times the standing crop (Table 11.1). (Compare values on the continents of 1837×10^9 t and 115×10^9 t, respectively.)

Most of this photosynthetic production is consumed in the same surface waters: the phytoplankton by herbivorous zooplankton and, in shallow waters, by the suspension-feeding benthos, and the littoral benthic algae by grazing herbivores (gastropod molluscs, etc.). A proportion of the gross primary production is also leaked into the surrounding water in the form of dissolved organic matter. Estimates of the exudation of dissolved organic carbon (DOC) compounds by living phytoplankton cells range from 7% to 62% of the photosynthetically fixed carbon. Estimates of the leaching of DOC by macroalgae are variable and lie between 1% and 40% of the fixed carbon (Valiela 1984). This dissolved organic material is consumed by bacteriaplankton (< 2 µm in size), which are in turn consumed by heterotrophic microflagellates (in the nanoplankton size range 2–20 µm), which appear to be the principal, if not the only, consumers of free planktonic bacteria in aquatic ecosystems (Azam *et al.* 1983; Ducklow 1983). Microzooplankton (in the 20–200 µm size range, including taxa such as tintinnid ciliates, rotifers, sarcodinians, copepod nauplii stages) are known to feed most effectively on particles in the nanoplankton size range (Sheldon *et al.* 1986), and

Table 11.1 World marine photosynthetic biomass and productivity.

	Biomass (10^9 t)	Productivity (10^9 t year^{-1})
Oceanic	1	41.5
Neritic and upwelling	0.3	9.8
Littoral	2.6	3.7

whose diet then probably consists of heterotrophic microflagellates, small phytoplankton and suspended detritus particles in that size range. Mesozooplankton (> 200 µm in size), such as copepods, feed to a large degree on microzooplankton, and on phytoplankton and suspended detritus, thus completing the shunt of energy via the so-called 'microbial loop' in aquatic ecosystems. This 'loop' is a parallel food chain to the conventional 'grazing' chain of phytoplankton–zooplankton–fish. The excretory products, mainly faecal pellets, of the water column organisms contain much material of potential food value. These faeces, together with detrital substances derived from the fringing seaweeds (and in some areas from unconsumed phytoplankton), collectively known as particulate organic matter (POM), form the substrate for bacterial growth and, ultimately, for bacterial-dependent populations of heterotrophic protists and small metazoans. As these aggregations sink through the water, they are consumed by members of the zooplankton, which digest the living microbiota and microfauna, until eventually the recycled aggregations reach the sea bed. There they support a deposit- and suspension-feeding benthos which, in regions below the photic zone, is ultimately similarly dependent on bacterial production. In some areas of sea bed, genuinely primarily producing bacteria can fix carbon dioxide by chemosynthesis and these, too, will pass organic matter into the benthic food web, as will those bacteria that chemosynthetically can regenerate organic matter using the end products of previous photosynthetic activity. Some of these various chemosynthetic bacteria are symbiotic in gutless animals and supply all of their host's nutritional requirements (as indeed can symbiotic photosynthetic algae in their shallow-water flatworm, cnidarian, mollusc or sea-squirt partners).

Herbivorous and detritivorous planktonic organisms are consumed by the carnivorous zooplankton, and equivalent benthic carnivores take the deposit and suspension feeders. All these planktonic and benthic consumers then form the food of the nekton, which occupy the uppermost portions of the marine food web over all but the deepest regions of the abyssal trenches. The large majority of the nekton are carnivores, although a limited number of species can consume the primary producers and even fewer appear able to take the detrital–bacterial aggregates.

As a consequence of primary production being a surface-water phenomenon and because shallow-water zones are the most productive regions of the sea per unit area, secondary production declines with both distance away from the coast and depth in the sea. Where deep waters lie immediately adjacent to the coast, however, detrital bonanzas may occur at great depths, and debris from coastal tracheophytes is found over large areas of the abyssal plain; nevertheless, the abyssal benthos is the slowest growing, longest lived and least productive of all marine systems. Relative food shortage places a selective premium on efficiency of utilization, and hence passage of materials through the food web is most efficient in the open ocean, on coral reefs and probably in the deep sea, and least efficient in areas of upwelling, in estuaries and in shallow waters generally.

Total animal biomass in the sea has been estimated at 1000×10^6 t (dry weight), a comparable value to that occurring on the continents. However, this biomass probably produces more than three times more organic matter per year than does the terrestrial/fresh-water fauna; the marine biomass produces 3025×10^6 t (dry weight) as compared with the continental production of 900×10^6 t (dry weight). Thus, although marine primary production is much lower than that on land, marine secondary production is much greater (primary to secondary production ratios of $1 : 0.008$ on land and $1 : 0.06$ in the sea). This is a result of the greater digestibility of marine primary production, the greater efficiency of trophic relationships in the sea and, in part, the ectothermic nature of marine consumers (less of the food energy has to be devoted to fuelling a temperature differential between body and environment). Man currently exploits about 100×10^6 t (wet weight) of marine production, with the world catch consisting of a wide diversity of animals including crustaceans, molluscs, echinoderms, fish and marine mammals.

So far in this synopsis, we have been following the passage of food materials from their fixation to the top consumers, but essential elements such as nitrogen and phosphorus are present in finite amount and unless these are cycled through the ecosystem back to the primary producers, productivity will cease. Supply of nutrients is probably the most frequent rate-limiting process in the sea, both

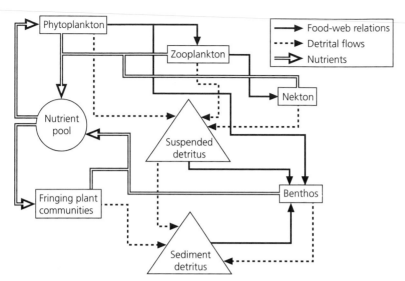

Fig. 11.1 Simplified diagram showing major interactions of an aquatic ecosystem.

to the photosynthesizers and to the bacteria. Nutrients are released from their organic binding by the animal consumers during their metabolism; ammonia and phosphates are then excreted or diffuse back to the water mass from where they may be withdrawn again by bacteria, algae, etc. Nevertheless, some nutrients will be incorporated in the detrital and faecal materials that sink towards the sea bed, so that although most (in the tropics, 90–95%) nutrient regeneration probably occurs *in situ* in the photic zone, there is a continual draining of these essential substances from the surface waters.

In shallow regions, tidal currents and wind-induced mixing will suffice to reinject those nutrients remineralized by benthic animals into the photic zone (if indeed the photic zone does not extend all the way to the sea bed), but nutrient loss by sinking towards the bottom is particularly important in waters overlying a thermocline, which separates permanently the photic zone from deeper waters and from the benthos. Nutrients regenerated below the thermocline will tend to remain at depth, except where deep water (i.e. down to 600 m) upwells to the surface. Hence nutrient cycling will be most unconstrained in shallow waters and in zones of upwelling, and, all things being equal, these areas will therefore be maximally productive, although often still nutrient limited. In waters with small stocks of nutrients, high productivity can be maintained only by a very rapid recycling (aided, in coral reefs, by

symbiotic associations) or by the presence of moneran species capable of fixing atmospheric nitrogen.

Some nutrients are incorporated into organic substances that are highly resistant to decomposition. Although these may go into solution, their half-life as dissolved organic compounds is likely to be very long, and although a huge pool of these substances is present in the ocean it appears largely to be inert. Other dissolved organic materials released by both algae and the animal consumers may be cycled, but much more slowly than the inorganic nutrients required by most photosynthetic organisms. Productivity may then be dependent on the rate of re-mineralization of this second pool. Standing stocks of nutrients are often of very little interest in marine ecology; it is the flux rates that govern photosynthetic production. At any one time, the finite pool of, say, nitrogen will be divided between living tissues, inorganic forms dissolved in the water, dissolved organically bound nitrogen, and particulate organic nitrogen in detritus and faeces, and exchanges between these compartments may be fast or slow. Unfortunately, we know much more about the size of these pools than of their dynamic interaction.

This, then, is a summary of the patterns with which carbon (or energy) flows through the marine ecosystem and in which nutrients must cycle (see Fig. 11.1), and we will now explain these processes in more detail and provide quantitative information on some of the fluxes.

11.2 Energy flow

Although we have a good general understanding of the major pathways of energy flow in the marine environment, the great diversity and complexity of localized regions and sub-systems (e.g. estuaries, shallow seas, bays, upwelling regions, coral reefs, and rocky and sandy substrates) make the construction of a generalized, representative model of energy flow of the marine environment virtually impossible. However, a number of recent studies provided most useful information on the flow and behaviour of energy in marine ecosystems. In these studies complex networks of energy and material flow have been constructed, depicting the various biotic and abiotic components of the particular system, the biomass (or standing crop) of each of these, and the rate of energy or material flux between them. In this chapter we present network models of energy (or carbon, a surrogate for energy) and nitrogen flow in the Chesapeake Bay, the largest estuarine ecosystem in the continental USA, and of the northern Benguela upwelling ecosystem, situated along the west coast of southern Africa off Namibia.

These descriptive flow models were assembled assuming that (1) for primary producers gross primary production (GPP) is equal to net primary production (NPP) plus respiration, and (2) the carbon budget for heterotrophic organisms is according to the balanced energy budget equation $C = P + R + E$, where C is consumption or total energy intake, P is secondary production, R is respiration, and E is egestion of assimilated and metabolized products such as faecal material, exoskeletons, urine, etc. The NPP of plants and the P of animals represent the energy available for other consumers in the ecosystem. In most cases, not all of the primary or secondary produced energy is utilized; the excess production is generally either exported from the system by currents, or broken down and decomposed to detrital material and reused within the system. The rate and magnitude of energy flow in temperate ecosystems, such as the Chesapeake Bay, vary significantly, as do the productivity and biotic diversity, between seasons, as a result of differences in temperature, salinity, and other environmental parameters. Figures 11.2 and 11.3 illustrate the food web networks of the Chesapeake Bay during aggregated summer and winter, respectively. (Networks for all four seasons have been constructed (see Baird & Ulanowicz (1989) for details.) The summer model consists of 10 living compartments and three non-living ones (i.e. compartments 11, 12 and 13 in Fig. 11.2). The arrows indicate the flow from prey to consumer, and the rate of flow, or the amount, associated with each exchange between the compartments is also given. During summer, when productivity in the system is maximal and the full complement of species is present, the total biomass and biological production of plants and animals amounts to about 14.8 g C m^{-2} and 377.6 g C m^{-2} per season (or 4.2 g C m^{-2} day^{-1}), respectively. Plant production (NPP of compartments 1 and 2) contributes about 36% of the total production in the system during summer, with a primary to secondary ratio of 1 : 1.81, which is much higher than the global ratios given above in Section 11.1. This ratio varies considerably between different marine ecosystems globally. For example, this ratio is 1 : 0.33 in the Ems–Dollard estuary in the Netherlands, 1 : 0.38 in the Baltic Sea, and 1 : 0.27 and 1 : 0.59 in the Peruvian and Benguela upwelling systems, respectively. Another important ratio in ecology is the production : biomass (or $P : B$ ratio), which gives the rate of turnover of the biomass of a particular component of the system. For example, the turnover rate, and thus the $P : B$ ratio, for phytoplankton in summer in the Chesapeake Bay is 0.4 day^{-1} or about 32 times during the season. In contrast, the $P : B$ ratio for carnivorous fish (compartment 9, Fig. 11.2) is only 0.004 day^{-1} (or about 0.24 for the entire season). The $P : B$ ratio is an important indicator of the rate of production in relation to the standing biomass, and because of the high turnover rates of phytoplankton, its relatively low biomass can fuel and sustain virtually the entire system. This principle is even more clearly illustrated in upwelling systems, where a relatively small phytoplankton biomass can sustain a vast production of planktivorous fish over a short period of time. This ratio is, for example, 1 : 0.2 day^{-1} in both the upwelling regions mentioned above, but only 1 : 0.08 day^{-1} in the Baltic Sea.

The food web during winter in the Chesapeake Bay (Fig. 11.3) is very different from the summer one. Some of the species have emigrated out of the system, resulting in a network with only 9 living compartments, as opposed to the summer situation

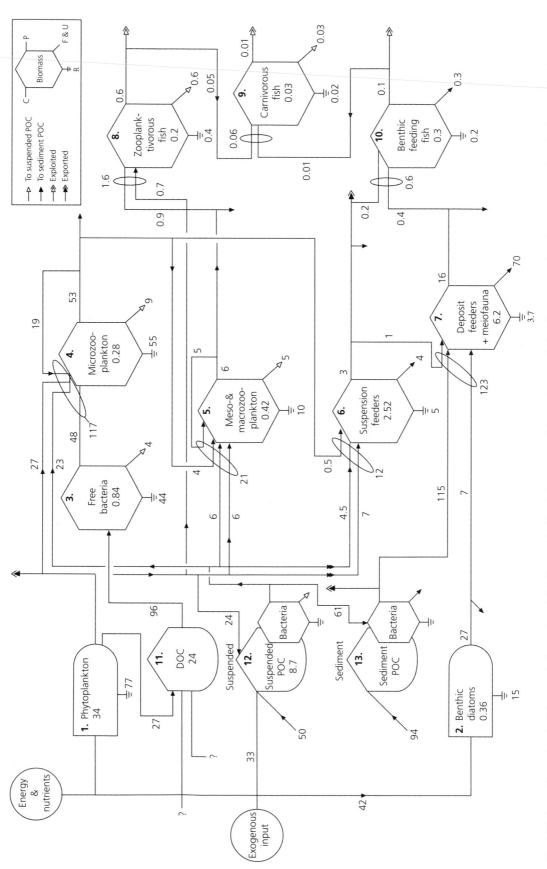

Fig. 11.2 Energy flow network of the mesohaline region of the Chesapeake Bay during summer (biomass in g C m^{-2}, flow between numbered compartments in g C m^{-2} per summer). (Adapted from Baird & Ulanowicz 1989.)

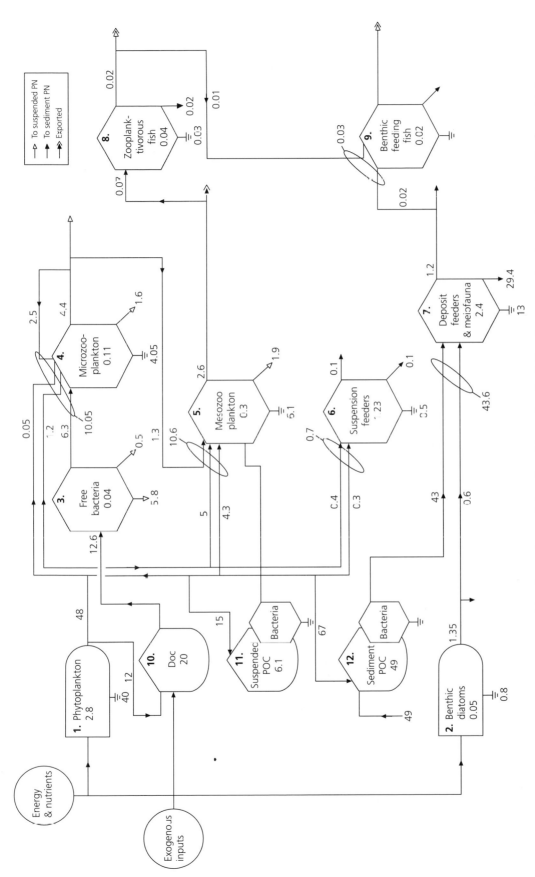

Fig. 11.3 Energy flow network of the mesohaline region of the Chesapeake Bay during winter (biomass in g C m⁻²; flows between numbered compartments in g C m⁻² per winter).

when 10 are present. Of the 13 fish species in summer, only seven remained through winter (see Baird & Ulanowicz 1989).

Upwelling regions are some of the most productive ecosystems on Earth. They are usually situated on the eastern boundary of large oceanic circulation systems, on continental shelves, and it is in these regions where most of the world's fish are caught. One such system is the northern Benguela upwelling system, situated between 15°S and 29°S and covering an area of about 179 000 km² off the coast of Namibia on the west coast of southern Africa. This upwelling region is very productive and large catches of fish and crustaceans have been made there since early this century. Catches of pelagic shoaling fish (e.g. sardines, anchovies, horse-mackerel; compartment 5 in Fig. 11.4) by purse seine boats exceeded 1.5×10^6 t in 1968, but declined in subsequent years to about 150 000 t in 1995. Bottom trawl catches for demersal fish species, such as the hake (compartment 7 in Fig. 11.4), reached a peak at about 820 000 t in 1972, and the commercial and recreational line fishery lands considerable numbers of carnivorous fish (compartment 8 in Fig. 11.4) annually. Rock-lobsters and the deep-sea red crab are two crustacean species that contribute significantly to the annual yield of the system.

The biomass of the major living components and the rates of exchange between them are illustrated in Fig. 11.4. The aggregated flow model consists of two non-living and 9 living components. Because of the monetary importance of the commercially exploited fish, cephalopod and shellfish species, much more information is available on their diet and general ecology. The figure clearly shows the increase in detailed information if the flow model is viewed from the lower to the higher trophic levels, a typical situation in regions of commercial fishing.

In the above two examples energy, or carbon, was used as the currency to quantitatively express standing stocks and flows. We know, however, that other elements are also imported, taken up, chemically transformed, recycled, transferred through the food web, and exported from marine ecosystems. Nitrogen (N) is one of the macro-elements essential for life and of great importance in ecological systems. Not only is N an essential component of animal tissue, but it is often the controlling factor in plant nutrition and productivity. A wide variety of nitrogenous compounds occurs in marine ecosystems and there is a complex pattern and process of N recycling that can be regulated by a number of physical, chemical and biological factors. The seasonal dynamics of nitrogen flow and recycling has recently been studied in some detail by Baird et al. (1995), who have presented a general concept of nitrogen behaviour in the Chesapeake Bay. In doing so they quantified the input of 'new' N (i.e. imported N from adjacent systems), the regeneration of N within the system by all components, the uptake of N, and the export of N from the system. All these processes are common to marine systems, whether upwelling regions or shallow embayments.

In general terms, N is taken up by plants and animals, converted into tissue or excreted, or passed on to other consumers in the food web. Figures 11.5 and 11.6 represent the aggregated summer and winter food webs in the Chesapeake Bay, respectively, in terms of N. They depict the standing biomass of the 13 compartments in summer and the 12 in winter, as well as the flow of nitrogen between them. The generalized N budget for the heterotrophic compartments is given by the equation $C_n = P_n + F_n + U_n$ where C_n (total nitrogen consumption) equals P_n (secondary production in terms of N) plus F_n (particulate faecal material) plus U_n (dissolved ammonium N excretion). In the absence of any clear guidelines from the literature on the release of photosynthate nitrogen, Baird et al. (1995) assumed nitrogen production equivalent to N uptake by plants. The flow models illustrate the N produced (P_n) by animals as a fraction of the amount consumed, those fractions excreted as dissolved (U_n) and particulate (F_n) N, as well as those fractions exploited or exported. The black arrows indicate the particulate N deposited on the sediment, while the open arrows indicate excreted dissolved N compounds contributing to the dissolved N pool (DN, compartment 10). The dissolved N from compartment 10 is directly taken up by phytoplankton, benthic algae and free living bacteria. Suspended particulate N (in compartment 11) may form aggregates and sink to the bottom (compartment 12).

The dynamics of N flow in marine systems is driven by the magnitude of the input from adjacent systems (e.g. river flow) and from deeper waters (as in upwelling regions), and by the heterotrophic

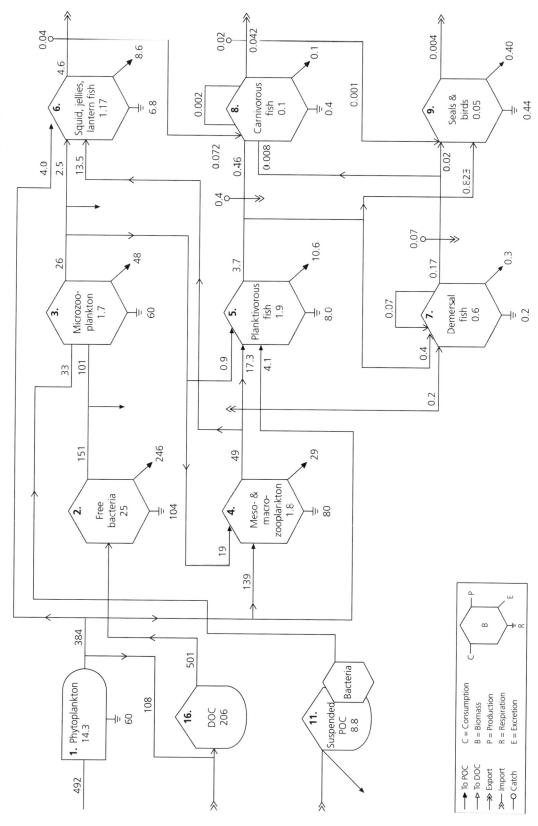

Fig. 11.4 Energy flow network of the northern Benguela upwelling region, Namibia (biomass in g C m^{-2}, flows between numbered compartments in g C m^{-2} day^{-1}).

regeneration of dissolved and particulate N in the water column and sediments. Both sources (i.e. 'new' N and recycled N) provide the necessary N to sustain primary production and bacterial growth, which fuel the food web to a very large extent. In the Chesapeake Bay the inputs of dissolved and particulate N vary seasonally; a phenomenon universal to most marine ecosystems. In the Chesapeake Bay the highest inputs of 'new' N are in winter (166 mg N m^{-2} day^{-1}) and spring (211 mg N m^{-2} day^{-1}), with lower inputs in summer (53 mg N m^{-2} day^{-1}) and autumn (74 mg N m^{-2} day^{-1}) (Baird *et al.* 1995). The regeneration of dissolved N within the system is mainly mediated by the excretion of N (mainly as ammonium-N) by microbes and other animals in the water column and top few centimetres of the sediment. It was found by Baird *et al.* (1995) that the largest proportion of regenerated N originates from the sediment when dissolved N efflux occurs from the sediment to the overlying water, a phenomenon most likely to dominate in marine ecosystems. In the Chesapeake Bay, for example, about 126 mg N m^{-2} day^{-1} is released from the sediments to the overlying waters during summer, and about 48, 24 and 40 mg N m^{-2} day^{-1} during autumn, winter and spring, respectively. Of these amounts about 78%, 68%, 42% and 44%, respectively, are regenerated through micobial activity in the sediments during the same seasons; the balance is due to benthic macro- and meiofauna. Benthic regenerated dissolved N contributed about 30% in summer, 22% in autumn, 26% in winter and 18% in spring to the collective seasonal nitrogen demand of phytoplankton, benthic algae and bacteria. Water column organisms < 200 µm in size (compartments 3 and 4, Figs 11.5 and 11.6) contribute between 14% (in summer) and 7% (in winter) to the seasonal demand of the components mentioned above.

From the information available on N fluxes in the Chesapeake Bay a seasonal supply and demand budget has been proposed by Baird *et al.* (1995), and is given in Table 11.2.

The table shows, for example, that the total supply of 'new' and regenerated N exceeds the demand in only two seasons, namely spring and winter, whereas the demand exceeds the supply in summer and autumn. One could infer that in situations such as these nitrogen could be in limited supply for

Table 11.2 Seasonal nitrogen demand and supply budget in the mesohaline Chesapeake Bay (values in mg N m^{-2} day^{-1}). (Adapted from Baird *et al.* 1995.)

	Season			
	Spring	Summer	Autumn	Winter
Demand				
Phytoplankton	130.7	232.5	175.3	71.0
Benthic algae	31.3	46.5	7.6	2.4
Water column bacteria	66.6	139.3	35.6	17.0
Total	228.6	418.3	218.5	90.4
Supply				
Regenerated N	78.4	201.5	79.8	35.6
'New' N	211.0	53.2	73.6	166.2
Total	289.4	254.7	153.4	201.8

primary production, leading to the 'winding down' of the system in late summer and early autumn.

The amount of nitrogen regenerated varies between systems at different geographical locations. The comparison of regeneration values between systems is, however, tenuous because of their inherent differences. It is interesting to note that about 80% of the total N supply is generated *in situ* in the Chesapeake Bay during summer, and this percentage fluctuates from 52% in autumn to 18% in winter and 27% in spring. Studies of some other systems reported that approximately 35% of the N supply is regenerated *in situ* in Carmarthen Bay (Mantoura *et al.* 1988), 72% in the English Channel (Newell & Linley 1984), 48% on the outer shelf of the middle Atlantic Bight (Harrison *et al.* 1983), and 41% in Narangansett Bay, Rhode Island, USA (Nixon & Pilson 1983). The highest percentage of recycled N was reported by Christian *et al.* (1992) in a study on the Neuse River estuary in North Carolina. They estimated that about 98% and 86% of the total N throughput was generated within the system during summer and spring, respectively.

The carbon flow models of the Chesapeake and the Benguela upwelling systems are undoubtedly incomplete, yet they reflect the complexity of energy flow in natural ecosystems. The food web networks reveal interesting aspects and information on the structure of ecosystems, and the differences in the dynamics of systems on temporal and spatial scales.

When nitrogen is used as a currency (of biomass and fluxes), a whole new approach to how a system functions is revealed, and illustrates that different kinds of information and interpretations are required. It is also clear from these examples on the Chesapeake Bay that annual averages of biomass and flows do not adequately reflect the dynamic nature of an ecosystem over time. However, the constraints on the collection and analysis of data sets representing seasons or shorter units of time are considerable. We will have to make do for the foreseeable future with generalized, annual averaged flow models of natural ecosystems.

11.3 Nutrient cycling

One of the universal processes inherent in all ecosystems is the recycling of matter. Without this process life on Earth cannot be sustained, as most of the 30–40 elements necessary for the growth and development of living organisms are in finite supply. We alluded above to the rates of nitrogen cycling and its importance in the functioning of the Chesapeake Bay. Other elements, such as C, P, Si, S, Fe, Mn, etc., are also essential for organic life and are also recycled at different rates in marine ecosystems. Despite the chemical differences between elements, the underlying principles of recycling for different elements are essentially the same. For example, all elements undergo chemical change during this process, and living organisms play a crucial role in their remineralization and recycling. All biogeochemical cycles thus involve living organisms so that the biotic and abiotic components are intertwined with one another. Biogeochemical cycles can generally be classified into two basic types: (1) gaseous nutrient cycles, where the atmosphere is the reservoir where the element collects—typical cycles of this kind include the oxygen, carbon and nitrogen cycles; (2) sedimentary nutrient cycles and those where the reservoir is mainly sedimentary deposits; these cycles are slower than the gaseous ones, and exert a more limiting influence on living organisms —the phosphorus and sulphur cycles are typical of sedimentary cycles.

Here we discuss the N and P cycles briefly. We have seen above that N is washed into the coastal seas and this is one way by which N is imported into the marine environment. N, however, occurs mainly in the gaseous form and is as such unavailable to all organisms except to various moneran blue–green algae and bacteria. In aquatic ecosystems N is recycled in the water column and the sediment at various rates and quantities as discussed in Section 11.2. Ammonium-nitrogen (NH_4^+) is the preferred N species for uptake by phytoplankton and bacteria. Phytoplankton can also assimilate nitrite-N (NO_2^-) and nitrate-N (NO_3^-), but at a cost, as they have to shed the oxygen molecules before uptake.

In the water column NH_4^+ is cycled mainly between a dissolved NH_4^+ pool, phytoplankton, bacteria and zooplankton, as illustrated in Fig. 11.7. In the Chesapeake Bay, for example, about 10% of the mean annual N demand is recycled in the water column, whereas approximately 25% is remineralized in the sediments. It would appear from recently published research that the sediments are the major site of N regeneration in coastal seas. The process of N cycling in the sediments is illustrated in Fig. 11.8 and consists essentially of three interlinked processes, namely ammonification, nitrification and denitrification. NH_4^+ is first produced by the decomposition and deamination of organic particulate N through ammonification. NH_4^+ is also produced

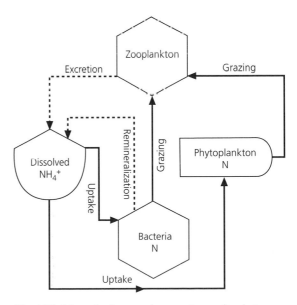

Fig. 11.7 Schematic diagram of ammonium cycling between phytoplankton, bacteria and zooplankton in the water column.

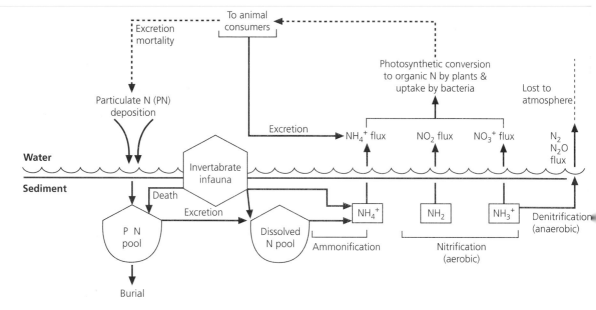

Fig. 11.8 Conceptual diagram showing nitrogen pools, transformation and fluxes across the sediment–water interface in marine ecosystems.

by the excretion of benthic infauna. The free NH_4^+ is oxidized to NO_2^- and NO_3^- by a process called nitrification, which takes place in the aerobic zone of the sediment. Aerobic bacteria play an important role in the nitrification process, with bacteria such as *Nitrosomonas* sp. and *Nitrobacter* sp. being the principal organisms involved. Nitrification is also a sink of N in that it prevents the flux of NH_4^+ to the overlying waters. The third process, denitrification, is mediated by denitrifying anaerobic bacteria (e.g. *Pseudomonas denitrificans*). In this phase NO_3^- is reduced to the gaseous forms of molecular N gas (N_2) and nitrous oxide (N_2O), both of which escape through the water column to the atmosphere. These processes occur in the top few centimetres of the sediment, and the rate of nitrogen metabolism and the flux of N across the sediment–water interface depend on the differential concentrations, nature and abundance of animals in the sediment, temperature and other environmental variables. In strongly seasonal marine systems, such as shallow coastal seas in temperate areas, upwelling systems, etc., large differences in flux rates can be observed. For example, in the Chesapeake Bay the flux of

regenerated NH_4^+ from the sediment to the water column ranges from 126 mg N m^{-2} day^{-1} in summer at water temperatures of 21–29°C to 24 mg N m^{-2} day^{-1} at 2–6°C in winter. Complete budgets of N demand, regeneration, inputs and exports are not readily available, and much research needs to be done in this field. Nevertheless, a substantial body of information exists on the remineralization of N in aquatic systems, and the reader is referred to Blackburn & Sorensen (1988) for more detail.

Phosphorus is another macro-nutrient whose availability in marine ecosystems significantly affects biological production. The rate of P recycling has been found to vary significantly, with the turnover time of PO_4^{3-} ranging from days in temperate seas to hours in coral reefs. The biological removal of PO_4^{3-} in the sea involves uptake by both phytoplankton and bacteria. Classical ecological schemes cast bacteria mainly as remineralizers of organic matter, but more recent research has revealed that bacteria account for a substantial proportion of PO_4^{3-} uptake in aquatic systems, and that active competition occurs between phytoplankton and bacteria for available phosphate. In the marine environment dissolved P is consumed during phytoplankton and bacterial growth, and is regenerated as a consequence of heterotrophic consumption of these organisms. Much of the regeneration takes place in the water column, but in relatively shallow environments, such

Table 11.3 Mass balance of particulate and dissolved phophorus in the Chesapeake Bay (values in mg P m^{-2} day^{-1}). (Adapted from Baird 1998.)

	Inputs			Losses			
Seasons	Inputs	Deposition	Sum	Burial	Exports	Sum	+ or −
Summer	10.7	9.4	20.1	31.0	17.5	38.5	−18.4
Autumn	12.8	23.9	36.7	15.8	9.0	25.0	+11.70
Winter	16.2	12.5	28.7	7.2	3.9	11.2	+17.5
Spring	21.0	23.5	44.5	14.5	6.2	21.0	+23.5

as estuaries, shallow seas and continental shelves, sediments may also play an important role in the regeneration of P.

The cycle of P utilization, transformation and regeneration is relatively simple compared with that of nitrogen. P enters the sea mainly from the weathering of rocks, from bone and guano deposits, and from river run-off into coastal seas. Three major forms of P occur in the marine environment: (1) dissolved inorganic phosphorus (DIP), which is assimilated by algae and bacteria into cellular organic matter; (2) excreted particulate organic phosphorus (POP); (3) dissolved organic phosphorus (DOP), which can be decomposed by bacteria releasing DIP. The relative importance of bacteria in the regeneration of DIP is larger in the sediments than in the water column. These substances occur in the sediment pore water and are exchanged via diffusion and some advective pumping by the infauna with the overlying water.

DIP, and possibly DOP, is involved in complex reactions with metal (such as iron and manganese) oxides and hydroxides to form insoluble precipitates under aerobic conditions, but is redissolved when redox becomes sufficiently reduced. This is a crucial chemical interaction, which strongly influences the seasonal availability of DIP for algal assimilation. At a much slower, but still significant rate, DIP participates in precipitation reactions to form various minerals such as apatite, which is important in some animals' skeletons. The highly charged phosphate anion PO_4^{3-} is also readily sorbed onto the surfaces of charged clay (and detrital) particles.

Phosphorus is supplied to the sediment via the sinking of particulate matter from the overlying waters, from decaying organic material (dead plant and animal tissue), surrounding land masses, and from the weathering of phosphorus deposits. A portion of the P that reaches the sediment–water interface will be buried; another fraction will decompose

or dissolve and release the phosphate to the sediment pore water. The regenerated P may be released to the overlying water, reprecipitated within the sediment or adsorbed to other constituents of the sediment. The portion of the sedimenting P that is not buried must be solubilized and returned to the overlying water; this portion is called mobilizable phosphorus. The reservoir of mobilizable phosphorus is the top few centimetres of the sediment, where it occurs in the pore water, and from where it is released to the overlying water. The concentration of dissolved phosphorus in the pore water decreases with depth. Adsorption on the oxidized surface sediment affects the flux of P from the sediment to the overlying water, hence the oxidized surface layer acts as a trap for phosphate.

Wright (1977–78) provided information on the gross annual flux of P through the world's oceans. About $130\,350 \times 10^6$ kg of particulate organic P sink to the sediments, of which approximately 98% is remineralized. Inputs of P from rivers amount to about 1350×10^6 kg year^{-1}. Few detailed studies on P dynamics in marine ecosystems have been carried out to date. However, Baird (1998) and Ulanowicz & Baird (1998) estimated that the regeneration rate of P in the Chesapeake Bay ranges from 54 mg P m^{-2} day^{-1} in summer to about 7 mg P m^{-2} day^{-1} in winter. The largest proportion of P is also regenerated in the sediments as opposed to P regenerated by organisms in the water column. Those workers provided a mass balance model of particulate and dissolved P for the Chesapeake Bay, given above in Table 11.3. These results show the seasonal input of total P from external sources and deposition rates of P from the water column. The table also shows the amounts buried and the export of P from the system including emigration and fisheries yield. Of interest is the apparent deficit of P towards the end of summer, indicating possible

P limitation in the Chesapeake Bay at that time of the year.

11.4 Ecosystem analysis

Ecosystems are by their very nature difficult entities to study, and it is difficult to fully understand their inherent functional processes, or to predict their responses to natural or anthropogenic disturbances. The development of computer technology in recent years has provided the opportunity for ecologists and mathematicians to attempt to mimic ecosystem behaviour and to predict the state of the system, or components thereof, in the future. Although this approach has yielded some interesting and valuable insights into the probable behaviour of systems, simulation models are rarely able to make meaningful predictions about the future state of an ecosystem. The main reasons for the inability of ecological simulation models to make good predictions are the inherent variability in the biological factors, the hierarchical structure of ecosystems, and the capacity of self-organization of ecosystems. In search for alternatives to simulation modelling, ecologists have recently developed techniques such as network analysis to quantify attributes believed to characterize ecosystems (Mann *et al.* 1989). These 'system attributes' can then be used to compare systems on temporal and spatial scales.

Questions often asked about ecosystems are, for example, how 'active' is the system, how efficient is energy or material processed within the system, or how much material is recycled by the system. Network analysis provides some answers, and we will discuss some of the results from these analyses below. The extent of the activity of a system can be measured by the sum of all flows through all the compartments of a system; this is called the 'total systems throughput' (TST), and is considered to be a surrogate for the total power generated within a system. In our examples above, the annual TST for the Chesapeake Bay amounts to about 4.23×10^6 mg C m^{-2}. The TST values vary dramatically between seasons so that winter and summer represent 13% and 39% of the annual TST, respectively (Baird & Ulanowicz 1989). These values clearly show the differences in the dynamics of a system at different times of the year, a phenomenon that would certainly occur in most temperate marine ecosystems.

In regions where the temperature and other environmental factors remain fairly constant throughout time, such as in the tropics or deep ocean basins, it is expected that the system's activity would also remain fairly constant.

The efficiency by which energy is transferred between the various components of an ecosystem can be determined from the construction of the so-called 'Lindeman Spine', an output from network analysis. The concept of trophic energy levels introduced by Lindeman (1942) held great intrinsic appeal to ecologists to assign organisms to specific trophic levels. The problem with applying Lindeman's concept to real ecosystems is that heterotrophic organisms can rarely be assigned to a single level, as they obtain their energy from more than one level. However, a routine from network analysis using matrix algebra yields information by which all organisms in a system can be allocated to a trophic level based on their dietary requirements (Ulanowicz & Kemp 1979). From these data a Lindeman Spine can be constructed, and Fig. 11.9 illustrates the Spine for the Chesapeake Bay. The first trophic level, indicated by the symbols I + D, combines primary producers and detritus (a common practice adopted during the International Biological Programme). The Spine shows the return of detrital energy from each discrete trophic level to first level (I + D), external inputs (888 723 mg C m^{-2} year^{-1}), the amount of energy transferred from one level to the next, exports and respiration at each level, and the efficiency of transfer between levels (indicated by the per cent value given in each box of the Spine). Of interest is the decline in the trophic efficiency from the lower to the higher levels; for example, about 35% of the energy entering level II is transferred to level III, whereas only 3.4% of the input to VI is passed on to VII. The mean efficiency for the system as a whole, incorporating all levels, can be calculated from the logarithmic mean of the individual values from each level, which is approximately 9.6% for the Chesapeake Bay. This efficiency differs, however, between seasons and varies between a high of 11.4% in summer to a low of 8.1% in winter in the Chesapeake Bay (Baird & Ulanowicz 1989).

As discussed above, the recycling of matter is one of the universal processes present in all ecosystems. Network analysis provides a routine that expresses the ratio of the amount of material recycled to the

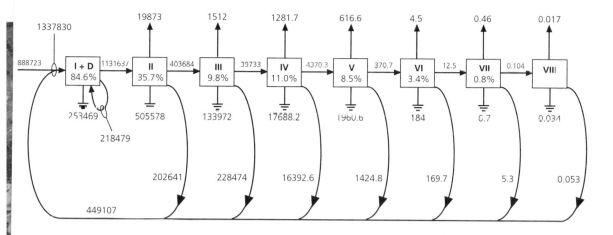

Fig. 11.9 The Lindeman Spine (or 'trophic chain') of the Chesapeake Bay, with primary producers and detritus merged in the first trophic level (1 + D). All values are given in mg C m^{-2} year^{-1}. The percentages in the boxes represent annual trophic efficiencies. (Adapted from Baird & Ulanowicz 1989.)

TST in the Finn Cycling Index (FCI). This index tells us how much of a system's activity is devoted to recycling; a high value would indicate a system self-reliant to a greater degree than one that recycles only a small fraction of its throughput. It has also been postulated that 'mature' systems may have a higher FCI than a young system, which may well be more dependent on the import of elements from adjacent systems.

The mean annual FCI for carbon cycling in the Chesapeake Bay is approximately 23%, and remains remarkably constant throughout the year. This index obviously differs between ecosystems from different geographical positions. For example, the FCI for the Baltic Sea is 22.8%, but is only 3.4% for the Peruvian upwelling system (Baird et al. 1991). It is also possible that different elements may be recycled at greater or lesser quantities than others. In the

Chesapeake Bay the FCI for nitrogen peaks at 70% in summer, compared with an FCI of 23% in terms of C during the same season. The FCI for N varies between 34% and 46% for the other seasons. FCI values for phosphorus in the Chesapeake Bay range between 75% in spring and 53% in autumn. In the case of P, maximum recycling occurs in spring, with FCI values of 60% in summer and 66% in winter. These data from the Chesapeake Bay illustrate two important points about the magnitude of recycling in marine ecosystems: (1) different elements are recycled at different rates and quantities; (2) large differences may occur between seasons within the same system.

Network analysis provides furthermore a wealth of information on the structure and function of an ecosystem. For example, information is given on the direct and indirect pathways of energy flow by means of dependence coefficients, on the structure of recycling pathways, and on other global ecosystem properties such as system specialization and organization. The reader is referred to Wulff et al. (1989) for further information on the methods and application of network analysis in marine ecology.

sideways, unharmed. Much of the southeastern North Sea is similar to this type of habitat and is also subjected to the most intense bottom-fishing effort. Not surprisingly, this is the ideal habitat for sole and plaice, which prefer shallow water with a sandy sea bed in which they find their main food, i.e. worms and bivalves. In fact, there is now evidence suggesting that far from degrading the habitat of these species, trawling enhances the food supply of flatfish by maintaining the system in a state of high productivity (Millner & Whiting 1996; Rijnsdorp & van Leeuwen 1996).

Whereas bottom trawling might be sustainable in an unstable habitat subject to frequent natural disturbance, its effects are amplified in more stable environments. As part of the same study, Kaiser & Spencer (1996) found that trawling caused a short-term decrease in the number of species (50%) and abundance of fauna (33%) in a more stable coarse sand, gravel and shell sediment in slightly deeper (40 m) water. Although the benthic community had recovered 6 months later (Kaiser et al. 1998) it is important to note that experimental studies may not truly represent the chronic effects of fishing disturbance created by an entire fleet of beam trawlers.

An alternative to the experimental approach of examining the effects of bottom trawls is to compare areas that have been exposed to different levels of fishing intensity. Collie et al. (1997) examined areas of similar substrata in the deep waters of the Georges Bank in the northwestern Atlantic. In areas that had been heavily fished by scallop dredgers, the cobble sea bed was relatively bare with few epifauna apart from crabs and starfish. This contrasted sharply with an area that was fished infrequently, where the sea bed was overgrown with foliose bryozoans, hydrozoans and tubeworms (Fig. 12.2a, b and Plates 32 and 33, facing p. 208). The heavily fished areas were then closed to mobile fishing gear and after 2 years the epibenthic community began to show signs of recovery.

These findings are important because they suggest that the relative magnitude of the effects of fishing disturbance within a particular environment can be related to the stability of the habitat and the local natural disturbance regime. Thus, in general, the consequences of fishing disturbance will become more severe with increasing habitat stability and a lower frequency of natural disturbance (Fig. 12.3). The sandy sediments of a surf beach are a good example of a highly disturbed environment. At the other extreme, the soft sediments of the abyssal plain, or slow-growing coral reefs (Chapter 6) are examples of stable and therefore sensitive habitats. However, it would be erroneous to assume that all sandy habitats are immune to significant change caused by fishing activity. Thrush et al. (1996) demonstrated the importance of dense beds of spionid worms to the stability of mudflats in New Zealand. The tube-building activities of the worms bind sand particles together, thereby consolidating the substratum. When the worms were experimentally removed from the sediment, it became unstable such that other species were less likely to recolonize the affected areas and recovery of the habitat was delayed.

Many of the species affected by bottom fishing have no commercial value and are often relatively common. So, does it matter that they are removed by fishing activity? Sainsbury (1987) recorded a gradual decrease in the bycatch of sponges in the Australian northwest shelf pair-trawl fishery between 1967 and 1985. This fishery targeted snappers, *Lutjanus* spp., and emperors, *Lethrinus* spp., which fed and sheltered from predators among the complex three-dimensional structure of the sponge community. As their habitat was reduced, so the catches of these fish also declined. Structurally complex habitats, such as those formed by sponges, corals, worm tubes and bryozoans, shelter a diverse community of prey species that are fed upon by small and juvenile commercial fish species (Peattie & Hoare 1981; Bowman & Michaels 1984). In addition, Walters & Juanes (1993) found that juvenile cod suffered much higher mortality when they were unable to hide within a structured habitat.

These and other studies demonstrate that whereas fishing may be relatively benign (even beneficial) in relatively featureless, highly disturbed habitats, its effects are more severe in structured habitats that fulfil a variety of functions on which commercial and other species depend.

12.1.2 Incidental catches and 'ghost fishing'

Advances in technology and the advent of man-made materials such as nylon have led to a revolution in some sectors of the fishing industry. The use of sonar and echo-sounders allows entire shoals of

(a)

(b)

Fig. 12.2 Photographs of the sea bed in a heavily fished scallop ground (a) and a similar habitat in a lightly fished area (b) on the Georges Bank. The fauna in the lightly fished area is typically dominated by foliose bryozoan and hydrozoan species that add to the structural complexity of the habitat. Photos by Dann Blackwood and Page Valentine, U.S. Geological Survey (Collie *et al.*, 1997).

Fig. 12.3 A conceptual model demonstrating the relationship between the relative effects of fishing disturbance, natural disturbance and habitat stability. The effects of fishing become more important as disturbance is reduced and habitat stability increases. (Adapted from Jennings & Kaiser 1998.)

Fig. 12.5 Hermit crabs and starfish respond to the odours of animals damaged by bottom trawls. These scavengers can aggregate within minutes to take advantage of an easily accessible meal like this damaged whelk. Photo by James Perrins.

landing on the sea bed (Fig. 12.5). These animals also tend to be resistant to the damaging effects of trawling, either because they inhabit a strong shell or because they are able to regenerate damaged body parts. So far, there is little evidence to suggest that populations of benthic invertebrate scavengers have expanded in response to carrion generated by fishing activities. One of the main differences between seabirds and benthic invertebrate scavenging behaviour is the mechanism of carrion acquisition. Seabirds actively seek out and target fishing vessels as a source of food, migrating from distances in excess of 10 km (Camphuysen *et al.* 1993). Consequently, seabirds are able to obtain a relatively constant food source by feeding at trawlers. Conversely, benthic invertebrates rely on the chance occurrence of food-falls of fisheries carrion, or a trawling disturbance occurring on the sea bed within the range of their sensory and ambulatory capabilities.

12.2 Dredging and mineral extraction

As humans have depleted terrestrial mineral resources, it has become economically feasible to begin mining the marine habitat. Large scars in the landscape in the case of open-cast mining, piles of spoil and contaminated run-off from subterranean mining are obvious environmental impacts of terrestrial mining. These effects are also evident in the marine environment, where mineral extraction leads to habitat alteration, faunal removal and contaminant resuspension. We highlight two examples, dredging of aggregates (gravel) used by the construction industry and manganese nodule extraction.

Extraction of marine aggregate in the UK has increased from 7×10^6 t per annum in 1969 to 25×10^6 t per annum in 1989. Although demand for aggregate was reduced during the recession of the early 1990s, Charlier & Charlier (1992) predicted that production will reach 200×10^6 t in the next century. Disturbance of the sea bed by dredging and bottom fishing are similar in that both perturb benthic communities. However, whereas fishing affects only the superficial layers of the sediment (0–20 cm deep), the process of aggregate extraction completely removes the substratum to depths in excess of 1 m. In this case, the issue is not recovery of the fauna, rather recovery of the habitat. Kenny & Rees (1996) undertook an experiment in which they used a commercial gravel dredger to remove 50 000 t of aggregate off the east coast of England. Two years later, the sediment composition at this site was still dominated by much finer particles than at a nearby reference area and was apparently much less stable. Whereas the majority of the smaller invertebrates had recolonized the dredged site, fauna with a higher biomass (horse mussels and foliose bryozoans) were still absent. Habitat modifications such as these may alter the suitability of the environment for commercial species such as flatfish and crustaceans that previously used it as a feeding or breeding ground. These potential environmental effects on local fisheries are addressed by the requirement for aggregate companies to conduct environmental impact assessments when applying for permission to dredge new areas.

Manganese nodules were first dredged from the deep sea during the H.M.S. *Challenger* expedition of 1872–76. These nodules contain numerous metals, including iron, manganese, copper, nickel, cobalt, zinc and silver. As the value of metals on world markets has risen, it has become economically feasible to mine them in the extreme environmental conditions found at depths exceeding 4000 m. Manganese nodules provide a hard substratum in a mainly soft muddy habitat. Dredging for manganese nodules has several environmental effects: it reduces both habitat and species diversity within the dredge's track, creates a trench in the sediment and produces a sediment plume. Sedimentation at these depths occurs at the rate of a few millimetres every 1000 years. As a result, dredge tracks will persist for many years and filter-feeding organisms may be adversely affected by the unnaturally large pulse of sediment that settles out of the plume (Theil & Schriever 1990). Recovery of habitats and benthic communities is likely to be much slower in the deep-sea environment compared with that on the continental shelf (Chapter 7). Consequently, the issues of extraction and the dumping of waste and contaminated material in the deep sea are highly contentious.

12.3 Pollution

A wide array of objects and substances are discharged either deliberately or accidentally into the sea: oil and other petrochemicals, solvents, insecticides and herbicides from agricultural run-off, heavy metals, domestic waste, radionuclides, sewage and other organic materials, plus a host of others. Any issue of the *Marine Pollution Bulletin* or *Marine Environmental Research* gives an idea of the range of polluting substances that are the subject of past and present research. Pollution causes shifts in the species composition of communities, affects breeding success and can reduce DNA integrity (Everaarts 1997). The realization that many substances have cumulative and long-lasting effects in the marine ecosystem, such as PCBs (polychlorinated biphenyls), has led to a stricter attitude towards monitoring the environment. For example, in Europe, legislation now exists to protect public health on bathing beaches by monitoring coliform counts. Some nations now restrict or have banned the dumping of wastes such as sewage sludge and spent nuclear fuel. We give two examples of pollution incidents: oil spillage and tributyl tin (TBT) poisoning. Oil spillage incidents tend to release large quantities of toxic substances into the sea, whereas the more subtle effects of TBT poisoning are only manifested in certain species.

12.3.1 Oil spillage

The list of specific pollution incidents is almost endless, but large-scale oil spills usually attract the greatest media attention. Recent examples include the grounding of the *Exxon Valdez* off Alaska, and the *Braer* off the Shetland Islands, or the destruction of oil pipelines and refineries during the Gulf war. Technology for dealing with these incidents and our understanding of the ecology of the biological recovery processes that occur after such incidents is much improved. For example, in the early 1970s the oil dispersants used to clean beaches after the *Torrey Canyon* oil spill were more harmful to organisms than the oil itself (Southward & Southward 1978). Modern dispersants are much less toxic, and bioremediation techniques using bacterial digestion to speed the breakdown of the oil are environmentally safer.

In coastal areas, the effects of pollution are most severe in any body of water that has a restricted or slow rate of exchange relative to its volume, e.g. estuaries and semi-enclosed bays or harbours. For example, discharges from the coastal areas surrounding the Baltic Sea resulted in a 50% decrease in the saturated oxygen content of the bottom water after 1960. Although the incidence and effects of pollution in coastal shelf waters are well studied, little is known about the likely consequences of polluting the deep sea. This was highlighted by the abandonment by Shell™ of plans to dispose of a redundant oil storage platform, the *Brent Spar*, by sinking it in water 2200 m deep in the north eastern Atlantic. Environmental groups protested that toxic residues would leak from the sunken wreck and cause contamination throughout the food chain. Greenpeace boarded the wreck in a bid to prevent its sinking. Eventually the *Brent Spar* was towed to a fjord in Norway, where it was moored until it could be dismantled. This episode sparked disagreement among leading experts on deep-sea biology. Whereas some considered that too little is known about the deep ocean to risk such waste disposal (Gage & Gordon 1995), others contended that the size and assimilative capacity of the oceans provided a *prima facie* case for using them for the disposal of some kinds of waste (Angel & Rice 1996). Furthermore, the environmental risks associated with mooring the storage platform in inshore waters, where marine mammals, seabirds, commercial fish and shellfish stocks could be affected in the event of an accident, may have exceeded the potential risks of sinking it offshore.

References

Åberg P. (1992) A demographic study of two populations of the seaweed *Ascophyllum nodosum*. *Ecology*, **73**, 1473–87.

Achituv Y. & Barnes H. (1978) Some observations on *Tetraclita squamosa rufotincta* Pilsbry. *J. exp. mar. Biol. Ecol.*, **31**, 315–24.

Adam P. (1990) *Saltmarsh Ecology*. Cambridge University Press, Cambridge.

Alverson D., Freeberg M., Pope J. & Murawski S. (1994) A global assessment of fisheries bycatch and discards. *FAO Fisheries Technical Paper*, **339**. FAO, Rome.

American Zoologist (1982) The role of uptake of organic solutes in the nutrition of marine organisms. *Am. Zool.*, **22**(3), 611–733.

Angel M.V. (1994) Long-term, large-scale patterns in marine pelagic systems. In Giller P.S., Hildrew A.G. & Raffaelli D.G. (eds), *Aquatic Ecology. Scale, Pattern and Process*, pp. 403–39. Blackwell Science, Oxford.

Angel M.V. & Rice A.L. (1996) The ecology of the deep ocean and its relevance to global waste management. *J. appl. Ecol.*, **33**, 915–26.

Au W. & Jones L. (1991) Acoustic selectivity of nets: implications concerning incidental take of dolphins. *Mar. Mammal Sci.*, **7**, 258–73.

Azam F., Fenchel T.L., Field J.G., Gray J.S., Meyer-Reil L.A. & Thingstad F. (1983) The role of water column microbes in the sea. *Mar. Ecol. Prog. Ser.* **10**, 257–63.

Bainbridge R. (1953) Studies on the inter-relationships of zooplankton and phytoplankton. *J. mar. biol. Ass. U.K.*, **32**, 385–447.

Baird D. (1998) Seasonal phosphorus dynamics in the Chesapeake Bay. University of Maryland Center for Environmental Science, Chesapeake Biological Laboratory. Ref. 98-015. Solomons, MD.

Baird D. & Ulanowicz R.E. (1989) The seasonal dynamics of the Chesapeake Bay ecosystem. *Ecol. Monogr.*, **59**, 329–64.

Baird D., McGlade J.M. & Ulanowicz R.E. (1991) The comparative ecology of six marine ecosystems. *Phil. Trans. Roy. Soc. Land. B*, **333**, 15–29.

Baird D., Ulanowicz R.E. & Boynton W.R. (1995) Seasonal nitrogen dynamics in the Chesapeake Bay, a network approach. *Estuar coast. shelf Sci.*, **41**, 37–62.

Barnes D.J. (1973) Growth in colonial scleractinians. *Bull. mar. Sci.*, **23**, 280–98.

Barnes R.S.K. (1980) *Coastal Lagoons*. Cambridge University Press, Cambridge.

Barnes R.S.K. (1984) *Estuarine Biology*, 2nd edn. Edward Arnold, London.

Barnes R.S.K. (1994) Macrofaunal community structure and life histories in coastal lagoons. In Kjerfve B. (ed.), *Coastal Lagoon Processes*, pp. 311–62. Elsevier, Amsterdam.

Barnes R.S.K. (ed.) (1998) *The Diversity of Living Organisms*. Blackwell Science, Oxford.

Barnes R.S.K. & Hughes R.N. (1988) *An Introduction to Marine Ecology*, 2nd edn. Blackwell Scientific Publications, Oxford.

Barnes R.S.K. & Mann K.H. (eds) (1980) *Fundamentals of Aquatic Ecosystems*. Blackwell Scientific Publications, Oxford.

Barnes R.S.K. & Mann K.H. (eds) (1991) *Fundamentals of Aquatic Ecology*. Blackwell Scientific Publications, Oxford.

Basford A., Eleftheriou A. & Raffaelli D. (1990) The infauna and epifauna of the northern North Sea. *Neth. J. Sea Res.*, **25**, 165–73.

Battaglia B., Bisol P.M. & Fava G. (1978) Genetic variability in relation to the environment in some marine invertebrates. In Battaglia B. & Beardmore J.A. (eds), *Marine Organisms: Genetics, Ecology and Evolution*, pp. 53–70. Plenum, New York.

Bearman G. (ed.) (1989) *Oceanography*. Open University Press, Milton Keynes.

Bech G. (1995) Retrieval of lost gillnets at Ilulissat Kangia. NAFO Science Council Research Document 95/67, pp. 1–5.

Behrens-Yamada S. (1977) Geographic range limitation of the intertidal gastropods *Littorina sitkana* and *L. planaxis*. *Mar. Biol.*, **39**, 61–5.

BELYAEV G.M. (1966) Bottom fauna of the ultra-abyssal depths of the world ocean (in Russian). *Akad. Nauk SSSR*, **591**(9), 1–248.

BEVERTON R.J.H. & HOLT S.J. (1957) On the dynamics of exploited fish populations. *Fish. Invest. (Lond.)*, Ser. 2, **19**, 533 pp.

BLACK K., MORAN P., BURRAGE D. & DE'ATH G. (1995) Association of low-frequency currents and crown-of-thorns starfish outbreaks. *Mar. Ecol. Prog. Ser.*, **125**, 185–94.

BLACKBURN T.H. & SORENSEN T. (eds) (1988) *Nitrogen Cycling in Coastal Marine Environments*. John Wiley, New York.

BOLD H.C. & WYNNE M.J. (1978) *Introduction to the Algae*. Prentice–Hall, Englewood Cliffs, NJ.

BOWERBANK J.S. (1874) *A Monograph of British Spongiidae, Vol. 3*. Ray Society, London.

BOWMAN R.E. & MICHAELS W.L. (1984) Food of seventeen species of northwest Atlantic fish. NOAA Technical Memorandum NMFS-F/NEC 28.

BRANCH G.M. (1975) Intraspecific competition in *Patella cochlear* Born. *J. Anim. Ecol.*, **44**, 263–82.

BRAWLEY S.H. (1992) Mesoherbivores. *Systematics Association Special Volume* **46**, 235–63.

BRIGGS J.C. (1974) *Marine Zoogeography*. McGraw–Hill, London.

BUCHANAN J.B., SHEADER M. & KINGSTON P.R. (1978) Sources of variability in the benthic macrofauna off the south Northumberland coast, 1971–6. *J. mar. biol. Ass. U.K.*, **58**, 191–210.

BUSS L.W. & JACKSON J.B.C. (1979) Competitive networks: nontransitive competitive relationships in cryptic coral reef environments. *Am. Nat.*, **113**, 233–4.

BUSTAMENTE R.H., BRANCH G.M. & EEKHOUT S. (1995) Maintenance of an exceptional intertidal grazer biomass in South Africa: subsidy by subtidal kelps. *Ecology*, **76**, 2314–29.

CAMERON A.M., ENDEAN R. & DE VANTIER L.M. (1991) Predation on massive corals: are devastating population outbreaks of *Acanthaster planci* novel events? *Mar. Ecol. Prog. Ser.*, **75**, 251–8.

CAMPHUYSEN C.J., ENSOR K., FURNESS R.W. *et al.* (1993) Seabirds feeding on discards in winter in the North Sea. Netherlands Institute for Sea Research, Den Burg, Texel.

CHAPMAN V.J. (ed.) (1977) *Wet Coastal Ecosystems*. Elsevier, Amsterdam.

CHARLIER R. & CHARLIER C. (1992) Environmental, economic and social aspects of marine aggregates' exploitation. *Environ. Cons.*, **19**, 29–37.

CHILDRESS J.J. & FISHER C.R. (1992) The biology of hydrothermal vent animals: physiology, biochemistry, and autotrophic symbioses. *Oceanogr. mar. Biol., Ann. Rev.*, **30**, 337–441.

CHRISTIAN R.R., BOYER J.N., STANLEY D.W. & RIZZO W.M. (1992) Network analysis of nitrogen cycling in an estuary. In Hurst C. (ed.), *Modelling the Metabolic and Physiology Activities of Microorganisms*, pp. 217–47. John Wiley, New York.

CLARK R.B. (1964) *Dynamics in Metazoan Evolution*. Clarendon Press, Oxford.

CLARKE K.R. & WARWICK R.M. (1994) Change in marine communities: an approach to statistical analysis and interpretation. Natural Environment Research Council, Plymouth Marine Laboratory, Plymouth.

COLE J.J., FINDLEY, S. & PACE, M.L. (1988) Bacterial production in fresh and salt water ecosystems: a cross-system overview. *Mar. Ecol. Prog. Ser.*, **43**, 1–10.

COLLIE J.S., ESCANERO G.A. & VALENTINE P.C. (1997) Effects of bottom fishing on the benthic megafauna of George's Bank. *Mar. Ecol. Prog. Ser.*, **155**, 159–72.

CONNELL J.H. (1972) Community interactions on marine rocky intertidal shores. *Ann. Rev. Ecol. Syst.*, **3**, 169–92.

CONNELL J.H. (1973) Population ecology of reef-building corals. In Jones O.A. & Endean R. (eds), *Biology and Geology of Coral Reefs, Vol. II, Biology 1*, pp. 271–324. Academic Press, New York.

CONNELL J.H. (1978) Diversity in tropical rainforests and coral reefs. *Science, N.Y.*, **199**, 1302–10.

CONNELL J.H. (1980) Diversity and the coevolution of competitors, or the ghost of competition past. *Oikos*, **35**, 131–8.

CONNELL J.H. (1985) Variation and persistence of rocky shore populations. In Moore P.G. & Seed R. (eds), *The Ecology of Rocky Coasts*, pp. 57–69. Hodder & Stoughton, London.

CRISP D.J. (1962) *Grazing in Terrestrial and Marine Environments*. (Brit. Ecol. Soc. Symp. 4.) Blackwell Scientific Publications, Oxford.

CRISP D.J. (1976) The role of the pelagic larva. In Spencer Davies P. (ed.), *Perspectives in Experimental Zoology*, pp. 145–55. Pergamon Press, Oxford.

CRISP D.J. (1978) Genetic consequences of different reproductive strategies in marine invertebrates. In Battaglia B. & Beardmore J.A. (eds), *Marine Organisms: Genetics, Ecology and Evolution*, pp. 257–73. Plenum, New York.

CROMIE W.J. (1988) Grappling with coupled systems. *Mosaic*, **19**, 12–23.

CUNDELL A.M., BROWN M.S., STANFORD R. & MITCHELL R. (1979) Microbial degradation of *Rhizophora mangle* leaves immersed in the sea. *Estuar. coast. mar. Sci.*, **9**, 281–6.

CUNNINGHAM P.N. & HUGHES R.N. (1984) Learning of predatory skills by shore-crabs *Carcinus maenas* feeding on mussels and dogwhelks. *Mar. Ecol. Prog. Ser.* **16**, 21–6.

CUSHING D.H. (1959) The seasonal variation in oceanic production as a problem in population dynamics. *J. Cons.*, 24, 455–64.

CUSHING D.H. (1968) Fisheries biology: a study in population dynamics. Univ. Wisconsin Press, Madison.

CUSHING D.H. (1971) Upwelling and the production of fish. *Adv. mar. Biol.*, 9, 255–334.

CUSHING D.H. (1975) *Marine Ecology and Fisheries.* Cambridge University Press, Cambridge.

CUSHING D.H. (1995) *Population Production and Regulation in the Sea.* Cambridge University Press, Cambridge.

CUSHING D.H. (1996) *Towards a Science of Recruitment in Fish Populations.* Ecology Institute, Oldendorf.

CUSHING D.H. & BRIDGER J.P. (1966) The stock of herring in the North Sea and changes due to fishing. *Fish. Invest. Land. Ser.*, 2, 20(11), 1–31.

DAIBER F.C. (1982) *Animals of the Tidal Marsh.* Van Nostrand Reinhold, New York.

DALY J.M. (1972) The maturation and breeding biology of *Harmothoë imbricata* (Polychaeta: Polynoidae). *Mar. Biol.*, 12, 53–66.

DAVIS G., HAAKER P. & RICHARDS D. (1996) Status and trends of white abalone at the California Channel Islands. *Trans. Am. Fish. Soc.*, 125, 42–8.

DAY J.W., SMITH W.G., WAGNER P.R. & STOWE W.C. (1973) Community structure and carbon budget of a salt-marsh and shallow bay estuarine system in Louisiana. *Center Wetland, Louisiana State Univ. Publ.*, LSU-SG-72-04.

DAYTON P.K. (1971) Competition, disturbance, and community organization: the provision and subsequent utilization of space in a rocky intertidal community. *Ecol. Monogr.*, 41, 351–89.

DAYTON P.K. (1975) Experimental studies of algal canopy interactions in a sea otter-dominated kelp community at Amchitka Island, Alaska. *Fish Bull.*, 73, 230–7.

DAYTON P.K. (1984) Processes structuring some marine communities: are they general? In Strong D.R.Jr, Simberloff D., Abele L.G. & Thistle A.B. (eds), *Ecological Communities: Conceptual Issues and the Evidence*, pp. 181–97. Princeton University Press, Princeton, NJ.

DAYTON P.K. & HESSLER R.R. (1972) Role of biological disturbance in maintaining diversity in the deep sea. *Deep-Sea Res.*, 19, 199–208.

DAYTON P.K., TEGNER P.K., PARNELL P.E. & EDWARDS P.B. (1992) Temporal and spatial patterns of disturbance and recovery in a kelp forest community. *Ecol. Monogr.*, 62, 421–45.

DAYTON P.K., THRUSH S.F., AGARDY M.T. & HOFMAN R.J. (1995) Environmental effects of marine fishing. *Aquat. Cons.*, 5, 205–32.

DE BEER G. (1958) *Embryos and Ancestors*, 3rd edn. Oxford University Press, Oxford.

DENMAN K.L. (1994) Scale-determining biological–physical interactions in oceanic food webs. In Giller P.S., Hildrew A.G. & Raffaelli D.G. (eds), *Aquatic Ecology. Scale, Pattern and Process*, pp. 377–402. Blackwell Science, Oxford.

DE VANTLIER L.M. (1992) Rafting of tropical marine organisms on buoyant coralla. *Mar. Ecol. Prog. Ser.*, 86, 301–2.

DÖRJES J. & HOWARD J.D. (1975) Estuaries of the Georgia coast, USA: Sedimentology and biology. IV Fluvial–marine transition indicators in an estuarine environment, Ogeechee River–Ossabaw Sound. *Senckenb. marit.*, 7, 137–9.

DOWSON P., BUBB J. & LESTER J. (1996) Persistence and degradation pathways of tributyltin in freshwater and estuarine sediments. *Estuar. coast. Shelf Sci.*, 42, 551–62.

DUCKLOW H.W. (1983) Production and fate of bacteria in the oceans. *BioScience* 33, 494–501.

EDGAR G.J. (1990) The use of size structure of benthic communities to estimate faunal biomass and secondary production. *J. exp. mar. Biol. Ecol.*, 137, 195–214.

EDWARDS P., MAY R. & WEBB N. (1994) *Large-scale Ecology and Conservation Biology.* Blackwell Science, Oxford.

EKMAN S. (1967) *Zoogeography of the Sea.* Sidgwick & Jackson, London.

ENDEAN R. (1973) Population explosions of *Acanthaster planci* and associated destruction of hermatypic corals in the Indo-West Pacific region. In Jones O.A. & Endean R. (eds), *Biology and Geology of Coral Reefs II: Biology I*, pp. 389–438. Academic Press, New York.

ENFIELD D.B. (1989) El Niño, past and present *Rev. Geophys.*, 27, 159–87.

EPPLEY R.W., ROGERS J.N. & McCARTHY J.J. (1969) Half saturation constants for uptake of nitrate and ammonium by marine phytoplankton. *Limnol. Oceanogr.*, 14, 912–20.

ERZINI K., MONTEIRO C., RIBEIRO J. *et al.* (1997) An experimental study of gill net and trammel net 'ghost-fishing' off the Algarve (southern Portugal). *Mar. Ecol. Prog. Ser.*, 158, 257–65.

EVANS S., LEKSONO T. & McKINNELL P. (1995) Tributyltin pollution: a diminishing problem following legislation limiting the use of TBT-based anti-fouling paints. *Mar. Poll. Bull.*, 30, 14–21.

EVERAARTS J. (1997) DNA integrity in *Asterias rubens*: a biomarker reflecting the pollution of the North Sea? *J. Sea Res.*, 37, 123–9.

FAO (1995) Yearbook of Fishery Statistics, 26, FAO, Rome.

FARRELL T.M., BRACHER D. & ROUGHGARDEN J. (1991) Cross-shelf transport causes recruitment to intertidal populations in central California. *Limnol. Oceanogr.,* **36,** 279–88.

FENCHEL T. (1975) Character displacement and coexistence in mud snails (Hydrobiidae). *Oecologia (Berl.),* **20,** 19–32.

FENCHEL T. (1978) The ecology of micro- and meiobenthos. *Ann. Rev. Ecol. Syst.,* **9,** 99–121.

FENCHEL T. (1988) Marine plankton food chains. *Ann. Rev. Ecol. Syst.,* **19,** 19–38.

FENCHEL T. & RIEDL R.J. (1970) The sulphide system: a new biotic community underneath the oxidised layer of marine sand bottoms. *Mar. Biol.,* **7,** 255–68.

FIELD J.G., JARMAN N.G., DIECKMANN G.S. *et al.* (1977) Sun, waves, seaweed and lobsters: the dynamics of a west coast kelp-bed. *S. Afr. J. Sci.,* **73,** 7–10.

FISHER R.A. (1930) *The Genetical Theory of Natural Selection.* Clarendon Press, Oxford.

FLACH E.C. (1996) The influence of the cockle, *Cerastoderma edule,* on the macrozoobenthic community of tidal flats in the Wadden Sea. *Mar. Ecol., Stn Zool. Napoli,* **17,** 87–98.

FOGG G.E. (1980) Phytoplanktonic primary production. In Barnes R.S.K. & Mann K.H. (eds), *Fundamentals of Aquatic Ecosystems,* pp. 24–45. Blackwell Scientific Publications, Oxford.

FRIEDRICH H. (1965) *Meeresbiologie.* Borntraeger, Berlin.

FURNESS R.W. (1996) A review of seabird responses to natural or fisheries-induced changes in food supply. In Greenstreet S.P.R. & Tasker M.L. (eds), *Aquatic Predators and their Prey,* pp. 168–73. Blackwell Science, Oxford.

GAGE J.D. (1996) Why are there so many species in deep-sea sediments? *J. exp. mar. Biol. Ecol.* **200,** 257–86.

GAGE J. & GORDON J. (1995) Sound bites, science and the Brent Spar: environmental considerations relevant to the deep-sea disposal option. *Mar. Poll. Bull.,* **30,** 772–9.

GAGE J.D. & TYLER P.A. (1991) *Deep-Sea Biology.* Cambridge University Press, Cambridge.

GAINES S., BROWN S. & ROUGHGARDEN J. (1985) Spatial variation in larval concentrations as a cause of spatial variation in settlement for the barnacle, *Balanus glandula. Oecologia (Berl.),* **67,** 267–72.

GAINES S.D. & ROUGHGARDEN J. (1987) Fish in offshore kelp forests affect recruitment to intertidal barnacle populations. *Science,* **235,** 479–81.

GASTON K. (1996) *Biodiversity.* Blackwell Science, Oxford.

GEORGE J.D. & GEORGE J.J. (1979) *Marine Life.* Harrap, London.

GERLACH S.A. (1978) Food chain relationships in subtidal silty sand marine sediments and the role of meiofauna in stimulating bacterial productivity. *Oecologia (Berl.),* **33,** 55–69.

GIBBS P., BRYAN G., PASCOE P. & BURT G. (1987) The use of dog-whelk, *Nucella lapillus,* as an indicator of tributyltin (TBT) contamination. *J. mar. biol. Ass. U.K.,* **67,** 507–23.

GLYNN P.W. (1988) El Niño–Southern Oscillation 1982–83: nearshore population, community, and ecosystem responses. *Ann. Rev. Ecol. Syst.,* **19,** 309–45.

GOHAR H.A.F. & SOLIMAN G.N. (1963) On the mytilid species boring in living corals. *Publ. Mar. Biol. Stn. Ghardaqua,* **12,** 65–92.

GOMEZ E., ALCALA A. & YAP H. (1987) Other fishing methods destructive to coral. In Salvat B. (ed.), *Human Impacts on Coral reefs: Facts and Recommendations,* pp. 67–75. Antenne Museum, French Polynesia.

GOREAU T.F. & GOREAU N.I. (1973) The ecology of Jamaican coral reefs. II. Geomorphology, zonation and sedimentary phases. *Bull. mar. Sci.,* **23,** 399–464.

GOULD S.J. (1977) *Ontogeny and Phylogeny.* Belknap, Cambridge, MA.

GRAHAM M. (1935) Modern theory of exploiting a fishery and application to North Sea trawling. *J. Cons. int. Explor. Mer.,* **10,** 264–74.

GRAHAM M. (1955) Effect of trawling on animals of the sea bed. *Deep-Sea Res.,* Supplement 3, 1–16.

GRASSLE J.F. (1986) The ecology of deep-sea hydrothermal vent communities. *Adv. mar. Biol.,* **23,** 301–62.

GRASSLE J.F. & GRASSLE J.P. (1977) Temporal adaptations in sibling species of *Capitella.* In Coull B.C. (ed.), *Ecology of Marine Benthos,* pp. 177–88. University of South Carolina Press, Columbia.

GRAY J. & BOUCOT A.J. (eds) (1979) *Historical Biogeography, Plate Tectonics and the Changing Environment.* Oregon State University Press, Corvallis.

GRAY J.S. (1981) *The Ecology of Marine Sediments: An Introduction to the Structure and Function of Benthic Communities.* Cambridge University Press, Cambridge.

GREENSTREET S.P.R. (1996) Estimation of the daily consumption of food by fish in the North Sea in each quarter of the year. Scottish Office Agriculture, Fisheries and Environment Department, Aberdeen.

GREENSTREET S.P.R. & HALL S.J. (1996) Fishing and ground-fish assemblage structure in the north-western North Sea: an analysis of long-term and spatial trends. *J. Anim. Ecol.,* **65,** 557–98.

GRICE G.D. & REEVE M.R. (1982) *Marine Mesocosms. Biological and Chemical Research in Experimental Ecosystems.* Springer, New York.

GRIFFITHS R.J. (1977) Reproductive cycles in littoral populations of *Choromytilus meridionalis* (Kr.) and *Aulacomya ater* (Molina) with a quantitative assessment of gamete production in the former. *J. exp. mar. Biol. Ecol.*, **30**, 53–71.

GRIGG R.W., POLOVINA J.J. & ATKINSON M.J. (1984) Model of a coral reef ecosystem. III. Resource limitation, community regulation, fisheries yield and resource management. *Coral Reefs*, **3**, 23–7.

GRIME J.P. (1979) *Plant Strategies and Vegetation Processes*. John Wiley, Chichester.

GROSHOLZ E.D. & RUIZ G.M. (1995) Spread and potential impact of the recently introduced European green crab, *Carcinus maenas*, in central California. *Mar. Biol.*, **122**, 239–47.

HALL S.J. (1994) Physical disturbance and marine benthic communities: life in unconsolidated sediments. *Oceanogr. mar. Biol., Ann. Rev.*, **32**, 179–239.

HALL S.J., RAFFAELLI D. & THRUSH S.F. (1994) Patchiness and disturbance in shallow water benthic assemblages. In Giller P.S., Hildrew A.G. & Raffaelli D.G. (eds), *Aquatic Ecology. Scale, Pattern and Process*, pp. 333–75. Blackwell Science, Oxford.

HARDEN JONES F.R. (1968) *Fish Migration*. Edward Arnold, London.

HARDEN JONES F.R. (1980) The nekton: production and migration patterns. In Barnes R.S.K. & Mann K.H. (eds), *Fundamentals of Aquatic Ecosystems*, pp. 119–42. Blackwell Scientific Publications, Oxford.

HARDY A.C. (1924) The herring in relation to its animate environment. Part 1. The food and feeding habits of the herring with special reference to the east coast of England. *Fish. Invest. (Lond.)*, (Ser, 2), **7**(3), 1–53.

HARDY A.C. (1962) *The Open Sea: Its Natural History. Part I: The World of Plankton*. Collins, London.

HARGRAVE B.T. (1991) Ecology of deep-water zones. In Barnes R.S.K. & Mann K.H. (eds), *Fundamentals of Aquatic Ecology*, pp. 77–90. Blackwell Scientific Publications, Oxford.

HARPER J.L. (1977) *Population Biology of Plants*. Academic Press, New York.

HARPER J.L., ROSEN B.R. & WHITE J. (1986) The growth and form of modular organisms. *Phil. Trans. Roy. Soc. Lond.* B, **373**, 1–250.

HARRISON W.G., DOUGLAS D., FALKOWSKI P., ROWE G. & VIDAL J. (1983) Summer nutrient dynamics of the middle Atlantic Bight: nitrogen uptake and regeneration. *J. plankt. Res.*, **5**, 539–56.

HARVEY P.H. & PAGEL M.D. (1991) *The Comparative Method in Evolutionary Biology*. Oxford University Press, Oxford.

HAWKINS S.J. & HARTNOLL R.G. (1983) Grazing of intertidal algae by marine invertebrates. *Oceanogr. mar. Biol. Ann. Rev.*, **21**, 195–282.

HAWKINS S.J., HARTNOLL R.G., KAIN J.M. & NORTON T.A. (1992) Plant–animal interactions on hard substrata in the north-east Atlantic. *Systematics Association Special Volume* **16**, 1–32.

HECKY R.E. & KILHAM P. (1988) Nutrient limitation of phytoplankton in freshwater and marine environments: a review of recent evidence on the effects of enrichment. *Limnol. Oceanogr.*, **33**, 796–822.

HEINRICH A.K. (1962) The life histories of plankton animals and seasonal cycles of plankton communities in the oceans. *J. Cons.*, **27**, 15–24.

HERON A.C. (1972) Population ecology of a colonizing species: the pelagic tunicate *Thalia democratica*. 1. Individual growth rate and generation time. *Oecologia (Berl.)*, **10**, 269–93.

HESSLER R.R. & SANDERS H.L. (1967) Faunal diversity in the deep sea. *Deep-Sea Res.*, **14**, 65–78.

HEUVELMANS B. (1968) *In the Wake of the Sea-serpents*. Hart-Davis, London.

HIGHSMITH R.C. (1979) Coral growth rates and environmental control of density banding. *J. exp. mar. Biol. Ecol.*, **37**, 105–25.

HIGHSMITH R.C., RIGGS A.C. & D'ATTONIO C.M. (1980) Survival of hurricane-generated coral fragments and a disturbance model of reef calcification/growth rates. *Oecologia (Berl.)*, **46**, 322–9.

HOLLAND A.F., MOUNTFORD N.K., HIEGEL M.H., KAUMEYER K.R. & MIHURSKY J.A. (1980) Influence of predation on infaunal abundance in Upper Chesapeake Bay, USA. *Mar. Biol.*, **57**, 221–35.

HOWARTH R.W. (1988) Nutrient limitation of net primary production in marine ecosystems. *Ann. Rev. Ecol. Syst.*, **19**, 89–110.

HUGHES R.N. & THOMAS M. (1971) The classification and ordination of shallow-water benthic samples from Prince Edward Island, Canada. *J. exp. mar. Biol. Ecol.*, **7**, 1–39.

HUGHES R.N. (1977) The biota of reef-flats and limestone cliffs near Jeddah, Saudi Arabia. *J. nat. Hist.*, **11**, 77–96.

HUGHES R.N. (1980a) Predation and community structure. In Price J.H., Irvine D.E.G. & Farnham W.F. (eds), *The Shore Environment, Vol. 2: Ecosystems*, pp. 699–728. Syst. Ass. Spec. Vol. 17 (6).

HUGHES R.N. (1980b) Strategies for survival of aquatic organisms. In Barnes R.S.K. & Mann K.H. (eds), *Fundamentals of Aquatic Ecosystems*, pp. 162–84. Blackwell Scientific Publications, Oxford.

HUGHES R.N. (1989) *A Functional Biology of Clonal Animals*. Chapman & Hall, London.

HUGHES R.N. (1995) Resource allocation, demography and the radiation of life histories in rough periwinkles (Gastropoda). *Hydrobiologia*, **309**, 1–14.

HUGHES R.N. & GRIFFITHS C.L. (1987) Self-thinning in barnacles and mussels: the geometry of packing. *Am. Nat.*, **132**, 484–91.

HUGHES R.N. & ROBERTS D.J. (1980) Reproductive effort of winkles (*Littorina* spp.) with contrasted methods of reproduction. *Oecologia (Berl.)*, **47**, 130–6.

HUSTON M. (1979) A general hypothesis of species diversity. *Am. Nat.*, **113**, 81–101.

HUSTON M.A. (1994) *Biological Diversity*. Cambridge University Press, Cambridge.

HUTCHINSON G.E. (1967) *A Treatise on Limnology. Vol. II. Introduction to Lake Biology and the Limnoplankton*. John Wiley, New York.

IRLANDI E.A. & PETERSON C.H. (1991) Modification of animal habitat by large plants: mechanisms by which seagrasses influence clam growth. *Oecologia (Berl.)*, **87**, 307–18.

IVLEV V.S. (1961) Experimental ecology of the feeding of fishes. Yale University Press, New Haven.

JACCARINI V. & MARTENS E. (1992) *The Ecology of Mangrove and Related Ecosystems*. Kluwer, Dordrecht.

JACKSON J.B.C. (1979) Morphological strategies of sessile animals. In Larwood G. & Rosen B.R. (eds), *Biology and Systematics of Colonial Organisms*, pp. 499–555. Syst. Ass. Spec. Vol. 11.

JACKSON J.B.C. (1986) Modes of dispersal of clonal benthic invertebrates: consequences for species' distributions and genetic structure of local populations. *Bull. mar. Sci.*, **39**, 588–606.

JACKSON J.B.C. (1994) Constancy and change of life in the sea. *Phil. Trans. Roy. Soc. Lond.* B, **344**, 55–60.

JACKSON J.B.C. & COATES A.G. (1986) Life cycles and evolution of clonal (modular) animals. *Phil. Trans. Roy. Soc. Lond.* B, **313**, 7–22.

JACKSON J.B.C., BUSS L.W. & COOK R.E. (1985) *Population Biology and Evolution of Clonal Organisms*. Yale University Press, New Haven, CT.

JAKLIN S. & GÜNTHER C.P. (1996) Macrobenthic driftfauna of the Gröningen Plate. *Senckenb. marit.*, **26**, 127–34.

JENNINGS S. & KAISER M.J. (1998) The effects of fishing on marine ecosystems. *Adv. Mar. Biol.*, **34**, 000–000.

JENNINGS S., REYNOLDS J.D. & MILLS S. (1998) Life history correlates of responses to fisheries exploitation. *Proc. Roy. Soc. Lond.* B, **265**, 333–9.

JENSEN K.T. (1992) Macrozoobenthos on an intertidal mudflat in the Danish Wadden Sea: comparisons of surveys made in the 1930s, 1940s and 1980s. *Helgol. wiss. Meeresunters.*, **46**, 363–76.

JENSEN K.T. & KRISTENSEN L.D. (1990) A field experiment on competition between *Corophium volutator* (Pallas) and *Corophium arenarium* Crawford (Crustacea: Amphipoda): effects on survival, reproduction and recruitment. *J. exp. mar. Biol. Ecol.*, **137**, 1–24.

JOHANNESSON K. (1988) The paradox of Rockall: why is a brooding gastropod (*Littorina saxatilis*) more widespread than one having a planktonic larval dispersal stage (*L. littorea*)? *Mar. Biol.*, **99**, 507–13.

JOHNSON W.S., GIGON A., GULMON S.L. & MOONEY H.A. (1974) Comparative photosynthetic capacities of intertidal algae under exposed and submerged conditions. *Ecology*, **55**, 450–3.

JOHNSTONE J., SCOTT A. & CHADWICK H.C. (1924) *The Marine Plankton*. Hodder & Stoughton, London.

JOKIEL P.L. (1990) Long-distance dispersal by rafting: reemergence of an old hypothesis. *Endeavour*, **14**, 66–73.

JONES M.L. (ed.) (1985) Hydrothermal vents of the eastern Pacific: an overview. *Bull. biol. Soc. Wash.*, **6**, 1–547.

JONES O.A. & ENDEAN R. (1973) *Biology and Geology of Coral Reefs. II: Biology 1*. Academic Press, New York.

JØRGENSEN C.B. (1990) *Bivalve Filter-feeding: Hydrodynamics, Bioenergetics, Physiology and Ecology*. Olsen & Olsen, Fredensborg.

KAISER M.J., EDWARDS D.B., ARMSTRONG P. et al. (1998) Changes in megafaunal benthic communities in different habitats after trawling disturbance. *ICES J. mar. Sci.*, **55**, 353–61.

KAISER M.J., BULLIMORE B., NEWMAN P., LOCK K. & GILBERT S. (1996) Catches in 'ghost-fishing' set nets. *Mar. Ecol. Prog. Ser.*, **145**, 11–16.

KAISER M.J. & SPENCER B.E. (1996) The effects of beam-trawl disturbance on infaunal communities in different habitats. *J. Anim. Ecol.*, **65**, 348–58.

KENNEDY A.D. (1993) Minimal predation upon meiofauna by endobenthic macrofauna in the Exe Estuary, south west England. *Mar. Biol. (Berl.)*, **117**, 311–19.

KENNY A.J. & REES H.L. (1996) The effects of marine gravel extraction on the macrobenthos: results 2 years post-dredging. *Mar. Poll. Bull.*, **32**, 615–22.

KING M. (1995) *Fisheries Biology Assessment and Management*. Fishing News Books, Oxford.

KINGSFORD M.J., UNDERWOOD A.J. & KENNELLY S.J. (1991) Humans as predators on rocky reefs in New South Wales, Australia. *Mar. Ecol. Prog. Ser.*, **72**, 1–14.

KIØRBOE T. (1993) Turbulence, phytoplankton cell size, and structure of pelagic food webs. *Adv. mar. Biol.*, **29**, 1–72.

KNEIB R.T. (1985) Predation and disturbance by grass shrimp, *Palaemonetes pugio* Holthuis, in soft-substratum benthic invertebrate assemblages. *J. exp. mar. Biol. Ecol.*, **93**, 91–102.

KNOWLTON N. (1993) Sibling species in the sea. *Ann. Rev. Ecol. Syst.*, **24**, 189–216.

KOBLENTZ-MISHKE O.J., VOLKOVISNY V.V. & KABANOVA J.G. (1970) Plankton primary production of the world ocean. In Wooster W.S. (ed.), *Scientific Exploration of the South Pacific*, pp. 183–93. National Academy of Science, Washington, DC.

KOCK K. & SHIMADZU Y. (1994) Trophic relationships and trends in population size and reproductive parameters in antarctic high-level predators. In El-Sayed S. (ed.), *Southern Ocean Ecology: the BIOMASS Perspective*, pp. 287–312. Cambridge University Press, Cambridge.

KOHN A.J. (1959) The ecology of *Conus* in Hawaii. *Ecol. Monogr.*, **29**, 47–90.

KOHN A.J. & PERRON F.E. (1993) *Life History and Biogeography: Patterns in Conus*. Oxford University Press, Oxford.

KOLDING S. (1986) Interspecific competition for mates and habitat selection in five species of *Gammarus* (Amphipoda: Crustacea). *Mar. Biol. (Berl.)*, **91**, 491–5.

KREBS J.R. & MCCLEERY R.H. (1984) Optimization in behavioural ecology. In Krebs. J.R. & Davies N.B. (eds), *Behavioural Ecology: An Evolutionary Approach*, pp. 91–121. Blackwell Scientific Publications, Oxford.

KRUSE G.H. & KIMBER A. (1993) Degradable escape mechanisms for pot gear: a summary report to the Alaska Board of Fisheries. Alaska Department of Fisheries and Game, Juneau.

KUIPERS B.R., DE WILDE P.A.W.J. & CREUTZBERG F. (1981) Energy flow in a tidal flat ecosystem. *Mar. Ecol. Progr. Ser.*, **5**, 215–21.

LAMBSHEAD P.J.D., PLATT H.M. & SHAW K.M. (1983) The detection of differences among assemblages of marine benthic species based on an assessment of dominance and diversity. *J. nat. Hist.*, **17**, 859–74.

LANG J. (1973) Interspecific aggression by scleractinian corals. 2. Why the race is not only to the swift. *Bull. mar. Sci.*, **23**, 260–79.

LASSEN H.H. & TURANO F.J. (1978) Clinal variation and heterozygote deficit at the Lap-locus in *Mytilus edulis*. *Mar. Biol.*, **49**, 245–54.

LASSIG B.R. (1977) Communication and coexistence in a coral community. *Mar. Biol.*, **42**, 85–92.

LAWS R. (1977) Seals and whales of the Southern Ocean. *Phil. Trans. Roy. Soc. Lond. B*, **279**, 81–96.

LE FÈVRE J. (1986) Aspects of the biology of frontal systems. *Adv. Mar. Biol.*, **23**, 163–299.

LEONARD J.L. & LUKOWIAK K. (1991) Sex and the simultaneous hermaphrodite: testing models of male–female conflict in a sea slug, *Navanax inermis* (Opisthobranchia). *Anim. Behav.*, **41**, 255–66.

LEVINTON J.S. (1979) Deposit-feeders, their resources, and the study of resource limitation. In Livingston R.J. (ed.), *Ecological Processes in Coastal and Marine Systems*, pp. 117–41. Plenum, New York.

LEVITAN D. & PETERSEN C. (1995) Sperm limitation in the sea. *Trends Ecol. Evol.*, **10**, 228–31.

LEWIS J.R. (1964) *The Ecology of Rocky Shores*. English Universities Press, London.

LINDEMAN R.L. (1942) The trophic-dynamic aspect of ecology. *Ecology*, **23**, 399–418.

LITTLER M.M. & LITTLER D.S. (1980) The evolution of thallus form and survival strategies in benthic macroalgae: field and laboratory tests of a functional form model. *Am. Nat.*, **116**, 25–44.

LLUCH-BELDA D., CRAWFORD R.J.M., KAWASAKI T. et al. (1989) World-wide fluctuations of sardine and anchovy stocks: the regime problem. *S. Afr. J. Mar. Sci.*, **8**, 195–205.

LONG S.P. & MASON C.F. (1983) *Saltmarsh Ecology*. Blackie, Glasgow.

LOPEZ G. & LEVINTON J.S. (1987) Ecology of deposit-feeding animals in marine sediments. *Quart. Rev. Biol.*, **62**, 235–60.

LUBCHENCO J. (1978) Plant species diversity in a marine intertidal community: importance of herbivore food preference and algal competitive abilities. *Am. Nat.*, **112**, 23–39.

LÜNING K. (1979) Growth strategies of three *Laminaria* species (Phaeophyceae) inhabiting different depth zones in the sublittoral region of Helgoland (North Sea). *Mar. Ecol. Prog. Ser.*, **1**, 195–207.

MCCLANAHAN T. (1995) A coral-reef ecosystem–fisheries model—impacts of fishing intensity and catch selection on reef structure and processes. *Ecol. Modelling*, **80**, 1–19.

MCCLANAHAN T.R. (1992) Resource utilization, competition and predation: a model and example from coral reef grazers. *Ecol. Modelling*, **61**, 195–215.

MCCLANAHAN T.R. & MUTHIGA N.A. (1988) Changes in Kenyan coral reef community structure and function due to exploitation. *Hydrobiologia*, **166**, 269–76.

MCCLANAHAN T.R. & OBURA D. (1995) Status of Kenyan coral reefs. *Coastal Management*, **23**, 57–76.

MACISAAC J.J. & DUGDALE R.C. (1969) The kinetics of nitrate and ammonia uptake by natural populations of marine phytoplankton. *Deep-Sea Res.*, **16**, 45–57.

MCLAREN I.A. (1965) Some relationships between temperature and egg size, body size, development rate and fecundity, of the copepod *Pseudocalanus*. *Limnol. Oceanogr.*, **10**, 528–38.

MANGUM C.P. (1976) Primitive respiratory adaptations. In Newell R.C. (ed.), *Adaptation to Environment*, pp. 191–278. Butterworth, London.

MANN K.H. (1973) Seaweeds: their productivity and strategy for growth. *Science (N.Y.)*, **182**, 975–81.

MANN K.H. & FIELD J.G. & WULFF F. (1989) Network analysis in marine ecology: an assessment. In Wulff F., Field J.G. & Mann K.H. (eds), *Network Analysis in Marine Ecology: Methods and Applications*, pp. 259–82. Springer-Verlag, Berlin.

MARTEL A. & CHIA F.-S. (1991) Drifting and dispersal of small bivalves and gastropods with direct development. *J. exp. mar. Biol. Ecol.*, **150**, 131–47.

MATTHIESSEN P., WALDOCK R., THAIN J., WAITE M. & SCROPE-HOWE S. (1995) Changes in periwinkle (*Littorina littorea*) populations following the ban on TBT-based antifoulings on small boats in the United Kingdom. *Ecotoxicol. environ. Safety*, **30**, 180–94.

MENGE B.A., DALEY B.A., WHEELER P.A. & STRUB P.T. (1997) Rocky intertidal oceanography: an association between community structure and nearshore phytoplankton concentration. *Limnol. Oceanogr.*, **42**, 57–66.

MENGE B.A. & SUTHERLAND J.P. (1976) Species diversity gradients: synthesis of the roles of predation, competition, and temporal heterogeneity. *Am. Nat.*, **110**, 351–69.

MENZIES R.J., GEORGE R.Y. & ROWE G.T. (1973) *Abyssal Environment and Ecology of the World Ocean*. John Wiley, New York.

MERGNER H. (1971) Structure ecology and zonation of Red Sea reefs (in comparison with South Indian and Jamaican reefs). In Stoddart D.R. & Yonge C.M. (eds), *Regional Variation in Indian Ocean Coral Reefs*, pp. 141–61. Symp. Zool. Soc. Lond. 28.

METAXAS A. & YOUNG C.M. (1998) Responses of echinoid larvae to food patches of different algal densities. *Mar. Biol. (Berl.)*, **130**, 433–45.

MILLAR R.H. (1970) *British Ascidians*. Academic Press, London.

MILLS E.L. (1969) The community concept in marine zoology, with comments on continua and instability in some marine communities: a review. *J. fish. Res. Bd Canada*, **26**, 1415–28.

MILNER R.S. & WHITING C.L. (1996) Long-term changes in growth and population abundance of sole in the North Sea from 1940 to the present. *ICES J. Mar. Sci.*, **53**, 1185–95.

MORROW C.C., THORPE J.P. & PICTON B.E. (1992) Genetic divergence and cryptic speciation in two morphs of the common subtidal nudibranch *Doto coronata* (Opisthobranchia: Dendronotacea: Dotoidae) from the northern Irish Sea. *Mar. Ecol. Prog. Ser.*, **84**, 53–61.

MYERS A.A. (1994) Biogeographic patterns in shallow-water marine systems and the controlling processes at different scales. In Giller P.S., Hildrew A.D. & Raffaelli D.G. (eds), *Aquatic Ecology, Scale, Pattern and Process*, pp. 547–74. Blackwell Scientific Publications, Oxford.

NAKAMARA K. & SEKIGUCHI K. (1980) Mating behaviour and oviposition in the pycnogonid *Propellane longiceps*. *Mar. Ecol. Prog. Ser.*, **2**, 163–8.

NAVARRETE S. & CASTILLA J.C. (1990) Barnacle walls as mediators of intertidal mussel recruitment: effects of patch size on the utilization of space. *Mar. Ecol. Prog. Ser.*, **68**, 113–19.

NELSON W.G. & CAPONE M.A. (1990) Experimental studies of predation on polychaetes associated with seagrass beds. *Estuaries*, **13**, 51–8.

NEWELL R.C. & LINLEY E.A.S. (1984) Significance of microheterotrophs in the decomposition of phytoplankton: estimates of carbon and nitrogen flow based on the biomass of plankton communities. *Mar. Ecol. Prog. Ser.*, **16**, 105–19.

NEWMAN G.G. (1970) Stock assessment of the pilchard *Sardinops ocellata* at Walvis Bay, South West Africa. *Invest. Rep. Div. Sea Fish. S. Afr.*, **85**, 1–13.

NIXON S.W. (1980) Between coastal marshes and coastal waters—a review of twenty years of speculation and research on the role of salt marshes in estuarine productivity and water chemistry. In Hamilton P. & Macdonald K.B. (eds), *Estuarine and Wetland Processes with Emphasis on Modeling*, pp. 437–525. Plenum, New York.

NIXON S.W., OVIATT C.A. & HALE S.S. (1976) Nitrogen regeneration and the metabolism of coastal marine bottom communities. In Anderson J.M. & Macfadyen A. (eds), *The Role of Terrestrial and Aquatic Organisms in Decomposition Processes*, pp. 269–83. Blackwell Scientific Publications, Oxford.

NIXON S.W. & PILSON M.Q. (1983) Nitrogen in estuarine and coastal environments. In Carpenter E.J. & Capone D.G. (eds), *Nitrogen in the Marine Environment*, pp. 565–648. Academic Press, New York.

NORTH W.J. (ed.) (1971) The biology of giant kelp beds (*Macrocystis*) in California. *Beih. Nova Hedwigia*, **32**.

OGDEN J.C. & EBERSOLE J.P. (1981) Scale and community structure of coral reef fishes: a long-term study of a large artificial reef. *Mar. Ecol. Progr. Ser.*, **4**, 97–103.

ÓLAFSSON E.B. (1986) Density dependence in suspension-feeding and deposit-feeding populations of the bivalve *Macoma balthica*: a field experiment. *J. Anim. Ecol.*, **55**, 517–26.

ÓLAFSSON E.B., PETERSON C.H. & AMBROSE W.G. Jr (1994) Does recruitment limitation structure populations and communities of macro-invertebrates in marine soft sediments: the relative significance of pre- and post-settlement processes. *Oceanogr. mar. Biol., Ann. Rev.*, **32**, 65–109.

ORMOND R.F.G., GAGE J.D. & ANGEL M.V. (eds) (1998) *Marine Biodiversity. Patterns and Processes.* Cambridge University Press, Cambridge.

OTA (US Congress Office of Technology Assessment) (1987) *Technologies to Maintain Biological Diversity.* US Government Printing Office, Washington, DC. (1.2).

PAINE R.T. (1966) Food web complexity and species diversity. *Am. Nat.*, **100**, 65–75.

PAINE R.T. & LEVIN S.A. (1981) Intertidal landscapes: disturbance and the dynamics of pattern. *Ecol. Monogr.*, **51**, 145–78.

PALMER M.A. (1988) Epibenthic predators and marine meiofauna: separating predation, disturbance and hydrodynamic effects. *Ecology*, **69**, 1251–9.

PALUMBI S.R. (1994) Genetic divergence, reproductive isolation, and marine speciation. *Ann. Rev. Ecol. Syst.*, **25**, 547–72.

PALUMBI S.R. (1996) What can molecular genetics contribute to marine biogeography? An urchin's tale. *J. Mar. Biol. Ass. U.K.*

PARSONS T.R. & LALLI C.M. (1988) Comparative oceanic ecology of the plankton communities of the subarctic Atlantic and Pacific Oceans. *Oceanogr. mar. Biol., Ann. Rev.*, **26**, 317–59.

PARSONS T.R., TAKAHASHI M. & HARGRAVE B. (1977) *Biological Oceanographic Processes*, 2nd edn. Pergamon Press, New York.

PARSONS T.R., TAKAHASHI M. & HARGRAVE B. (1984) *Biological Oceanographic Processes*, 3rd edn. Pergamon Press, Oxford.

PAULY D. & CHRISTENSEN V. (1995) Primary production required to sustain global fisheries. *Nature*, **374**, 255–7.

PAYNE M. (1977) Growth of a fur seal population. *Philos. Trans. Roy. Soc. Lond. B*, **279**, 67–79.

PEARSON T.H. (1975) The benthic ecology of Loch Linnhe and Loch Eil, a sea-loch system on the west coast of Scotland. IV. Changes in the benthic fauna attributable to organic enrichment. *J. exp. mar. Biol. Ecol.*, **20**, 1–41.

PEATTIE M. & HOARE R. (1981) The sublittoral ecology of the Menai Strait II. The sponge *Halichondria panicea* (Pallas) and its associated fauna. *Estuar. coast. Shelf Sci.*, **13**, 621–33.

PERRIN W., DONOVAN G. & BARLOW J. (1994) *Gillnets and Cetaceans.* International Whaling Commission, Cambridge.

PETERSON C.H. (1979) Predation, competitive exclusion, and diversity in the soft-sediment benthic communities of estuaries and lagoons. In Livingston R.J. (ed.), *Ecological Processes in Coastal and Marine Systems*, pp. 233–64. Plenum, New York.

PHILANDER G. (1990) *El Niño, la Niña, and the Southern Oscillation.* Academic Press, San Diego, CA.

PHILLIPS R.C. & MENEZ E.G. (1988) *Seagrasses. Smithsonian Contr. Mar. Sci.*, **34**.

PIANKA E.R. (1974) *Evolutionary Ecology.* Harper & Row, London.

PIKE G.C. (1962) Migration and feeding of the gray whale (*Eschrichtius gibbosus*). *J. fish. Res. Bd Canada*, **19**, 815–38.

POMEROY L.R. & WIEGERT R.G. (eds) (1981) *The Ecology of a Salt Marsh.* Springer, New York.

QASIM S.Z. (1970) Some problems related to the food chain in a tropical estuary. In Steele J.H. (ed.), *Marine Food Chains*, pp. 46–51. Oliver & Boyd, Edinburgh.

RAMSAY K., KAISER M.J., MOORE P.G. & HUGHES R.N. (1997) Consumption of fisheries discards by benthic scavengers: utilisation of energy subsidies in different marine habitats. *J. Anim. Ecol.*, **66**, 884–96.

RAMUS J., BEALE S.I. & MAUZERALL D. (1976) Correlation of changes in pigment content with photosynthetic capacity of seaweeds as a function of water depth. *Mar. Biol.*, **37**, 231–8.

RAYMONT J.E.G. (1983) *Plankton and Productivity in the Oceans*, 2nd Edn. Pergamon Press, Oxford.

RAYMONT J.E.G. (1983) *Plankton and Productivity in the Oceans*, 2nd edn. Vol. 2: Zooplankton. Pergamon Press, Oxford.

REEB C.A. & AVISE J.C. (1990) A genetic discontinuity in a continuously distributed species: mitochondrial DNA in the American oyster, *Crassostrea virginica*. *Genetics*, **124**, 397–406.

REID D.G. (1996) *Systematics and Evolution of Littorina.* Ray Society, London.

REID J.L. (1962) On circulation, phosphate phosphorus content, and zooplankton volumes in the upper part of the Pacific Ocean. *Limnol. Oceanogr.*, **7**, 287–306.

REISE K. & AX P. (1979) A meiofauna 'thiobios' limited to the anaerobic sulfide system of marine sand does not exist. *Mar. Biol. (Berl.)*, **54**, 225–37.

REISE K. (1985) *Tidal Flat Ecology: An Experimental Approach to Species Interactions.* Springer, Berlin.

REX M.A. (1981) Community structure in the deep-sea benthos. *Ann. Rev. Ecol. Syst.*, **12**, 331–53.

RICE A.L. & LAMBSHEAD P.J.D. (1994) Patch dynamics in the deep-sea benthos: the role of a heterogeneous supply of organic matter. In Giller P.S., Hildrew A.G. & Raffaelli D.G. (eds), *Aquatic Ecology. Scale, Pattern and Process*, pp. 469–97. Blackwell Science, Oxford.

RICKER W.E. (1969) Food from the sea. In Cloud (ed.), *Resources and Man*, pp. 87–108. Freeman, Chicago, IL.

RIESEN W. & REISE K. (1982) Macrobenthos of the subtidal Wadden Sea: revisited after 55 years. *Helgol. wiss. Meeresunters.*, **35**, 409–23.

RIJNSDORP A.D., Buijs A.M., Storbeck F. & VISSER E. (1998) Micro-scale distribution of beam trawl effort in the southern North Sea between 1993 and 1996 in relation to the trawling frequency of the sea bed and the impact on benthic organisms. *ICES J. mar. Sci.*, **55**.

RIJNSDORP A.D. & VAN LEEUWEN P.I. (1996) Changes in growth of North Sea plaice since 1950 in relation to density, eutrophication, beam-trawl effort, and temperature. *ICES J. mar. Sci.*, **53**, 1199–213.

RISK M.J., SAMMARCO P.W. & SCHWARCZ H.P. (1994) Cross-continental shelf trends in $\delta^{13}C$ in coral on the Great Barrier Reef. *Mar. Ecol. Prog. Ser.*, **106**, 121–30.

ROBERTSON A.I. (1979) The relationship between annual production: biomass ratios and lifespans for marine macrobenthos. *Oecologia (Berl.)*, **38**, 193–202.

RODRIGUEZ C. & STONER A.W. (1990) The epiphyte community of mangrove roots in a tropical estuary: distribution and biomass. *Aquat. Bot.*, **36**, 117–26.

RODRIGUEZ J. & MULLIN M.M. (1986) Relationship between biomass and body weight in a steady state oceanic ecosystem. *Limnol. Oceanogr.*, **31**, 361–70.

ROSEN B.R. (1984) Reef coral biogeography and climate through the late Cainozoic: just islands in the sun or a critical pattern of islands? In *Fossils and Climate*, pp. 25–30. *Geol. J.* Special Issue 11.

ROWE G.T. (1983) Biomass and production of deep-sea macrobenthos. In Rowe G.T. (ed.), *The Sea*, **8**, 97–121. John Wiley, New York.

RUSSELL-HUNTER W.D. (1970) *Aquatic Productivity*. Macmillan, London.

RYLAND J.S. (1962) Biology and identification of intertidal Bryozoa. *Field Stud.*, **1**, 1–19.

RYTHER J.H. (1956) Photosynthesis in the ocean as a function of light intensity. *Limnol. Oceanogr.*, **1**, 61–70.

SAINSBURY K.J. (1987) Assessment and management of the demersal fishery on the continental shelf of northwestern Australia. In Polovina J.J. & Ralston S. (eds), *Tropical Snappers and Groupers—Biology and Fisheries Management*, pp. 465–503. Westview Press, Boulder, CO.

SALE P.S. (1980) The ecology of fishes on coral reefs. *Oceanogr. mar. Biol. Ann. Rev.*, **18**, 367–421.

SALEMAA H. (1987) Herbivory and microhabitat preferences of *Idotea* spp. (Isopoda) in the northern Baltic Sea. *Ophelia*, **27**, 1–15.

SANDERS H.L. (1968) Marine benthic diversity: a comparative study. *Am. Nat.*, **102**, 243–82.

SANTELICES B. (1992) Digestion survival in seaweeds: an overview. *Systematics Association Special Volume*, **46**, 363–84.

SAVIDGE G. (1976) A preliminary study of the distribution of chlorophyll *a* in the vicinity of fronts in the Celtic and western Irish Seas. *Estuar. coast. mar. Sci.*, **4**, 617–25.

SHAEFER M.B. (1954) Some aspects of the dynamics of populations important to the management of the commercial fish populations. *Inter. Amer. Trop. Tuna Comm. Bull.*, **1**, 27–56.

SCHAEFER M.B. (1957) A study of the dynamics of the fishery for yellowfin tuna in the eastern tropical Pacific Ocean. *Bull. Int. Am. trop. Tuna Comm.*, **2**, 245–85.

SCHÄFER W. (1972) *Ecology and Palaeoecology of Marine Environments*. (Translated by I. Oertel.) Oliver & Boyd, Edinburgh.

SCHEFFER V.B. (1973) The last days of the sea cow. *Smithsonian*, **3**(10), 64–7.

SCHELTEMA R.S. (1978) On the relationship between dispersal of pelagic veliger larvae and the evolution of marine prosobranch gastropods. In Battaglia B. & Beardmore J.A. (eds), *Marine Organisms: Genetics, Ecology and Evolution*, pp. 303–22. Plenum, New York.

SCHONBECK M. & NORTON T.A. (1978) Factors controlling the upper limits of fucoid algae on the shore. *J. exp. mar. Biol. Ecol.*, **31**, 303–13.

SCHONBECK M. & NORTON T.A. (1980) Factors controlling the lower limits of fucoid algae on the shore. *J. exp. mar. Biol. Ecol.*, **43**, 131–50.

SCHOPF T.J.M., FISHER J.B. & SMITH III C.A.F. (1978) Is the marine latitudinal diversity gradient merely another example of the species area curve? In Battaglia B. & Beardmore J.A. (eds), *Marine Organisms: Genetics, Ecology and Evolution*, pp. 365–86. Plenum, New York.

SCHOPF T.J.M. & GOOCH J.L. (1971) Gene frequencies in a marine ectoproct: a cline in natural populations related to sea temperature. *Evolution*, **25**, 286–9.

SCHUMACHER H. (1976) *Korallenriffe*. BLV, Munich.

SCHWINGHAMER P. (1983) Generating ecological hypotheses from biomass spectra using causal analysis: a benthic example. *Mar. Ecol. Prog. Ser.*, **13**, 151–66.

SEARLES R.B. (1980) The strategy of the red algal life history. *Am. Nat.*, **115**, 113–20.

SEED R. & O'CONNOR R.J. (1981) Community organization in marine algal epifaunas. *Ann. Rev. Ecol. Syst.*, **12**, 49–74.

SHELDON R.W., NIVAL P. & RASSOULZADEGAN F. (1986) An experimental investigation of a flagellate–ciliate–copepod food chain with some observations relevant to the linear biomass hypothesis. *Limnol. Oceanogr.*, **31**, 184–88.

SHEPPARD C.R.C. (1980) Coral cover, zonation and diversity on reef slopes of Chagos atolls, and population structures of the major species. *Mar. Ecol. Progr. Ser.*, **2**, 193–205.

SHERR F., & SHERR B. (1988) Role of microbes in pelagic food webs: a revised concept. *Limnol. Oceanogr.*, **33**, 1225–7.

SHORT F.T. (1980) A simulation model of the seagrass production system. In Phillips R.C. & McRoy C.P. (eds), *Handbook of Seagrass Biology: An Ecosystem Perspective*, pp. 277–95. Garland Press, New York.

SIBLY R.M. & CALOW P. (1986) *Physiological Ecology of Animals: an Evolutionary Approach.* Blackwell Scientific Publications, Oxford.

SIEBURTH J.M. (1979) *Sea Microbes.* Oxford University Press, New York.

SIMON J.L. & DAUER D.M. (1977) Re-establishment of a benthic community following natural defaunation. In Coull B.C. (ed.), *Ecology of Marine Benthos*, pp. 139–54. University of South Carolina Press, Columbia.

SIMPSON J.H. (1976) A boundary front in the summer regime of the Celtic Sea. *Estuar. coast. mar. Sci.*, **4**, 71–81.

SLADEN W. (1964) The distribution of the Adélie and chinstrap penguins. In Carrick R., Holdgate M. & Prevost J. (eds), *Biologie Antarctique*, pp. 359–65. Hermann, Paris.

SMALDON G. (1979) *British Coastal Shrimps and Prawns.* Academic Press, London.

SMITH T.D. (1983) Changes in size of three dolphin (Stenella spp.) populations in the eastern tropical Pacific. *Fisheries Bulletin*, **81**, 1–13.

SMITH G., READ A. & GASKIN D. (1983) Incidental catch of harbour porpoise, *Phocoena phocoena* (L.) in herring weirs in Charlotte county, New Brunswick. *Can. fish. Bull.*, **81**, 660–2.

SNELGROVE P.V.R. & BUTMAN C.A. (1994) Animal–sediment relationships revisited: cause versus effect. *Oceanogr. mar. Biol., Ann. Rev.*, **32**, 179–239.

SOMERFIELD P., OLSGARD F. & CARR M. (1997) A further examination of two new taxonomic distinctness measures. *Mar. Ecol. Prog. Ser.*, **154**, 303–6.

SOUTHWARD A.J., HAWKINS S.J. & BURROWS M.T. (1995) 70 years observations of changes in distribution and abundance of zooplankton and intertidal organisms in the western English Channel in relation to rising sea temperature. *J. therm. Biol.*, **20**, 127–55.

SOUTHWARD A. & SOUTHWARD E. (1978) Recolonization of rocky shores in Cornwall after use of toxic dispersants to clean up the *Torrey Canyon* spill. *J. fish. Res. Bd Canada*, **35**, 682–706.

SPENCER B.E., KAISER M.J. & EDWARDS D.B. (1998) Intertidal clam harvesting: benthic community change and recovery. *Aquaculture Research*, **29**, 429–37.

SPENCER-DAVIES P., STODDART D.R. & SIGEE D.C. (1971) Reef forms of Addu Atoll, Maldive Islands. In Stoddart D.R. & Yonge C.M. (eds), *Regional Variation in Indian Ocean Coral Reefs*, pp. 217–59. Symp. Zool. Soc. Lond. 28.

STAFFORD-DEITSCH J. (1996) *Mangrove. The Forgotten Habitat.* Immel, London.

STANWELL-SMITH D., PECK L.S., CLARKE A., MURRAY A.W.A. & TODD C.D. (1998) The distribution, abundance and seasonality of pelagic marine invertebrate larvae in the maritime Antarctic. *Proc. R. Soc. Lond. B.*, **353**, 1–14.

STEBBING A.R.D. (1973) Competition for space between the epiphytes of *Fucus serratus. J. mar. biol. Ass. U.K.*, **53**, 247–61.

STEHLI F.G. & WELLS J.W. (1971) Diversity and age patterns in hermatypic corals. *Systematic Zool.*, **20**, 115–26.

STEPHENSON T.A. & STEPHENSON A. (1972) *Life between Tide Marks on Rocky Shores.* W.H. Freeman, San Francisco, CA.

STODDART D.R. (1971) Environment and history in Indian Ocean reef morphology. In Stoddart D.R. & Yonge C.M. (eds), *Regional Variation in Indian Ocean Coral Reefs.* pp. 3–38. Symp. Zool. Soc. Lond. 28.

STODDART D.R. & YONGE C.M. (eds) (1971) *Regional Variation in Indian Ocean Coral Reefs.* Symp. Zool. Soc. Lond. 28.

STRONG K.W. & DABORN G.R. (1979) Growth and energy utilization of the intertidal isopod *Idotea baltica* (Pallas) (Crustacea: Isopoda). *J. exp. mar. Biol. Ecol.*, **41**, 101–23.

SUESS E. (1980) Particulate carbon flux in the ocean-surface productivity and oxygen utilization. *Nature (Lond.)*, **288**, 260–3.

SVERDRUP H.U. (1953) On conditions for the vernal blooming of phytoplankton. *J. Cons.*, **18**, 287–95.

SVERDRUP H.U., JOHNSON M.W. & FLEMING R.H. (1942) *The Oceans.* Prentice–Hall, New York.

TAYLOR J.D. & TAYLOR C.N. (1977) Latitudinal distribution of predatory gastropods on the eastern Atlantic Shelf. *J. Biogeogr.*, **4**, 73–81.

TAYLOR W.R. (1964) Light and photosynthesis in intertidal benthic diatoms. *Helgol. wiss. Meeresunters.*, **10**, 29–37.

TEAL J.M. (1980) Primary production of benthic and fringing plant communities. In Barnes R.S.K. & Mann K.H. (eds), *Fundamentals of Aquatic Ecosystems*, pp. 67–83. Blackwell Scientific Publications, Oxford.

THEIL H. & SCHRIEVER G. (1990) Deep-sea mining, environmental impact and the DISCOL project. *Ambio*, **19**, 245–50.

THORPE J.P., BEARDMORE J.A. & RYLAND J.S. (1978) Genetic evidence for cryptic speciation in the marine bryozoan *Alcyonidium hirsutum*. *Mar. Biol.*, **49**, 27–32.

THORSON G. (1946) Reproduction and larval development of Danish marine bottom invertebrates. Meddr. Komm. Danm. Fish.,-OG Havunders. Plankton 4, 1–523.

THORSON G. (1957) Bottom communities. In Hedgpeth J.W. (ed.), *Treatise on Marine Ecology and Paleoecology, 1 Ecology*, pp. 461–534. Geological Society of America, New York.

THORSON G. (1971) *Life in the Sea*. Weidenfeld & Nicolson, London.

THRUSH S.F., WHITLATCH R.B., PRIDMORE R.D. *et al.* (1996) Scale-dependent recolonization: the role of sediment stability in a dynamic sandflat habitat. *Ecology*, **77**, 2472–87.

TODD C.D. & DOYLE R.W. (1981) Reproductive strategies of marine benthic invertebrates: a settlement-time hypothesis. *Mar. Ecol. Prog. Ser.*, **4**, 75–83.

TOWNSEND C.R. & HUGHES R.N. (1981) Maximizing net energy returns from foraging. In Townsend C.R. & Calow P. (eds), *Physiological Ecology: An Evolutionary Approach to Resource Use*, pp. 86–108. Blackwell Scientific Publications, Oxford.

TREGENZA N., BERROW S., HAMMOND P. & LEAPER R. (1997) Harbour porpoise (*Phocoena phocoena* L.) by-catch in set gillnets in the Celtic Sea. *ICES J. mar. Sci.*, **54**, 896–904.

TUNNICLIFFE V. (1991) The biology of hydrothermal vents: ecology and evolution. *Oceanogr. mar. Biol.*, *Ann. Rev.*, **29**, 319–407.

ULANOWICZ R.E. & KEMP W.M. (1979) Toward canonical trophic aggregations. *Am. Nat.*, **114**, 871–83.

ULANOWICZ R.E. & BAIRD D. (1999) Nutrient controls on ecosystem dynamics: the Chesapeake mesohaline community. *J. mar. Syst.* (in press).

UNDERWOOD A.J. (1979) The ecology of intertidal gastropods. *Adv. mar. Biol.*, **16**, 111–210.

UNDERWOOD A.J. (1980) The effects of grazing by gastropods and physical factors on the upper limit of distribution of intertidal macroalgae. *Oecologia (Berl.)*, **46**, 201–13.

UNDERWOOD A.J. & DENLEY E.J. (1984) Paradigms, explanations and generalizations in models for the structure of intertidal communities on rocky shores. In Strong D.R., Simberloff D., Abele L.G. & Thistle A. (eds), *Ecological Communities: Conceptual Issues and the Evidence*, pp. 151–80. Princeton University Press, Princeton, NJ.

VALENTINE J.W. (1973) *Evolutionary Paleoecology of the Marine Biosphere*. Prentice–Hall, Englewood Cliffs, NJ.

VALENTINE J.W. & MOORES E.M. (1974) Plate tectonics and the history of life in the oceans. *Sci. Am.*, **230**(4), 80–9.

VALIELA I. (1984) *Marine Ecological Processes*. Springer, New York.

VALIELA I. (1991) Ecology of water columns. In Barnes R.S.K. & Mann K.H. (eds), *Fundamentals of Aquatic Ecology*, pp. 29–56. Blackwell Scientific Publications, Oxford.

VAN ANDEL T.H. (1979) An eclectic overview of plate tectonics, paleogeography, and paleoecology. In Gray J. & Boucot A.J. (eds), *Historical Biogeography, Plate Tectonics and the Changing Environment*, pp. 9–25. Oregon State University Press, Corvallis.

VAN DOVER C.L. (1990) Biogeography of hydrothermal vent communities along seafloor spreading centers. *Trends Ecol. Evol.*, **5**, 242–6.

VAN WOESIK R., DE VANTIER L.M. & GLAZEBROOK J.S. (1995) Effects of cyclone 'Joy' on nearshore coral communities of the Great Barrier Reef. *Mar. Ecol. Prog. Ser.*, **128**, 261–70.

VAUGHAN D. & DOAKE C. (1996) Recent atmospheric warming and retreat of ice shelves on the Antarctic Peninsula. *Nature*, **379**, 328–31.

VELIMIROV B., FIELD J.G., GRIFFITHS C.L. & ZOUTENDYK P. (1977) The ecology of kelp bed communities in the Benguela upwelling system. *Helgol. wiss. Meeresunters.*, **30**, 495–518.

VINOGRADOVA N.G. (1962) Vertical zonation in the distribution of the deep sea benthic fauna in the ocean. *Deep-Sea Res.*, **8**, 245–50.

WAFAR S., UNTAWALE A.G. & WAFAR M. (1997) Litter fall and energy flux in a mangrove ecosystem. *Estuar. coast. shelf Sci.*, **44**, 111–24.

WALDOCK M. (1986) TBT in UK estuaries, 1982–1986. Evaluation of the environmental problem. In *Organotin Symposium Oceans '86 'Science–Engineering–Adventure'*, Vol. 4, pp. 1324–30. Marine Technology Society, Washington, DC.

WALTERS C.J. & JUANES F. (1993) Recruitment limitation as a consequence of natural selection for use of restricted feeding habitats and predation risk taking by juvenile fishes. *Can. J. fish. aquat. Sci.*, **50**, 2058–70.

WARD J.E., MACDONALD B.A., THOMPSON R.J. & BENINGER P.G. (1993) Mechanisms of suspension feeding in bivalves: resolution of current controversies by means of endoscopy. *Limnol. Oceanogr.*, **38**, 265–72.

WARWICK R.M. (1986) A new method for detecting pollution effects on marine macrobenthic communities. *Mar. Biol.*, **92**, 557–62.

WARWICK R.M. (1989) The role of meiofauna in the marine ecosystem: evolutionary considerations. *Zool. J. Linn. Soc.*, **96**, 229–41.

WARWICK R.M. & CLARKE K.R. (1995) New 'biodiversity' measures reveal a decrease in taxanomic distinctness with increasing stress. *Mar. Ecol. Prog. Ser.*, **129**, 301–5.

WASSENBERG T. & HILL B.J. (1990) Partitioning of material discarded by prawn trawlers in Moreton Bay. *Aust. J. mar. Freshwater Res.*, **41**, 27–36.

WEIGERT R.G. (1979) Ecological processes characteristic of coastal *Spartina* marshes of the south-eastern USA. In Jefferies R.L. & Davy A.J. (eds), *Ecological Processes in Coastal Environments*, pp. 467–90. Blackwell Scientific Publications, Oxford.

WICKLER W. (1968) *Mimicry in Plants and Animals*. World University Library, McGraw–Hill, New York.

WIEBE P.H. (1970) Small-scale spatial distribution in oceanic zooplankton. *Limnol. Oceanogr.*, **15**, 205–17.

WILLIAMS G.C. (1977) *Sex and Evolution*. Princeton University Press, Princeton, NJ.

WILSON G.D.F. & HESSLER R.R. (1987) Speciation in the deep sea. *Ann. Rev. Ecol. Syst.*, **18**, 185–207.

WOLCOTT T.G. (1973) Physiological ecology and intertidal zonation in limpets (*Acmaea*): a critical look at 'limiting factors'. *Biol. Bull.*, **145**, 389–422.

WOLFF T. (1977) Diversity and faunal composition of the deep-sea benthos. *Nature (Lond.)*, **267**, 780–5.

WOLFF W.J. (1977) A benthic food budget for the Grevelingen Estuary, The Netherlands, and a consideration of the mechanisms causing high benthic secondary production in estuaries. In Coull B.C. (ed.), *Ecology of Marine Benthos*, pp. 267–80. University of South Carolina Press, Columbia.

WOOD E.J.F., ODUM W.E. & ZIEMAN J.C. (1969) Influence of sea-grasses on the productivity of coastal lagoons. In Castañares A.A. & Phleger F.B. (eds), *Lagunas Costeras: un Simposio*, pp. 495–502. Univ. Nac. Auton. Mexico, Mexico City.

WOODIN S.A. (1974) Polychaete abundance patterns in a marine soft-sediment environment: the importance of biological interactions. *Ecol. Monogr.*, **44**, 171–87.

WOODWELL G.M., HOUGHTON R.A., HALL C.A.S. *et al.* (1979) The Flax Pond ecosystem study: the annual metabolism and nutrient budgets of a salt marsh. In Jefferies R.L. & Davy A.J. (eds), *Ecological Processes in Coastal Environments*, pp. 491–511. Blackwell Scientific Publications, Oxford.

WRIGHT J.B. (ed.) (1977–78) *Oceanography*. Open University Press, Milton Keynes.

WULFF F., FIELD J.G. & MANN K.H. (eds) (1989) *Network Analysis in Marine Ecology: Methods and Applications*. Springer, Berlin.

YONGE C.M. (1963) The biology of coral reefs. In Russell F.S. (ed.), *Advances in Marine Biology. Vol. 1*, pp. 209–60. Academic Press, New York.

ZENKEVITCH L.A. & BIRSTEIN J.A. (1956) Studies of the deep-water fauna and related problems. *Deep-Sea Res.*, **4**, 54–64.

ZIEMAN J.C., THAYER G.W., RUBBLEE M.B. & ZIEMAN R.T. (1979) Production and export of sea-grasses from a tropical bay. In Livingston R.J. (ed.), *Ecological Processes in Coastal and Marine Systems*, pp. 21–23. Plenum, New York.

Index

Note: page numbers in *italics* refer to figures, those in **bold** refer to tables